PENGU

VOYAG

CHARLES DARWIN was born in ~~~~~~~~~~~~~~~~~~~~~~~~~~~~~~~~~ed at
Shrewsbury School, Edinburgh University and Christ's College, Cambridge. He took his degree in 1831 and in the same year embarked on a
five-year voyage on HMS *Beagle* as a companion to the captain; the
purpose of the voyage was to chart the coasts of Patagonia and Tierra del
Fuego, and to carry a chain of chronometric readings round the world.
While he was away some of his letters on scientific matters were pri-
vately published, and on his return he at once took his place among the
leading men of science. In 1839 he was elected a Fellow of the Royal
Society. Most of the rest of his life was occupied in publishing the find-
ings of the voyage and in documenting his theory of the transmutation
of species. *On the origin of species by means of natural selection* appeared
in 1859. Darwin spent many years with his wife – his cousin Emma
Wedgwood, whom he had married in 1839 – and their children at Down
House in Kent. He died in 1882, and was buried in Westminster
Abbey.

DR MICHAEL NEVE is a lecturer in the history of medicine at the
Wellcome Institute, London and University College, London. His
courses include histories of the life sciences and psychiatry.

DR JANET BROWNE teaches history of the life sciences at University
College, London. She has been an editor of Charles Darwin's correspondence, and is currently engaged on a biography of Charles Darwin.

VOYAGE

OF THE

BEAGLE

CHARLES DARWIN

CHARLES DARWIN'S

Journal of Researches

EDITED AND ABRIDGED
WITH AN INTRODUCTION BY

JANET BROWNE and
MICHAEL NEVE

PENGUIN BOOKS

PENGUIN BOOKS

Published by the Penguin Group
Penguin Books Ltd, 27 Wrights Lane, London W8 5TZ, England
Penguin Books USA Inc., 375 Hudson Street, New York, New York 10014, USA
Penguin Books Australia Ltd, Ringwood, Victoria, Australia
Penguin Books Canada Ltd, 10 Alcorn Avenue, Toronto, Ontario, Canada M4V 3B2
Penguin Books (NZ) Ltd, 182–190 Wairau Road, Auckland 10, New Zealand

Penguin Books Ltd, Registered Offices: Harmondsworth, Middlesex, England

Text first published by Henry Colburn 1839
Appendices first published 1839
This abridged text, incorporating Fitzroy's appendices and with new introduction
and notes, first published in Penguin Books 1989
7 9 10 8

Printed in England by Clays Ltd, St Ives plc
Filmset in Linotron Bembo

CONTENTS

MAPS AND ILLUSTRATIONS

Cover illustration: the *Rhea darwinii*, a second type of Rhea found only in the southern parts of the Argentine pampas, collected by Darwin and named in his honour by the London zoologist John Gould.

ACKNOWLEDGEMENTS

The editors would like to express their thanks to Kate Griffin, Nick Browne, Ben Barkow, our colleagues at the Wellcome Institute for the History of Medicine and those who worked on the *Correspondence of Charles Darwin* for their help and support. It gives us equal pleasure to be able to thank all the students who, over the years, have helped us clarify and expand our views about Darwin and the *Beagle* voyage. We are also very grateful to Paul Keegan of Penguin Books for his loyalty and patience during the preparation of this edition, and to Joanna Swinnerton for her helpful copy-editing. For permission to use the *Correspondence of Charles Darwin* in compiling the introduction, grateful acknowledgement is made to Cambridge University Press, and for permission to use the facsimile page on p. 29, grateful acknowledgement is made to the British Library.

CHARLES DARWIN: A CHRONOLOGY FROM 1809 TO 1839

1809 Born at Shrewsbury on 12 February, in the parish of St Chad, the son of Robert Waring Darwin and Susannah, née Wedgwood.

1813 In the summer, goes to Gros near Abegele for sea-bathing, some of his earliest recollections coming from this.

1817 Attends Dr G. Case's day-school at Shrewsbury, aged eight. His mother dies.

1818 At midsummer, goes to Samuel Butler's school at Shrewsbury. Butler (1744–1839) was the grandfather of Samuel Butler (1835–1902), science writer and the author of *Erewhon*. In September, Darwin is ill with scarlet fever.

1822 In June, with his sister Caroline (1800–1888), Darwin goes touring and recollects taking pleasure in scenery.

1825 In October, matriculates at the University of Edinburgh. In the winter of 1826/1827 he examines marine animals on the shore of the Firth of Forth with Dr R. E. Grant (1793–1874), Lamarckian and naturalist.

1827 On 15 October Darwin is admitted to Christ's College, Cambridge, after ending his medical education in Edinburgh. He becomes friendly with his second cousin W. D. Fox (1805–1880) and they share an interest in entomology.

1828 Becomes a keen collector of insects, as well as spending time either hunting, shooting or being idle.

1831 Passes his BA on 22 January without Honours, and remains at Cambridge for a further two terms. Spends much time with J. S. Henslow (1796–1861), clergyman, botanist and mineralogist. In his *Autobiography*, Darwin describes Henslow as 'free from every tinge of vanity or other petty feeling'. In the spring of 1831 Darwin plans a visit to Tenerife and learns Spanish for that reason. In the summer he geologizes in North Wales with Adam Sedgwick (1785–1873), Woodwardian Professor of Geology at Cambridge. In August Henslow informs Darwin that Robert FitzRoy is seeking a gentleman companion and a naturalist aboard HMS *Beagle* for a journey to South America to 'survey the S. extremity'. Initially Darwin refuses the offer (30 August) because his father has strong objections, the first of which is 'the unfitting' of Darwin to become a clergyman. Robert Darwin is persuaded to drop his objections by Josiah Wedgwood II, Charles's uncle, who wrote in late August: 'The undertaking would be useless as regards his profession, but looking upon him as a man of enlarged curiosity, it affords him such an opportunity of seeing men and things as happens to few.' On 11 September, Darwin travels with Captain FitzRoy to Plymouth to see the *Beagle*, taking the steamer on a journey that lasts three days. He returns to Shrewsbury, via Cambridge, on 22 September and departs from home ten days later for the journey to South America, reaching Plymouth on 24 October. The *Beagle*'s departure is delayed during October and November, and two attempts to depart in December, on 10th and 21st, are abandoned because of bad weather. On 27 December the *Beagle* starts on the voyage of circumnavigation of the globe.

1832   On 16 January, Darwin makes his first landing 'on a tropical shore' at St Jago, Cape Verde Islands.

1836   On 2 October the *Beagle* drops anchor at Falmouth, and on the 4th Darwin returns home to Shrewsbury after an absence of five years and two days.

1837   In March Darwin takes up lodgings at 36 Great

Marlborough Street, in London. In May he gives a
paper to the Geological Society on coral formations. In
July, he opens his first notebook on 'transmutation of
Species', having for some months pondered the
character of South American fossils and the fauna of the
Galapagos Archipelago. Throughout October and
November he prepares the scheme for *The zoology of the
voyage of HMS* Beagle.

1838   Works on a variety of geological and natural historical
topics, and in September finishes his paper on the
Scottish site Glen Roy and its famous 'roads',
attempting to prove their marine origin. Throughout
September and October he ponders transmutation
questions and matters of religion. On 11 November, a
Sunday (Darwin called it 'the day of days'), his cousin
Emma Wedgwood (1808–1896), a daughter of Josiah
Wedgwood II, agrees to marry him. In December
Darwin is busy house-hunting and often feels unwell. In
the last days of the year they take possession of their
London home at 12 Upper Gower Street. On 24 January
1839 Darwin is elected a Fellow of the Royal Society,
and on the 29th, he and Emma are married in
Staffordshire.

INTRODUCTION

THE VOYAGE of the *Beagle* was, without a doubt, the most formative and influential event in Charles Darwin's life. Begun in 1831 and concluded towards the end of 1836, it transformed him from an amiable and somewhat aimless young man into an acknowledged expert in natural history and geology. It gave him an unrivalled opportunity to make scientific observations, to collect animals and plants and to travel through a succession of countries then little-known to European naturalists, and it convinced him that he should give up the idea of taking holy orders and that he was capable of joining the élite world of London science on an equal footing. Thereafter the expedition made him a scientific celebrity as he supervised the distribution and description of his specimens and produced several books and numerous articles about many of the things that he had seen. Moreover, and more famously, the intellectual currents set into motion by this voyage swirled through Darwin's later life until, precipitated by Alfred Russel Wallace, they burst into Victorian drawing-rooms in 1859 in the form of *On the origin of species by means of natural selection, or the preservation of favoured races in the struggle for life.*

Without Darwin's time on the *Beagle* there would not have been any of the important collections now distributed throughout the major museums of Great Britain; no *Journal of Researches*; no published accounts of the zoology and geology of the *Beagle*'s travels round the world, and, of course, no evolutionary theory, although here it would be unfair to forget that Wallace independently forged the idea of natural selection and set out the main themes of transmutation just as clearly as Darwin. But their joint formulation of the concept of evolution notwithstanding, even the most cursory glance at Darwin's achievements shows that the

Beagle period was the great time of his life. Even at the grand old age of sixty-seven, long after *Origin of Species* had been published and the biological revolution had slowly turned its circle, Darwin still insisted in his autobiography that 'the voyage of the *Beagle* has been by far the most important event in my life and has determined my whole career.'

Yet Darwin was not an evolutionist during his time on board the *Beagle*, and it would be incorrect to say that Darwin first thought of evolution as he explored the Galapagos Islands or as he travelled home across the Pacific. Recent research into Darwin's work on board, his notebooks and other writings, the busy round of meetings, library researches and scientific societies into which he threw himself after returning to England, and the more considered inspection of his specimens under the eyes of experienced naturalists in London and Cambridge all add up to the conclusion that Darwin became intrigued by the idea of transmutation only around the middle of March 1837, some five months after the *Beagle* had docked in Falmouth. His travels were, of course, in some important sense preparatory to arriving at this momentous decision, but the point remains that the received image of Darwin voyaging alone through vast, turbulent seas of thought as he paced the deck of the *Beagle* is a fantasy: reality was rather different.

The events and constantly changing scenery of the voyage must, therefore, be seen more as Darwin saw them. The fresh, youthful prose of Darwin's daily record-book and his letters to friends and family back home testify to the uncomplicated enthusiasm that he brought to his task. The voyage was an extraordinary opportunity for him to get out of Shropshire, to do things that he and fellow undergraduates had yearned to do, and to see something of the world and to make something of himself.

However, the best place to find a more detailed and personal account of Darwin's experiences is in the book published here, entitled *Journal of Researches*, in which he describes not just the mere facts of the *Beagle*'s travels, the ports of call and inland expeditions, but also his emotions, his intense feelings on first arriving in the tropics, his dangerous overland excursions with the Gauchos, the stars over the Cordillera de los Andes, rain

and depression on the island of Chiloe, the sight of half-naked
Fuegians standing on a headland at what seemed to Darwin like
the very edge of the world. The text is vivid and immediate,
giving us even today a very clear idea of what those five years
must have been like for Darwin. It was written during the
voyage, in journal form, most of it being composed at the time of,
or very soon after, the events it described. Bit by bit, the pages
were sent home for his father and sisters to read aloud and then
pass around the other branches of the extensive Darwin family.
Ever mindful of the ocean, Darwin kept other records on the
Beagle. From these and his journal he drew up a book almost
as soon as he arrived back in England. It was completed by
the following September, although publication was delayed for
another eighteen months while Robert FitzRoy, the captain of the
expedition, who was writing his own account of the voyage and
also editing material provided by Phillip Parker King, the previous captain of the *Beagle*, desperately tried to find the time to
finish his share of the work. The four volumes (FitzRoy's two,
plus a lengthy appendix, and Darwin's) eventually came out in
May 1839.

Rather to Darwin's surprise, his book was a huge success. Later
in the same year it was reissued independently by the original
publisher, Henry Colburn, with a different title page and,
apparently, without seeking Darwin's permission. The trans-
action was made also without a fee. Yet Darwin always felt a
special affection for this particular work, however much money
may have been forfeited; he spoke truly when he wrote at the end
of his life that 'the success of this my first literary child always
tickles my vanity more than that of any of my other books.'

Though the events leading up to Darwin's acceptance on to the
Beagle are well known, some of them are worth recapitulating
here. It was not the first time that the *Beagle* had been on the high
seas. Under the overall command of Phillip Parker King, HMS
Adventure and HMS *Beagle* had already surveyed a substantial
portion of the eastern coast of South America, significant in naval
and commercial terms for Great Britain since 1825, when George
Canning had signed a commercial treaty with a newly independent federation of Argentinian states. Robert FitzRoy had joined

this earlier expedition from another posting halfway through the voyage, and was then catapulted into temporary captaincy of the *Beagle* because of the dramatic and unforeseen suicide of her captain, Pringle Stokes. When the expedition eventually returned home in 1830, and King expressed his desire to retire and settle in Australia, FitzRoy was appointed overall commander for a second surveying voyage to set out as soon as practicable to complete the work.

Taking his new duties very seriously, FitzRoy refitted the *Beagle* with all the latest Admiralty surveying devices and some expensive equipment of his own purchasing, including twenty-two chronometers, which, when calibrated against those held at Greenwich, would enable him to determine the ship's longitude far more accurately than had previously been attempted. His instruments were first class and his maps were the best that could be procured. His crew was largely drawn from the previous expedition with some judicious promotions plus some others known officially as 'supernumeraries', a term that meant that the Admiralty was responsible for neither their conduct nor their salaries. Among these were three natives of Tierra del Fuego, who had been brought to England by FitzRoy during the previous expedition (a fourth had died of smallpox on arriving in England) and were now being returned to their country in the company of a missionary, Richard Matthews, in order to set up an Anglican mission. There was also an artist, Augustus Earle, and an instrument-maker to tend the chronometers, both in FitzRoy's private employment.

However, the captain was in some sense alone at the head of his seventy-four officers, men and passengers. FitzRoy was acutely aware of the loneliness and despair that could overtake senior officers, and can hardly have welcomed sailing in the same cabin that had seen Stokes's suicide, a death evidently brought on by the pressures of command. Moreover, FitzRoy's own pedigree indicated that there might be mental instability in the family, for he was the nephew of the aristocratic Tory foreign secretary, Viscount Castlereagh (Lord Londonderry), who in the last years of his life had teetered on the edge of insanity before finally cutting his throat. FitzRoy, therefore, approached a friend, Harry

Chester – possibly the Henry Chester, younger son of Sir Robert Chester, who in 1831 was a clerk in the Privy Council Office – asking if he would care to accompany him for part or the whole of the voyage. It seems that Chester had other things to do, for in August 1831 FitzRoy wrote in haste to the Admiralty (probably to Francis Beaufort, Hydrographer to the Navy, who had dealt with most of FitzRoy's arrangements for the expedition) asking if a suitable gentleman companion might be found.

Here the Cambridge University network came into full play. Beaufort notified George Peacock, a fellow of Trinity College and lecturer in mathematics, who in turn wrote to his colleague at St John's, the Professor of Botany and curate of Little St Mary's Church, John Stevens Henslow. Henslow, himself only thirty-five years of age, thought briefly of taking up the offer himself, but a new baby and other family, collegiate and parochial affairs combined to persuade him to pass the invitation on to his friend and brother-in-law, Leonard Jenyns. Jenyns also felt that he ought not to leave his parish after only a few years at his post. So, on 24 August 1831, Henslow wrote to Darwin.

Henslow had known Darwin since 1828, after Robert Waring Darwin had decided to withdraw his son from medical studies at Edinburgh University and had sent him to Cambridge in order to take the ordinary degree that was a prerequisite for entering the church. At Cambridge, as at Edinburgh, Darwin soon became friendly with other young men with serious interests in natural history, his cousin William Darwin Fox among them, and began attending Henslow's informal natural history soirées held on Friday evenings during term. Together Henslow's young naturalists would make excursions to local villages to collect plants and insects; at his Friday evenings they were given an opportunity, as Darwin later put it, to become 'slightly acquainted with several of the learned men in Cambridge – which much quickened the little zeal, which dinner parties & hunting had not destroyed'. By the spring of 1831 Henslow had seen Darwin blossom into an enthusiastic and knowledgeable amateur entomologist, a young man with easy manners and a cheerful disposition who could ride and shoot, happy with the thought that he would soon be occupying a small country parsonage where he could follow the traditional natural history pursuits of the British clergyman, and yet capable of informed conversation with the savants of the University. He was, in many ways, just as Henslow had been fifteen years earlier, and each of the two men probably saw something important of himself in the other, something that fostered great affection and respect between them.

During that same spring, Darwin's last at Cambridge, for he had passed his BA examination in January, Henslow persuaded Darwin to take up the study of geology, at this time taught by the great Adam Sedgwick, and watched indulgently as Darwin and two friends planned a natural history expedition to the island of Tenerife. Suddenly enthused about geology, Darwin was therefore pleased to be allowed to accompany Sedgwick on his customary working tour through the mountains of Wales before a summer vacation in which he planned to do little except enjoy himself at home in Shrewsbury.

It was at Shrewsbury that Darwin received Henslow's letter about a proposed 'trip to Terra del Fuego & home by the East Indies'. The voyage as Henslow understood it was to last two years, and as he put it in his letter, 'if you take plenty of Books with you, any thing you please may be done – You will have ample opportunities at command – In short I suppose there never was a finer chance for a man of zeal & spirit.' He was, however, careful to explain that the position was 'more as a companion than a mere collector', and emphasized that he recommended Darwin 'not on the supposition of yr. being a *finished* Naturalist, but as amply qualified for collecting, observing, & noting any thing worthy to be noted in Natural History . . . Don't put on any modest doubts or fears about your disqualifications for I assure you I think you are the very man they are in search of.'

A few days later, Henslow's invitation was reiterated by George Peacock, who had been to see Francis Beaufort and could relay official information about the voyage and the situation offered. However, Darwin was already writing out his reluctant refusal: his father, almost despairing of his son ever settling down to a career and foreseeing all kinds of shipwreck, death and disease, as would any parent, had given Darwin such strong advice about not going that he declared he should not be comfortable if he did not follow it.

And yet when Darwin visited his uncle Josiah Wedgwood at his home in Staffordshire, in the heart of the pottery districts where the first Josiah Wedgwood had built up the china company that still bears his name, he found that this branch of the family were so much in favour of his going that he tentatively reopened the case with his father. Wedgwood wrote to Robert Waring Darwin, setting out all the objections that had been raised and giving his opinion on each. Just in case this failed to have the desired effect, Darwin and his uncle returned together to Shrewsbury to argue their point in person. The net result was that Robert Waring Darwin graciously and kindly offered to give his son 'all the assistance in my power'.

Robert Waring Darwin's offer of assistance was not an empty one: the Admiralty expected Darwin to pay for his expedition, and his father was here agreeing not only to send him around the world for some two years (it eventually stretched to five) but also to foot the bill.

These negotiations took seven days. On 1 September 1831 Darwin wrote accepting the offer, and by 5 September he had been to Cambridge to consult Henslow and thence up to London to meet FitzRoy, who quite understandably had had second thoughts about sharing his cabin with a complete stranger, even if he were a natural philosopher. He gave Darwin the opportunity to stand down and explained how easy it would be for him to leave the voyage at any major staging-post. But both young men evidently made themselves as agreeable as possible, for Darwin wrote to his elder sister Susan that 'it is no use attempting to praise him as much as I feel inclined to do, for you would not believe me.'

The deal was set, and Darwin busied himself in London, visiting savants at the British Museum and elsewhere, gathering information and getting his equipment in order before going to see the *Beagle*. A short coastal trip from Woolwich to the naval base of Devonport in Plymouth convinced him that he would have no problem with seasickness: 'as for sickness I utterly scorn the very name of it.' And after what seemed like interminable official and administrative delays coupled with prolonged stormy weather, which forced the *Beagle* to return to port twice in ten days, FitzRoy and Darwin at last sailed out of Plymouth on 27 December 1831.

It is worth emphasizing that this voyage was one undertaken by young men ready to experience all that the world could offer. Charles Darwin was twenty-two when the *Beagle* sailed and Robert FitzRoy was twenty-six. They were therefore confident of their ability to withstand all the privations of nineteenth-century sea-travel. Moreover, they shared a young man's conviction that these privations should be accepted as a necessary feature of starting out on the road to becoming an acknowledged expert, each in their own field. FitzRoy knew that his promotion had come early, perhaps even fortuitously, to a talented surveyor at a time when few naval captains were particularly interested in the minutiae of coastal work. He wished to succeed: to demonstrate that he could produce the best surveys, keep the best ship, complete the exercise as required. On Darwin's part, though he had few such clear objectives at the start, he, too, saw the voyage as a tremendous opportunity to use his skills to his best advantage. Why, for example, did he stay with the *Beagle* when the expedition lengthened from two to three and then to five years even when he could ask – and as a paying customer could afford – to leave whenever he chose? Desperately seasick, grumbling in private to his family about cramped conditions, ceaseless travelling and continued delays in turning the *Beagle* northwards towards Britain, why did he not return after a couple of years afloat? Why did any young man, for that matter, voluntarily undergo the rigours of contemporary travel without scuttling back home after the first port of call?

Looking beyond the personal factors that undoubtedly came into play here, not least the intellectual resilience shown by Darwin at this time and in his later career, the persistence displayed by Darwin, FitzRoy and others has something to tell us about the role that 'travel' may have had in the burgeoning professional world of the nineteenth century and the functions it may have carried out in the making of a career. Persistence, it seems, usually paid off, in that rewards were many and various, scientific and practical, individual and national. Britain, France, Portugal and the United States of America were predominantly seafaring nations held together by the great trading routes of the oceans. For each country, colonization and rapid industrialization were proceeding hand-in-hand as the complex logistics of supply and demand required that new and fruitful sources of basic commodities be located and made accessible to the manufacturers of Europe. Governments needed information about natural products, transport routes, the possibility of new harbours, staging-posts available in mid-ocean and so forth, just as much as they needed to encourage the development of railway systems, dockyards and inland navigation routes to provide the vital arteries of commerce. Over and above these obvious needs there was competition between European nations to make commercial links with undeveloped countries. The whole point of the Admiralty's desire to chart the South American coast was to enable informed decisions to be made on naval, military and commercial operations along the stretch from Bahia (now Salvador) in Brazil to Bahia Blanca in Argentina and into the unexplored coastline beyond, and to enable Britain to establish a stronger foothold in these areas, so recently released from their commitment to trade only with Spain and Portugal.

Exercises such as these were not always peaceful or even intended to be peaceful. As far as the *Beagle* was concerned, Captain FitzRoy was a government representative engaged on official business. But that was not necessarily how other nations might see it. FitzRoy was responsible for his crew and for the protection of other British nationals at such ports as they called at, and indeed the *Beagle* was involved in several incidents, including military action in Montevideo and being caught in a major naval blockade off Buenos Aires. Nor was it coincidental that the Falkland Islands and other disputed territories were listed by the Admiralty as necessary stopping-places at which the flag should be shown. FitzRoy, therefore, had to think politics and negotiate his ship through government policy as well as fulfilling the more practical tasks he had been set. The official Admiralty instructions for the *Beagle*, individually tailored for this particular surveying expedition and including all such government requirements as to stopping-places and the aim to reinforce British imperialism, are, therefore, given in an appendix at the end of this volume.

Similarly many of Darwin's excursions into the hinterland around Buenos Aires and Maldonado were marked by politics and the sporadic military activity of General Juan Manuel de Rosas, whose troops were then relentlessly hunting down his political opponents. Darwin had arrived right in the middle of Rosas's fiercest and most brutal campaigns; by 1835, just two years after the *Beagle* had left the area, Rosas had succeeded in concentrating all public authority and power into his own hands and had assumed the position of a popular saviour and dictator. Darwin describes several occasions when he and his party of Gauchos were forced to talk their way out of trouble or to flee for their lives. Natural history and coastal surveys were, therefore, heavily politicized nationalistic pursuits. For this reason, the version of the *Journal of Researches* that is printed here includes all of Darwin's stories of his experiences of the 1833 revolution in Argentina, his meeting with General Rosas, bandits and the blockade. For much the same reason, all Darwin's accounts of the social mores of the different groups of people that he met are also included, such as the details that he gives of the nomadic lifestyle of the Gauchos, the use of the bolas, the hospitality of the slave-holder on his plantation, the missionaries and their missions, the sorry life of expatriate Cornish miners in the bowels of the Andes. Darwin and FitzRoy were never neutral observers, and this text shows just how far Darwin, at least, saw foreign society through English eyes and endorsed the social order and parliamentary structure of his native country.

Moreover the nationalistic web that Britain threw over the activities of its travellers also covered the kind of work that was done overseas and the social relations of those who attempted to carry it out. Government and commercial aims could be fulfilled by the same person. Consequently, most of the careers offered by the army and navy during the first half of the nineteenth century permitted, sometimes even encouraged and expected, an interest in science and natural history – the subject area in which obser-vations about natural resources were most likely to be made. Positions in the forces were, therefore, for many people a perfectly acceptable way of combining philosophical inclinations with salary and prospects. The job of naval surgeon, in particular, carried with it the opportunity of taking advantage of a voyage into some distant part of the world to study or collect specimens, be they plant, animal, mineral or ethnographic. On many occasions the ship's doctor was the only member of the company, besides the captain, who had received a scientific training of sorts (although the Admiralty was never fussy about actual medical qualifications, issuing its own diploma whenever necessary) and was expected to know about man and the natural world. Benjamin Bynoe, for example, who was the surgeon on board the *Beagle* in Darwin's time, was an able naturalist. He supervised the gathering of wild celery and cranberries in Tierra del Fuego in order to replenish the ship's antiscorbutics: he went on to make substantial botanical collections, both on the *Beagle* voyage and elsewhere, which displayed genuine interest and a knowledge of plants. Bynoe, however, was a surgeon first and botanist second. Other more famous figures, such as Joseph Dalton Hooker or Thomas Henry Huxley, used their medical qualifications as a mere lever to get a place on a voyage, for their primary aim was to pursue natural history rather than a naval career. Yet the point, surely, is that there was a whole spectrum of motives and actualities that the post of naval surgeon could accommodate: an ocean-going medical post was a marvellous chance to practise natural history and get paid. Certainly Robert McCormick, the appointed senior surgeon on the *Beagle*, had every intention of making a fine natural history collection and he very much resented FitzRoy's carefree endorsement of Darwin's collecting activities; McCormick rightly believed that custom decreed that he was the expedition's official naturalist, and he left the *Beagle* in high dudgeon at the first major port of Rio de Janeiro, only four months out of Plymouth.

Thereafter Darwin assumed the quasi-official responsibilities of naturalist to the voyage, along with the nickname 'Philos' (ship's philosopher), although it should be emphasized that his collections were always held to be his own personal property, his to dispose of as he saw fit on his return to England. Others on board were equally keen to take on McCormick's customary task,

and both Bynoe (now promoted to acting surgeon) and Lieutenant Bartholomew James Sulivan, Darwin's closest friend on
board, made competent collections for the Crown, to be deposited
in the British Museum by the Admiralty. Even FitzRoy and lesser
mortals such as the midshipman, Philip Gidley King, made
sizeable collections. FitzRoy's bird-skins, for example, were
essential to Darwin after his return when he discovered that the
various islands of the Galapagos supported different species of
mocking-birds and finches: Darwin had mostly ignored the precise geographic locale for his own specimens. And, like Darwin,
FitzRoy had a special interest in geology, having read many
of the major texts in this field, including the first volume of
Charles Lyell's celebrated *Principles of Geology*, which had been
issued in 1830 (the next two volumes were published while the
Beagle was at sea, although Darwin managed to obtain these fairly
soon after publication). It was FitzRoy who had given Darwin a
copy of Lyell's first volume before they left England, and we can
be sure that Darwin lent the subsequent volumes to this 'geological' captain during the course of the voyage. FitzRoy also made a
useful collection of mineralogical specimens and evidently
thought deeply about the scientific interpretation of the land-
forms and geological features that were seen.

This mutual interest in geology was probably one of the areas
that drew Darwin and FitzRoy closely together during their time
on board. From a short essay on the recent geological history of
the earth that FitzRoy appended to his own account of the *Beagle*
voyage, it appears that he was, for some time at least, in favour of
Lyell's arguments for very slow and gradual geological changes
and had questioned the Biblical account of earth history sufficiently to register his doubts about the existence of a former
(Noachian) flood. It was FitzRoy, not Darwin, who had been
approached by Lyell before the *Beagle* sailed in order to ask
that observations on specific geological features such as erratic
boulders be recorded. However, and presumably while he was
travelling with Darwin, FitzRoy changed his mind and positively
reaffirmed his faith in the Biblical story in his essay entitled 'A
very few remarks with reference to the Deluge'. The essay is
reprinted at the end of this volume.

Darwin was evidently not alone in his natural historical interests. Nor was he alone in the customary sense of being a solitary explorer, a picturesque and romantic figure or an intrepid Victorian hero. He was part of an extensive and predominantly British network that was activated by the arrival of the *Beagle* at every port of call. He followed the life of an ordinary English gentleman on shore as much as was possible: he and FitzRoy paid social calls to local governers, dined out, visited representatives and agents for various British concerns, harbour authorities and so forth. They stayed in the homes of cultivated expatriate families when in port, even attending a musical evening in Tasmania, of which Darwin reported: 'I dined yesterday at the Attorneys General, where, amongst a small party of his most intimate friends he got up an excellent concert of first rate Italian Music. The house large, beautifully furnished; dinner most elegant with *respectable*! (although of course all Convicts) Servants –' Darwin himself stayed for several months in the Valparaiso home of an old schoolfriend, Richard Corfield, and subsequently paid other visits as the *Beagle*'s schedule allowed. Corfield was just one of an extended British support system composed of shipping agents, commercial entrepreneurs, retired gentlefolk, government officials, even the previous commander of the *Beagle*, who then resided at Bathurst, Australia, and the great astronomer, John Frederick William Herschel, who was living in Cape Town while making his southern observations.

This kind of overseas society was hardly provincial. Even the smaller towns of South America possessed their local gentry who pursued activities appropriate to their social standing. Larger cities had libraries and, occasionally, theatres and opera houses. There were, of course, local newspapers at every port, but what is perhaps not quite so well known is that FitzRoy and his crew received regular and bulky parcels of English papers and journals, just as the full-time residents of these areas did. Letters also arrived more or less regularly, depending on the steamer routes and Admiralty movements in the area. Darwin's three sisters took it in turns to write every month for nearly five years, and others on board probably also received news at every station. Although Darwin naturally complained of being out of touch with his family and country, the reality was that both he and FitzRoy were remarkably well informed about home activities, and conversations with local residents were as likely to be in-depth discussions about the passing of the Reform Bill as they were about the location of interesting scientific sites to visit. The overseas network held a very real and constantly replenished link with Britain: it was far more metropolitan in tone and up-to-date than is often nowadays supposed, and enabled Darwin to carry with him the latest scientific theories, the ideas of his friends and his developing personal aspirations, while maintaining social links that extended beyond the *Beagle* and contacts made during onshore visits, right back to London and Cambridge. Darwin may indeed have been alone in some literal sense, a solitary naturalist pursuing his individual researches, but he travelled along the strands of a well-organized web of Empire where the whole world of British natural history and English upper-class society constantly accompanied him.

One particular aspect of this close connection with Britain, which eventually stood Darwin in great stead, was his relationship with Henslow, his friend and professor. Though Henslow barely wrote to Darwin at all during the voyage, it is clear that Darwin saw him as his intellectual and scientific mentor and endeavoured to do those things that he believed Henslow would approve.

Darwin collected carefully and with much thought: he concentrated on complete suites of specimens of insects and plants, small vertebrates, birds, spiders, corals, molluscs and other invertebrates (his collections in this department were much esteemed by colleagues) and fossils when he could get hold of them, having no scruples about buying bones if necessary. Along with all this, he made extensive geological collections to supplement his researches in that field and to provide the requisite information about deposits where fossils were found. Everything was meticulously tagged, recorded in at least two different lists or catalogues along with notes on location, colour, behaviour, etc., wrapped or bottled, skinned or dried, and carefully packed away in barrels to be sent on the next Admiralty ship back to England. Shipping his boxes through the Admiralty was a considerable perk for

Darwin, for in his capacity as private gentleman he was expected to pay all his own expenses. Nevertheless, and fully aware that the specimens were Darwin's private property, FitzRoy generously made it possible for the collection to be treated as official cargo.

The boxes were addressed to Henslow in Cambridge. He had agreed to act as a receiving officer and to perform the necessary function of unpacking on receipt in order to check that bottles had not broken or needed topping up, or that insects or decay had not disposed of Darwin's trophies. Henslow's interest in the collections did not stop there. Before Darwin returned in 1836, he had already indicated to the right London people that there were some magnificent specimens waiting to be distributed as Darwin saw fit: in particular, he could not resist directing the fossilized skull of a Megatherium to William Clift at the Royal College of Surgeons, where eminent naturalists like William Buckland were eager to examine it. Nor could Henslow resist bringing the fruits of his protégé's labours to the attention of Cambridge savants by publishing extracts from Darwin's letters to him about natural history topics. The extracts were read before the Cambridge Philosophical Society meeting on 16 November 1835 and brought Darwin's name before some of the most notable natural scientists of the day, serving to generate a great deal of interest in his results and a general buzz of anticipation about his then impending return. Darwin was, therefore, known as an enterprising and thoughtful naturalist well before he came back: through the judicious placing of specimens and some careful editorial work, Henslow had somehow managed to create a public figure who now had only to return – not always a certainty in the age of tall ships – to take his place in scientific society. In this sense it might well be said that Henslow acted almost as an agent or business manager for Darwin, and completed his self-appointed task with admirable skill. Never one to forget this timely assistance, Darwin always believed that Henslow 'had made him what he was'.

Darwin's scientific achievements on the *Beagle* voyage were many and various. Impossible to summarize adequately, everything that he saw in the natural history line and much from other fields such as ethnology and political economy was recorded.

Modelling his work and activities on the famous example of
Alexander von Humboldt, who had travelled with his friend
Aimé Bonpland through the rain forests of the Orinoco and River
Negro in the years 1799 to 1804, on a perilous and dramatic
expedition that was subsequently written up in a comprehensive
Personal Narrative, Darwin succeeded in mastering an equivalent
range of scientific and personal interests. His *Journal of Researches*
is a vivid and accurate account of all these observations and
experiences.

One rather obvious point that is rarely made and deserves full
emphasis here is that Darwin's voyage on the *Beagle* was not so
much a journey at sea but a voyage on land. Although he was
away from home for very nearly five years, he was at sea for only
533 days (eighteen months in all), the longest stretch at sea being
one of forty-seven consecutive days, but, as he noted in his daily
diary, many landings were made during that period, the usual
sailing-run being between eight and eighteen days. Robert
FitzRoy and his crew undoubtedly spent more time on board than
Darwin, who had a privileged position as FitzRoy's guest, and it
seems that Darwin quite naturally took advantage of this useful
situation. He was on land for a grand total of three years and three
months. Some of these periods ashore ran to four months, as
when he travelled in the Rio Negro area of Patagonia, or on his
major geological tour of the Andes, setting out from Richard
Corfield's house in Valparaiso. On another occasion he spent
three months in Rio de Janeiro, then a further three and a half
months again based at Corfield's home. Most visits were longer
than a fortnight, and only five out of a total of thirty-seven
landings were less than a week in length. So the colourful picture
of FitzRoy and Darwin being cooped up in a cabin for weeks on
end, or of Darwin delving into his psyche alone on deck out in the
middle of the ocean, may indeed be true, but only in part. The
more mundane but inescapable conclusion is that Darwin was
ashore, exploring the countryside and going about his business,
for roughly three-fifths of his voyage.

But it is also probably true to say that this division of his time
worked greatly to Darwin's advantage. Like Alcide d'Orbigny,
the many-talented geologist who had accompanied the French

government's expedition to South America in the years 1826 to 1833, Darwin was always seasick and – according to his letters and diary – often unable to do anything on board ship but lie in his hammock or on the Captain's sofa, feeling dreadful. He developed a hatred of the sea that was frankly expressed to his family: 'I hate every wave of the ocean,' he told his cousin William Darwin Fox, 'with a fervour, which you who have only seen the green waters of the shore can never understand.' 'I loathe, I abhor the sea and all ships which sail on it,' he wrote to Susan Darwin; 'not even the thrill of geology,' he said, 'makes up for the misery and vexation of spirit that comes with seasickness.' FitzRoy took Darwin's condition seriously, even writing to the Admiralty full of concern that he would have to leave at the first possible stop at the island of St Jago. Right to the end of the expedition, FitzRoy was writing home that 'Mr Darwin was a martyr to confinement and seasickness when under way.'

Robert Waring Darwin, a noted physician, recommended that his son should try eating raisins, and circumspectly forbore to suggest that he should come home. But Darwin evidently found that his own best remedy was to stay on land whenever possible. FitzRoy, again obliging and solicitous, kept Darwin informed about the *Beagle*'s whereabouts and usually arranged to pick up Darwin at some port further along the coast in three or six weeks' time. For his part Darwin seemed like a man possessed while on shore: he literally crammed most of his life into those three years that he spent safely on terra firma. Furthermore the timetable of the *Beagle* gave him a concrete timetable of his own to work to. The ship's itinerary and planned sequence of landings created a definite programme of visits that determined what Darwin could effectively do with his time. In other words, he had to get on with it before the *Beagle* left.

There was, of course, another more potent reason behind Darwin's intensely active days on shore. It was here, on land, pre-eminently in South America, that he could pursue his fascination with geology, fostered so recently in Cambridge and Wales by Adam Sedgwick. It was here that he first began to believe he might be able to do something worthwhile in natural history and where his zest for geology started to lead him towards some of the major theoretical achievements of the voyage.

The role of Charles Lyell and his *Principles of Geology* in this enthusiastic commitment to geology in Darwin's early work is perhaps now so well known that there is very little to add to the various accounts given by recent historians. Darwin read Lyell's famous textbook volume by volume as it was published, and was delighted by the grand theoretical scheme he found there. Lyell's theory of the gradual elevation of land out of the sea, for instance, could be used by Darwin to explain many of the things he saw. Thick alluvial deposits on the east coast of South America were consequently interpreted as an elevated estuary, and Darwin thought that the fossil mammalian bones he found there must have been swept into the sea by Tertiary rivers, only to be covered with sediment and eventually raised above the surface. Rounding Cape Horn, he thought the archipelago looked very like a row of submerged mountains and he was gratified to find that the west coast of South America seemed to have been uplifted, step by step, from a state similar to that now exhibited by Tierra del Fuego. Further cross-country traverses and coastal surveys left no doubt in Darwin's mind that elevation had taken place exactly as Lyell had surmised.

But Lyell had not been to South America, nor had he taken more than a handful of his illustrative case histories from that country. It was virtually unknown territory to British geologists. Darwin rapidly came to see that he could make genuinely useful observations of the kind that were itemized in detail by Lyell. Throughout the *Principles of Geology*, Lyell had stressed the value of first-hand observation. He also gave the novice a historically based theory of breathtaking scope and application: that the earth had never experienced any geological forces that were greater in intensity than those acting today, thereby suggesting that the earth had neither a beginning nor an end, and that the actual crust of the globe was composed of giant blocks or segments that constantly and gradually changed their level in relation to each other and the sea. Lyell suggested that these movements were generated by the activity of red-hot, molten rock constituting the earth's core.

As the *Beagle*, with both Darwin and FitzRoy on board, proceeded slowly up the coast of Chile, Darwin was soon able to draw practical conclusions about the causes behind such changes in the levels of land and sea. He and his colleagues saw a volcano erupt and soon afterwards experienced a major earthquake. Correlating this dramatic example of geological forces in action with his knowledge of the recent history of the east coast, as worked out in Patagonia, Darwin came to believe that earthquakes, volcanic eruptions and elevation of the coastline were all parts of the same Lyellian phenomenon. If the volcanoes of a mountain range were in some way connected deep below the surface, they would tend to erupt together or in sequence. Pressure not relieved by eruption might be dissipated through the fracturing and shifting of strata, causing earthquakes. At a less intense level of activity, molten rock would be constantly injected into the basal zones of mountain chains, thereby lifting the ground level 'slowly and by little starts'.

Darwin was evidently theorizing on an extraordinarily large scale, learning to look for grand phenomena, searching for the all-pervasive unifying causes of global geology that Lyell had intimated were still in action in the landscape around him. In this respect it is remarkable just how often Darwin found the very bit of geological evidence he was looking for, remarkable because often only the most sophisticated and highly experienced geologists can recognize the significant piece of information in a complicated traverse or section, and remarkable in that Darwin had no real reason to believe, other than a hunch, that the areas he examined were likely to reveal crucial materials. An important example of this collector's serendipity came in April 1835, when Darwin travelled through the Cordillera from Valparaiso to Mendoza and back again by a different route, taking in both the Portillo and Uspallata passes. He was searching for additional evidence that the coasts of South America had been elevated during the Tertiary and more recent geological periods, but hardly expected to find that one entire range of the Cordillera was composed of the same Tertiary rocks as the coastal flats of Chile. Yet, as he wrote in some of the more exciting passages in his *Journal of Researches*, this conclusion became impossible to ignore after two days spent geologizing in the Uspallata pass. Moreover, even as he was arriving at this theoretical position, he happened to find a small 'forest' of silicified trees still standing upright. To Darwin these clearly represented larger memorials of the same thin layers of silicified wood that he had found to be particularly characteristic of the western coastal strip. The overall conclusion of this active and fortuitous week in Darwin's geological life was that at least part of the huge mountainous backbone of South America had been pushed up from ground level after the ordinary sedimentary beds of the Pampas had been laid down – a time so recent in geological history that the fossils that were now buried in the eastern deposits, although the original animals were certainly extinct, were of the same general kind of animal that still lived in the area (armadillos, llamas and sloths, for example) and were embedded with species of shells that were still in existence. Darwin was left gasping at the power and sheer explanatory force of Lyell's general principles.

What goes up must come down, as Lyell never quite said. So another similar change of level could be visualized in the Pacific basin, but here it was subsidence that Darwin believed he saw rather than elevation; subsidence of the ocean floor that was matched by coral polyps building their reefs right up to the surface and there forming coral islets. Geological and biological elements combined over the passage of aeons of time to create all the various forms of fringing reefs, lagoon islands and atolls. This elegant scheme, which Darwin freely confessed he had dreamed up long before he saw a coral island, also served to support Lyell's contention that the course of geological history was marked by endless oscillations of the crust of the earth.

Obviously Darwin was 'seeing' land-forms as if he had the eyes of Lyell, as indeed he said in letters to Henslow. This special 'vision' was also plainly to the fore during Darwin's visit to the Galapagos archipelago.

In September 1835, with all these ideas of elevation and subsidence fresh in his mind, Darwin sailed for the Galapagos. His interest in the archipelago was intense because it promised a new kind of geological experience. The island group was of volcanic origin and, he thought, geologically recent: mountainous and studded with craters, the landscape was composed entirely of volcanic rocks in various stages of decomposition, and at least two of the fifteen islands were covered with immense sheets of naked lava, signifying continuous geological activity. For Darwin, phenomena such as these meant one thing only: new land was being brought into existence. The geological forces that had elevated the South American continent and fuelled the processes of mountain-building were manifested here as a series of volcanic eruptions. At first hidden on the sea floor, these lava flows must have accumulated steadily, eventually emerging above the surface of the water as volcanic cones. Perhaps the process was accelerated by a general elevation of the sea-bed, thought by Darwin often to accompany eruption. Where there had once been 'unbroken ocean' there was now a constellation of new islands ready to receive all that the organic realm could offer. Here in the Galapagos, Darwin could examine genuinely new land and judge the path of its subsequent history; he could study the effects of denudation and weathering and assess the rate at which base rock disintegrated into soil; he could virtually witness the transformation of the earth's raw materials into a fully diversified topography. This was the very stuff of Lyellian uniformitarian geology.

More important in the light of his later career, Darwin also got the chance to investigate the animal and plant life of the archipelago. Insular populations were fascinating objects at any time, and the Galapagos Islands were known to possess a rich variety of endemic species. Here again it would be possible to see how animals and plants colonized new lands, how bare rock was clothed and peopled with living beings. In what way had these geologically recent islands acquired their species? Was it by immigration or by creation? Did new land have new species? Darwin's interpretation of the geology of this region raised fundamental biological questions, the answers to which would surely reinforce his theory of crustal movements and expand its scope. Just as Lyell recommended and explained so fully in the second volume of *Principles of Geology*, new lands such as these could be seen as having received their fauna and flora from the neighbouring continent. And, apparently, Darwin did indeed see the island populations as straightforward imports from Ecuador and surrounding regions. According to recent studies of his notes, it is clear that Darwin believed that all the different kinds of mocking-thrushes and finches that he collected were merely variants of a single species. Only later, in London, when he was told that these birds were distinctly separate species did he begin to puzzle over their meaning for theoretical natural history. Indeed, Darwin's understanding of the fauna was so far from an evolutionary one that he failed to notice that there were several different species of giant tortoise, one species for each island. The vice-governor of the Galapagos therefore surprised Darwin when he informed him that such was the case.

Nevertheless Darwin was plainly deeply impressed by the tiny worlds he had found on the islands and carried that image with him across the Pacific. It was still vivid when he came to question the taxonomic status of the bird life of the archipelago, and thence his famous recollection, repeated in several forms in various publications: 'In July opened first note Book on "transmutation of Species". – Had been greatly struck from about month of previous March – on character of S. American fossils – & species on Galapagos Archipelago. – These facts origin (especially latter) of all my views.'

What Darwin omitted to record in his little pocket-book was that he had been deeply impressed also by the geographic distribution of the American 'ostrich', usually known today as the Rhea. The common ostrich (as he called it) was replaced in the southern parts of the continent by another, closely allied species, later named after Darwin as the *Rhea darwinii*; the two were taxonomically related yet mutually exclusive forms. Darwin tells a funny story in his *Journal of Researches* of how he had few expectations of seeing this reputed second species and inadvertently ate most of one before he noticed its identifying characteristics.

The point for him, and the theoretical umbrella under which he later collected these significant observations, was that the two forms of Rhea lived in different but adjacent areas. Much the same could be said about the fossils he had discovered in the Pampas, which were plainly much larger, extinct relatives of the local armadillos and sloths: the two sets of mammals were clearly related and yet mutually exclusive in a chronological sense. Though Darwin did not at the time put them together, he shortly came to see that the spatial distribution of the Rhea was comparable to the chronological relationships of past and present mammalians and that the processes that encouraged spatial differentiation might also have stimulated changes over time: whatever it was that made one modern species devolve from another would perhaps account for the existence of a succession of different species in the past. Something along these lines can be read in Darwin's last field notebook, written on the *Beagle* while sailing home towards Britain, together with his puzzled comments about the possible origin of insular faunas recorded in the ornithological notes he was writing up before finally disembarking in Falmouth.

Two other topics were of great significance for Darwin, one of which is made so plain in this first edition of his *Journal of Researches* that it deserves some comment. Darwin was intrigued by the problem of the causes of extinction. Many of his contemporaries argued that it could only have been brought about by major topographic changes or climatic catastrophes. But like Lyell, Darwin was not fully comfortable with explanations that rested on the occurrence of great geological revolutions and believed that the number of individuals within a species gradually declined until there was perhaps only a single pair left. Absolute extinction soon followed. Lyell's gradualism and uniformitarianism required that there were no sudden alterations in the landscape and so, in his and Darwin's view, the process of extinction was similarly paced and did not necessarily indicate that the history of the earth had been punctuated by geological catastrophes. The point on which this – apparently reasonable – assertion was fully stretched was in the case of the extinction of very large mammals, such as the animals whose bones he had found in the Pampas: the Megatherium, Toxodon, Glyptodon and others of huge proportions. Most naturalists would have assumed that animals of this size must have needed luxuriant forests in which to feed, and that some catastrophe that had wiped out the forests was the cause of the extinction of the giant mammals. Darwin tried to show that a meagre diet was always sufficient to feed them, arguing, by analogy with other large mammals, such as the rhinoceros or elephant in South Africa, seen by Darwin in 1836, that they could easily have existed on sparse, scrubby terrain. Equally the carcasses of giants like the Mastodon and Mammoth found preserved in the frozen subsoil of the Arctic Circle had not been rendered extinct by lack of food brought on by sudden cold. All previous suggestions on these subsistence lines, he felt, went against the Lyellian grain, for they implied that there were sudden and devastating catastrophes. Instead he argued that we can hardly begin to guess at the complex concatenation of causes that might lead to a gradual decline in numbers. Perhaps there was some inbuilt ageing device, which meant that a species came to the end of its 'lifespan': 'as with the individual, so with the species, the hour of life has run its course, and is spent.' Darwin later changed his mind about a possible inbuilt species senescence in favour of his better-known ideas about population pressure, competition and natural selection, and sections in this edition that deal with his views on the topic were consequently deleted from the 1845 version.

The other topic that should finally be mentioned is Darwin's discussion of the Fuegians – both of the trio on board, who fascinated and charmed Darwin, and of the relationship of these with the missionary and their own tribespeople after their return. FitzRoy and Darwin had a high opinion of Jemmy Button, in particular, and Darwin recounts his naïve amazement that after many years in English company Jemmy was now almost another 'species of man' compared to those who were his literal relatives. Both Englishmen were inexpressibly saddened by the eventual fate of the mission they had so hopefully set up in Tierra del Fuego, and dismayed at Jemmy's rapid reversion to the manners of his tribe. Of all of Darwin's varied experiences, this example of the temporary nature of civilization and the contrast of human habits and lifestyles moved him the most. 'I could not have believed,' he wrote, 'how wide was the difference, between savage and civilized man. It is greater than between a wild and domesticated animal, in as much as in man there is a greater power of improvement.'

FitzRoy took it upon himself to give the longer and fuller account of the Fuegians' experiences and the fate of the mission in his own volume about the *Beagle*'s surveying work. Being the captain, and for a long time closely involved with the three who had travelled to England, this seemed only proper to Darwin's eyes. But it is clear from Darwin's text that he and FitzRoy had discussed and thought deeply about these and similar untutored peoples, for they composed a joint letter to be sent to a South African newspaper about the successful operations of missionaries in Tahiti and the famous settlement of Waimate in New Zealand. Their letter was published in the *South African Christian Recorder* in 1836, and was both Darwin's and FitzRoy's first published paper. Many of Darwin's opinions – which reflect the emancipationist, liberal views of his father and grandfather before him – were endorsed for rather different reasons by FitzRoy. FitzRoy – like his father and grandfather before him – believed in the high-Tory, Christian ethic in which paternalism was the motive for providing direct economic and practical aid to primitive societies. United by their mutual respect and (perhaps disingenuous) concern for these aboriginal groups, Darwin and FitzRoy met on the common ground of British philanthropy at a time when optimism and high-minded ideals had not yet been submerged by the exploitations of late-Victorian times.

The accord that Darwin and FitzRoy plainly shared on this issue was more than simple accommodation between two men obliged to co-exist for a long time in one tiny cabin. FitzRoy was difficult, imperious and authoritarian: but he was also intelligent, fond of outdoor pursuits and natural philosophy, a good talker, thinker and companion. During the *Beagle* voyage the two men were friends, thrown upon each other's company to be sure, but nevertheless friends. Though there is, perhaps, little written material to support it, it seems very probable that Darwin's views were shaped as much by his close relationship with FitzRoy as they were by his enthusiasm for Lyell or his own private love-affair with nature. As time went on, Darwin tended to remember only the worst aspects of his captain's temper, but the remaining correspondence from the *Beagle* period indicates a frank, cheerful trust in each other animated by a marked community of tastes and boyish camaraderie. The long discussions, dangerous journeys, dust, dirt and shared enthusiasms of these two young men, perhaps even the voyage itself, could well be summarized in FitzRoy's affectionate astonishment at the news reported by a mutual friend, just after the *Beagle* had docked, that Darwin had actually been seen in a 'good hat!'

A NOTE ON THIS EDITION

THE TEXT printed here is a shortened version of the first edition of Charles Darwin's first published book. Though it is commonly known under the title *Journal of Researches* or sometimes as the *Voyage of the* Beagle, it was actually first issued in 1839 as the third volume of Robert FitzRoy's account of the *Beagle* voyage (*Narrative of the surveying voyages of H.M.S.* Adventure *and* Beagle) with the title *Journal and Remarks, 1832–1836*. A few weeks later, in August 1839, the same text was reissued with a different title-page, this time being called *Journal of researches into the geology and natural history of the various countries visited by H.M.S.* Beagle. Thereafter, it was released in many different editions and under several different titles, the most well known probably being the second edition of 1845. The editors thought that rather than offer these amended, later versions, it was important to give the text of the very first edition in order to get as close as we possibly can to the thoughts and adventures that Darwin experienced during his voyage. To bring Darwin's volume into a Penguin format it has, however, been necessary to shorten the text by about one-third. The policy has been to interfere as little as possible and to maintain the original sense and feel of his account by removing only those passages that the editors thought would be unlikely to be missed, and carefully balancing the range of examples and observations that Darwin recorded. The preference has been for removing entire paragraphs or sections rather than making piecemeal alterations. The editors have consequently been able to retain long, uninterrupted passages, as, for example, in the chapter on the Galapagos archipelago, in which no editorial cuts have been made at all. The passages that have been removed in other chapters are indicated by asterisks; Darwin's typically full table of contents is also given, in case readers wish to pursue certain sections further.

The footnotes are Darwin's own. As the bibliographical references contained in these notes are often very abbreviated or otherwise confusing, the editors have added a brief guide to the people and books mentioned in the text. There are also two items of additional interest: the first is the original British Admiralty instructions issued for the *Beagle* voyage; the other is an essay on the relationship between scientific geology and the Bible, written by Robert Fitzroy, the captain of the *Beagle*, which helps us to understand the interest in science and natural history that Darwin and Fitzroy shared and provides a counterpoint to the new geological views that Darwin put forward in his *Journal*.

JOURNAL OF RESEARCHES

INTO THE

GEOLOGY

AND

NATURAL HISTORY

OF THE

VARIOUS COUNTRIES
VISITED BY H. M. S. BEAGLE,
UNDER THE COMMAND OF CAPTAIN FITZROY, R.N.
FROM 1832 TO 1836.

BY

CHARLES DARWIN, Esq., M.A. F.R.S.

SECRETARY TO THE GEOLOGICAL SOCIETY.

LONDON:
HENRY COLBURN, GREAT MARLBOROUGH STREET.
1839.

AZORES

MADEIRA

CANARIES

CAPE DE VERD
ISLANDS

ST PAUL'S
ROCKS

FERNANDO
NORONHA

ASCENSION

ST HELENA

HAWAIIAN
ISLANDS

GALAPAGOS
ISLANDS

MARQUESAS

SOCIETY ISLANDS

TAHITI

Callao

Bahia

Rio de Janeiro

JUAN
FERNANDEZ

Valparaiso

Monte Video

Buenos Ayres

TRISTAN
DA CUNHA

CHONOS
ARCHIPELAGO

FALKLAND
ISLANDS

SOUTH
GEORGIA

CAPE
HORN

THE PRINCIPAL LOCATIONS MENTIONED
IN DARWIN'S TEXT

Though principally concerned with surveying the coastal waters of
South America for the British Admiralty, the *Beagle* also sailed round
the world, visiting Tahiti, New Zealand, Australia, Keeling Islands
and the Cape of Good Hope. The map shows the principal locations
discussed by Darwin in his text. The names are those in use during the
nineteenth century.

MALDIVES

SEYCHELLES

NEW
GUINEA

SOLOMON
ISLANDS

MAURITIUS

KEELING (COCOS)
ISLANDS

NEW HEBRIDES

BOURBON (RÉUNION)

NEW
CALEDONIA

MADAGASCAR

NORFOLK
ISLAND

PE OF
OD HOPE

King
George's Sound

Sydney

Waimate

TASMANIA   Hobart

CROZET
ISLANDS

KERGUELEN LAND

AUCKLAND
ISLAND

CAMPBELL
ISLAND

SOUTH AMERICA IN 1830

Darwin described his visits and expeditions into the interior of South America in great detail in his *Journal of Researches*. The principal locations are marked on the map with the names of Darwin. The political boundaries are those existing in 1830: many of the states altered their boundaries during the years that the *Beagle* travelled round the coast.

PREFACE

I HAVE stated in the preface to the Zoology of the Voyage of the *Beagle*, that it was in consequence of a wish expressed by Captain FitzRoy, of having some scientific person on board, accompanied by an offer from him, of giving up part of his own accommodations, that I volunteered my services, which received, through the kindness of the hydrographer, Captain Beaufort, the sanction of the Lords of the Admiralty. As I feel that the opportunities, which I enjoyed of studying the Natural History of the different countries we visited, have been wholly due to Captain FitzRoy, I hope I may here be permitted to express my gratitude to him; and to add that, during the five years we were together, I received from him the most cordial friendship and steady assistance. Both to Captain FitzRoy and to all the Officers of the *Beagle*,* I shall ever feel most thankful for the undeviating kindness with which I was treated, during our long voyage.

The present volume contains in the form of a journal, a sketch of those observations in Geology and Natural History, which I thought would possess some general interest. As it was originally intended to have preceded any more detailed account, and as its publication has been unavoidably delayed, the briefness and imperfection of several parts, I hope, will be excused. I have given a list of those errata (partly caused by my absence from town when some of the sheets were in the press) which affect the sense; and have added an Appendix, containing some additional facts (especially on the theory of the transportation of erratic blocks) which I have accidentally met with during the past year. I hope

---

* I must likewise take this opportunity of returning my sincere thanks to Mr Bynoe, the surgeon of the *Beagle*, for his very kind attention to me when I was ill at Valparaiso.

shortly to publish my geological observations; the first Part of which will be on the Volcanic Islands of the Atlantic and Pacific Oceans, and on Coral Formations; and the second Part will treat of South America. Several numbers of the Zoology of the Voyage of the *Beagle*, due to the disinterested zeal of several of our first naturalists, have already appeared. These works could not have been undertaken, had it not been for the liberality of the Lords Commissioners of Her Majesty's Treasury, who, through the representation of the Right Honourable the Chancellor of the Exchequer, have been pleased to grant a sum of £1,000 towards defraying part of the expenses of publication. I have repeated in this volume my account of the habits of some of the birds and quadrupeds of South America, as I thought such observations might interest those readers who would not, probably, consult the larger work. But I trust that naturalists will remember, that mere sketches are here given on several subjects, which will hereafter be more fully entered on, or have already been so:— for instance, the notices of the strange fossil quadrupeds of the eastern plains of South America are exceedingly imperfect, whilst an admirable account of them by Mr Owen now forms the first part of the Zoology of the Voyage of the *Beagle*.

I shall have the pleasure of acknowledging the great assistance I have received from several naturalists, in the course of this and the succeeding works; but I must be here allowed to return my most sincere thanks to the Reverend Professor Henslow, who, when I was an under-graduate at Cambridge, was one chief means of giving me a taste for Natural History, – who, during my absence, took charge of the collections I sent home, and by his correspondence directed my endeavours – and who, since my return, has constantly rendered me every assistance which the kindest friend could offer.

C.D.

CONTENTS

CHAPTER XIX

CHAPTER XX

CHAPTER XXI

CHAPTER XXII

CHAPTER XXIII

JOURNAL OF

CHARLES DARWIN, MA,

NATURALIST TO THE *BEAGLE*

CHAPTER I

Porto Praya – Ribeira Grande – Dry and clear atmosphere –
Effect of lava on calcareous beach – Habits of Aplysia and
Octopus – St Paul's rock non-volcanic – Incrustations and
stalactites of phosphate of lime – Insects first colonists –
Fernando Noronha – Bahia – Extent of granite – Burnished
rocks – Habits of Diodon – Pelagic confervæ, infusoria –
Causes of discoloured sea

ST JAGO — CAPE DE VERD ISLANDS

JAN. 16TH, 1832 – The neighbourhood of Porto Praya, viewed
from the sea, wears a desolate aspect. The volcanic fire of past
ages, and the scorching heat of a tropical sun, have in most places
rendered the soil sterile and unfit for vegetation. The country rises
in successive steps of table land, interspersed with some truncate
conical hills, and the horizon is bounded by an irregular chain of
more lofty mountains. The scene, as beheld through the hazy
atmosphere of this climate, is one of great interest; if, indeed, a
person, fresh from the sea, and who has just walked, for the first
time, in a grove of cocoa-nut trees, can be a judge of any thing but
his own happiness. The island would generally be considered as
very uninteresting; but to any one accustomed only to an English
landscape, the novel prospect of an utterly sterile land possesses a
grandeur which more vegetation might spoil. A single green leaf

can scarcely be discovered over wide tracts of the lava plains; yet flocks of goats, together with a few cows, contrive to exist. It rains very seldom, but during a short portion of the year heavy torrents fall, and immediately afterwards a light vegetation springs out of every crevice. This soon withers; and upon such naturally formed hay the animals live. At the present time it has not rained for an entire year. The broad, flat-bottomed valleys, many of which serve during a few days only in the season as a water-course, are clothed with thickets of leafless bushes. Few living creatures inhabit these valleys. The commonest bird is a kingfisher (*Dacelo jagoensis*), which tamely sits on the branches of the castor-oil plant, and thence darts on the grasshoppers and lizards. It is brightly coloured, but not so beautiful as the European species: in its flight, manners, and place of habitation, which is generally in the driest valleys, there is also a wide difference.

One day, two of the officers and myself rode to Ribeira Grande, a village a few miles to the eastward of Porto Praya. Until we reached the valley of St Martin, the country presented its usual dull brown appearance; but there, a very small rill of water produces a most refreshing margin of luxuriant vegetation. In the course of an hour we arrived at Ribeira Grande, and were surprised at the sight of a large ruined fort and cathedral. The little town, before its harbour was filled up, was the principal place in the island: it now presents a melancholy, but very picturesque appearance. Having procured a black Padre for a guide, and a Spaniard, who had served in the Peninsular war, as an interpreter, we visited a collection of buildings, of which an ancient church formed the principal part. It is here the governors and captain-generals of the islands have been buried. Some of the tombstones recorded dates of the sixteenth century.* The heraldic ornaments were the only things in this retired place that reminded us of Europe. The church or chapel formed one side of a quadrangle, in the middle of which a large clump of bananas were growing. On another side was a hospital, containing about a dozen miserable-looking inmates.

We returned to the 'Vênda'† to eat our dinners. A considerable

* The Cape de Verd Islands were discovered in 1449.
† Vênda, the Portuguese name for an inn.

number of men, women, and children, all as black as jet, were
collected to watch us. Our companions were extremely merry;
and every thing we said or did was followed by their hearty
laughter. Before leaving the town we visited the cathedral. It does
not appear so rich as the smaller church, but boasts of a little
organ, which sent forth most singularly inharmonious cries. We
presented the black priest with a few shillings, and the Spaniard,
patting him on the head, said, with much candour, he thought his
colour made no great difference. We then returned, as fast as the
ponies would go, to Porto Praya.

Another day we rode to the village of St Domingo, situated
near the centre of the island. On a small plain which we crossed, a
few stunted acacias were growing; their tops, by the action of the
steady trade wind, were bent in a singular manner – some of them
even at a right angle to the trunk. The direction of the branches
was exactly NE by N, and SW by S. These natural vanes must
indicate the prevailing direction of the force of the trade wind.
The travelling had made so little impression on the barren soil,
that we here missed our track, and took that to Fuentes. This we
did not find out till we arrived there; and we were afterwards very
glad of our mistake. Fuentes is a pretty village, with a small
stream; and every thing appeared to prosper well, excepting,
indeed, that which ought to do so most – its inhabitants. The
black children, completely naked, and looking very wretched,
were carrying bundles of firewood half as big as their own bodies.

Near Fuentes we saw a large flock of guinea-fowl – probably
fifty or sixty in number. They were extremely wary, and could
not be approached. They avoided us, like partridges on a rainy
day in September, running with their heads cocked up; and if
pursued, they readily took to the wing.

The scenery of St Domingo possesses a beauty totally unex-
pected, from the prevalent gloomy character of the rest of the
island. The village is situated at the bottom of a valley, bounded
by lofty and jagged walls of stratified lava. The black rocks afford
a most striking contrast with the bright green vegetation, which
follows the banks of a little stream of clear water. It happened to
be a grand feast-day, and the village was full of people. On our
return we overtook a party of about twenty young black girls,

dressed in most excellent taste; their black skins and snow-white linen being set off by their coloured turbans and large shawls. As soon as we approached near, they suddenly all turned round, and covering the path with their shawls, sung with great energy a wild song, beating time with their hands upon their legs. We threw them some *vintéms*, which were received with screams of laughter, and we left them redoubling the noise of their song.

It has already been remarked, that the atmosphere is generally very hazy; this appears chiefly due to an impalpable dust, which is constantly falling, even on vessels far out at sea. The dust is of a brown colour, and under the blowpipe easily fuses into a black enamel. It is produced, as I believe, from the wear and tear of volcanic rocks, and must come from the coast of Africa. One morning the view was singularly clear; the distant mountains being projected with the sharpest outline, on a heavy bank of dark blue clouds. Judging from the appearance, and from similar cases in England, I supposed that the air was saturated with moisture. The fact, however, turned out quite the contrary. The hygrometer gave a difference of 29.6 degrees, between the temperature of the air, and the point at which dew was precipitated. This difference was nearly double that which I had observed on the previous mornings. This unusual degree of atmospheric dryness was accompanied by continual flashes of lightning. Is it not an uncommon case, thus to find a remarkable degree of aerial transparency with such a state of weather?

The geology of this island is the most interesting part of its natural history. On entering the harbour, a perfectly horizontal white band, in the face of the sea cliff, may be seen running for some miles along the coast, and at the height of about 45 feet above the water. Upon examination, this white stratum is found to consist of calcareous matter, with numerous shells embedded, such as now exist on the neighbouring coast. It rests on ancient volcanic rocks, and has been covered by a stream of basalt, which must have entered the sea, when the white shelly bed was lying at the bottom. It is interesting to trace the changes, produced by the heat of the overlying lava, on the friable mass. For a thickness of several inches it is converted, in some parts, into a firm stone, as hard as the best freestone; and the earthy matter, originally mingled with the calcareous, has been separated into little spots, thus leaving the limestone white and pure. In other parts a highly crystalline marble has been formed, and so perfect are the crystals of carbonate of lime, that they can easily be measured by the reflecting goniometer. The change is even more extraordinary, where the lime has been caught up by the scoriaceous fragments of the lower surface of the stream; for it is there converted into groups of beautifully radiated fibres resembling arragonite. The beds of lava rise in successive gently sloping plains, towards the interior, whence the deluges of melted stone originally proceeded. Within historical times, no signs of volcanic activity have, I believe, been manifested in any part of St Jago. This state of quiescence is, probably, owing to the neighbouring island of Fogo being frequently in eruption. Even the form of a crater can but rarely be discovered on the summits of any of the red cindery hills; yet the more recent streams can be distinguished on the coast, forming a line of cliffs of less height, but stretching out in advance of those belonging to an older series: the height of the cliff thus affording a rude measure of the age.

During our stay, I observed the habits of some marine animals. A large Aplysia is very common. This sea-slug is about 5 inches long; and is of a dirty yellowish colour, veined with purple. At the anterior extremity, it has two pairs of feelers; the upper ones of which resemble in shape the ears of a quadruped. On each side of the lower surface, or foot, there is a broad membrane, which appears sometimes to act as a ventilator, in causing a current of water to flow over the dorsal branchiæ. It feeds on delicate seaweeds, which grow among the stones in muddy and shallow water; and I found in its stomach several small pebbles, as in the gizzards of birds. This slug, when disturbed, emits a very fine purplish-red fluid, which stains the water for the space of a foot around. Besides this means of defence, an acrid secretion, which is spread over its body, causes a sharp, stinging sensation, similar to that produced by the Physalia, or Portuguese man-of-war.

I was much interested, on several occasions, by watching the habits of an Octopus or cuttle-fish. Although common in the pools of water left by the retiring tide, these animals were not easily caught. By means of their long arms and suckers, they could drag their bodies into very narrow crevices; and when thus fixed, it required great force to remove them. At other times they darted tail first, with the rapidity of an arrow, from one side of the pool to the other, at the same instant discolouring the water with a dark chestnut-brown ink. These animals also escape detection by a very extraordinary, chameleon-like, power of changing their colour. They appear to vary the tints, according to the nature of the ground over which they pass: when in deep water, their general shade was brownish purple, but when placed on the land, or in shallow water, this dark tint changed into one of a yellowish green. The colour, examined more carefully, was a French gray, with numerous minute spots of bright yellow: the former of these varied in intensity; the latter entirely disappeared and appeared again by turns. These changes were effected in such a manner, that clouds, varying in tint between a hyacinth red and a chestnut brown,* were continually passing over the body. Any part being subjected to a slight shock of galvanism, became almost black: a similar effect, but in a less degree, was produced by scratching the skin with a needle. These clouds, or blushes, as they may be called, when examined under a glass, are described as being produced by the alternate expansions and contractions of minute vesicles, containing variously coloured fluids.†

This cuttle-fish displayed its chameleon-like power both during the act of swimming and whilst remaining stationary at the bottom. I was much amused by the various arts to escape detection used by one individual, which seemed fully aware that I was watching it. Remaining for a time motionless, it would then stealthily advance an inch or two, like a cat after a mouse; sometimes changing its colour: it thus proceeded, till having gained a deeper part, it darted away, leaving a dusky train of ink to hide the hole into which it had crawled.

While looking for marine animals, with my head about 2 feet above the rocky shore, I was more than once saluted by a jet of water, accompanied by a slight grating noise. At first I did not know what it was, but afterwards I found out that it was the

---

* So named according to Pat. Symes's nomenclature.
† See Encyclo. of Anat. and Physiol., article *Cephalopoda*.

cuttle-fish, which, though concealed in a hole, thus often led me to its discovery. That it possesses the power of ejecting water there is no doubt, and it appeared to me certain that it could, moreover, take good aim by directing the tube or siphon on the under side of its body. From the difficulty which these animals have in carrying their heads, they cannot crawl with ease when placed on the ground. I observed that one which I kept in the cabin was slightly phosphorescent in the dark.

ST PAUL'S ROCKS—In crossing the Atlantic we hove to, during the morning of February 16th, close to the island of St Paul. This cluster of rocks is situated in 0° 58' north latitude, and 29° 15' west longitude. It is 540 miles distant from the coast of America, and 350 from the island of Fernando Noronha. The highest point is only 50 feet above the level of the sea, and the entire circumference is under three-quarters of a mile. This small point rises abruptly out of the depths of the ocean. Its mineralogical constitution is not simple; in some parts, the rock is of a cherty, in others, of a felspathic nature; and in the latter case it contains thin veins of serpentine, mingled with calcareous matter.

The circumstance of these rocks not being of volcanic origin is of interest, because, with very few exceptions, the islands situated in the midst of the great oceans are thus constituted. As the highest pinnacles of the great mountain ranges probably once existed as islands distant from any continent, we are led to expect that they would frequently consist of volcanic rocks. It becomes, therefore, a curious point to speculate on the changes which many of the present islands would undergo, during the lapse of the countless ages, which would be required to elevate them into snow-clad summits. If we take the case of Ascension, or St Helena, both of which have long existed in an extinct condition, we may feel assured, before so vast a period could elapse, during the whole of which the surface would be exposed to constant wear and tear, that the mere nucleus or core of the island would remain; perhaps, every fragment of cellular rock having been decomposed, a mass of some compact stone, as phonolite or greenstone, would crown our new Chimborazo.

The rocks of St Paul appear from a distance of a brilliantly white colour. This is partly owing to the dung of a vast multitude of seafowl, and partly to a coating of a glossy white substance, which is intimately united to the surface of the rocks. This, when examined with a lens, is found to consist of numerous exceedingly thin layers, its total thickness being about the tenth of an inch. The surface is smooth and glossy, and has a pearly lustre; it is considerably harder than calcareous spar, although it can be scratched by a knife: under the blowpipe it decrepitates, slightly blackens, and emits a fetid odour. It consists of phosphate of lime, mingled with some impurities; and its origin without doubt is due to the action of the rain or spray on the bird's dung. I may here mention, that I found in some hollows in the lava rocks of Ascension considerable masses of the substance called *guano*, which on the west coast of the intertropical parts of South America occurs in great beds, some yards thick, on the islets frequented by seafowl. According to the analysis of Fourcroy and Vauquelin, it consists of the urates, phosphates, and oxalates of lime, ammonia, and potash, together with some other salts, and some fatty and earthy matter. I believe there is no doubt of its being the richest manure which has ever been discovered. At Ascension, close to the *guano*, stalactitic or botryoidal masses of impure phosphate of lime adhered to the basalt. The basal part of these had an earthy texture, but the extremities were smooth and glossy, and sufficiently hard to scratch common glass. These stalactites appeared to have shrunk, perhaps from the removal of some soluble matter, in the act of consolidation; and hence they had an irregular form. Similar stalactitic masses,* though I am not aware that they have ever been noticed, are, I believe, by no means of uncommon occurrence.

We only observed two kinds of birds – the booby and the noddy. The former is a species of gannet, and the latter a tern. Both are of a tame and stupid disposition, and are so unaccustomed to visitors, that I could have killed any number of them

---

* I may mention that I was shown, at Ascension, some very fine stalactites, composed of sulphate of lime, which had been taken out of a cavern. From their external appearance they would generally be mistaken for the ordinary calcareous kind. It was interesting to observe, in a fractured specimen, the double cleavage intersecting with its even planes, the irregular layers of successive deposition.

with my geological hammer. The booby lays her eggs on the bare rock; but the tern makes a very simple nest with seaweed. By the side of many of these nests a small flying-fish was placed; which, I suppose, had been brought by the male bird for its partner. It was amusing to watch how quickly a large and active crab (*Graspus*), which inhabits the crevices of the rock, stole the fish from the side of the nest, as soon as we had disturbed the birds. Not a single plant, not even a lichen, grows on this island; yet it is inhabited by several insects and spiders. The following list completes, I believe, the terrestrial fauna: a species of Feronia and an acarus, which must have come here as parasites on the birds; a small brown moth, belonging to a genus that feeds on feathers; a staphylinus (*Quedius*) and a woodlouse from beneath the dung; and lastly, numerous spiders, which I suppose prey on these small attendants on, and scavengers of the waterfowl. The often-repeated description of the first colonists of the coral islets in the South Sea, is not, probably, quite correct: I fear it destroys the poetry of the story to find, that these little vile insects should thus take possession before the cocoa-nut tree and other noble plants have appeared.

The smallest rock in the tropical seas, by giving a foundation, for the growth of innumerable kinds of sea-weed and compound animals, supports likewise a large number of fish. The sharks and the seamen in the boats maintained a constant struggle, who should secure the greater share of the prey caught by the lines. I have heard, that a rock near the Bermudas, lying many miles out at sea, and covered by a considerable depth of water, was first discovered by the circumstance of fish having been observed in the neighbourhood.

FERNANDO NORONHA, FEB. 20TH – As far as I was enabled to observe, during the few hours we stayed at this place, the constitution of the island is volcanic, but probably not of a recent date. The most remarkable feature is a conical hill, about 1,000 feet high, the upper part of which is exceedingly steep, and on one side overhangs its base. The rock is phonolite, and is divided into irregular columns. From the first impression, on viewing one of these isolated masses, one is inclined to believe, that the whole has been suddenly pushed up in a semi-fluid state. At St Helena, however, I ascertained that some pinnacles, of a nearly similar figure and constitution, had been formed by the injection of melted rock among the yielding strata; which thus formed the model for these gigantic obelisks. The whole island is covered with wood; but from the dryness of the climate there is no appearance of luxuriance. At some elevation great masses of the columnar rock, shaded by laurels, and ornamented by a tree covered by fine pink flowers like those of a foxglove, but without a single leaf, gave a pleasing effect to the nearer parts of the scenery.

BAHIA, OR SAN SALVADOR. BRAZIL, FEB. 29TH – The day has passed delightfully. Delight itself, however, is a weak term to express the feelings of a naturalist who, for the first time, has been wandering by himself in a Brazilian forest. Among the multitude of striking objects, the general luxuriance of the vegetation bears away the victory. The elegance of the grasses, the novelty of the parasitical plants, the beauty of the flowers, the glossy green of the foliage, all tend to this end. A most paradoxical mixture of sound and silence pervades the shady parts of the wood. The noise from the insects is so loud, that it may be heard even in a vessel anchored several hundred yards from the shore; yet within the recesses of the forest a universal silence appears to reign. To a person fond of natural history, such a day as this, brings with it a deeper pleasure than he ever can hope again to experience. After wandering about for some hours, I returned to the landing-place; but, before reaching it, I was overtaken by a tropical storm. I tried to find shelter under a tree which was so thick, that it would never have been penetrated by common English rain; but here, in a couple of minutes, a little torrent flowed down the trunk. It is to this violence of the rain we must attribute the verdure at the bottom of the thickest woods: if the showers were like those of a colder clime, the greater part would be absorbed or evaporated before it reached the ground. I will not at present attempt to describe the gaudy scenery of this noble bay, because, in our homeward voyage, we called here a second time, and I shall then have occasion to remark on it.

The geology of the surrounding country possesses little interest. Throughout the coast of Brazil, and certainly for a considerable space inland, from the Rio Plata to Cape St Roque, lat. 5° S, a distance of more than 2,000 geographical miles, wherever solid rock occurs, it belongs to a granitic formation. The circumstance of this enormous area being thus constituted of materials, which almost every geologist believes to have been crystallized by the action of heat under pressure, gives rise to many curious reflections. Was this effect produced beneath the depths of a profound ocean? or did a covering of strata formerly extend over it, which has since been removed? Can we believe that any power, acting for a time short of infinity, could have denuded the granite over so many thousand square leagues?

On a point not far from the city, where a rivulet entered the sea, I observed a fact connected with a subject discussed by Humboldt.* At the cataracts of the great rivers Orinoco, Nile, and Congo, the syenitic rocks are coated by a black substance, appearing as if they had been polished with plumbago. The layer is of extreme thinness; and on analysis by Berzelius it was found to consist of the oxides of manganese and iron. In the Orinoco it occurs on the rocks periodically washed by the floods, and in those parts alone, where the stream is rapid; or, as the Indians say, 'the rocks are black, where the waters are white'. The coating is here of a rich brown instead of a black colour, and seems to be composed of ferrugineous matter alone. Hand specimens fail to give a just idea of these brown, burnished, stones which glitter in the sun's rays. They occur only within the limits of tidal action; and as the rivulet slowly trickles down, the surf must supply the polishing power of the cataracts in the great rivers. In the same manner, the rise and fall of the tide probably answers to the periodical inundations; and thus the same causes are present under apparently very different circumstances. The real origin, however, of these coatings of metallic oxides, which seem as if cemented to the rocks, is not understood; and no reason, I believe, can be assigned for their thickness remaining constant.

One day I was amused by watching the habits of a Diodon, which was caught swimming near the shore. This fish is well known to possess the singular power of distending itself into a

* Pers. Narr., vol. v, pt. i, p. 18.

nearly spherical form. After having been taken out of water for a short time, and then again immersed in it, a considerable quantity both of water and air was absorbed by the mouth, and perhaps likewise by the branchial apertures. This process is effected by two methods; the air is swallowed, and is then forced into the cavity of the body, its return being prevented by a muscular contraction which is externally visible; but the water, I observed, entered in a stream through the mouth, which was wide open and motionless: this latter action must, therefore, depend on suction. The skin about the abdomen is much looser than that of the back; hence, during the inflation, the lower surface becomes far more distended than the upper; and the fish, in consequence, floats with its back downwards. Cuvier doubts whether the Diodon, in this position, is able to swim; but not only can it thus move forward in a straight line, but likewise it can turn round to either side. This latter movement is effected solely by the aid of the pectoral fins; the tail being collapsed, and not used. From the body being buoyed up with so much air, the branchial openings were out of water; but a stream drawn in by the mouth, constantly flowed through them.

The fish, having remained in this distended state for a short time, generally expelled the air and water with considerable force from the branchial apertures and mouth. It could emit, at will, a certain portion of the water; and it appears, therefore, probable, that this fluid is taken in partly for the sake of regulating its specific gravity. This Diodon possessed several means of defence. It could give a severe bite, and could eject water from its mouth to some distance, at the same time it made a curious noise by the movement of its jaws. By the inflation of its body, the papillæ, with which the skin is covered, became erect and pointed. But the most curious circumstance was, that it emitted from the skin of its belly, when handled, a most beautiful carmine red and fibrous secretion, which stained ivory and paper in so permanent a manner, that the tint is retained with all its brightness to the present day. I am quite ignorant of the nature and use of this secretion.

MARCH 18TH – We sailed from Bahia. A few days afterwards, when not far distant from the Abrolhos islets, my attention was called to a discoloured appearance in the sea. The whole surface of the water, as it appeared under a weak lens, seemed as if covered by chopped bits of hay, with their ends jagged. One of the larger particles measured .03 of an inch in length, and .009 in breadth. Examined more carefully, each is seen to consist of from twenty to sixty cylindrical filaments, which have perfectly rounded extremities, and are divided at regular intervals by transverse septa, containing a brownish-green flocculent matter. The filaments must be enveloped in some viscid fluid, for the bundles adhered together without actual contact. I do not know to what family these bodies properly belong, but they have a close general resemblance in structure with the confervæ which grow in every ditch. These simple vegetables, thus constituted for floating in the open ocean, must in certain places exist in countless numbers. The ship passed through several bands of them, one of which was about 10 yards wide, and, judging from the mud-like colour of the water, at least two and a half miles long. In almost every long voyage some account is given of these confervæ. They appear especially common in the sea near Australia. Off Cape Leeuwin, I found some very similar to those above described; they differed chiefly in the bundles being rather smaller, and being composed of fewer filaments. Captain Cook, in his third voyage, remarks, that the sailors gave to this appearance the name of sea-sawdust.

I may here mention that during two days preceding our arrival at the Keeling Islands in the Indian Ocean, I saw in many parts masses of flocculent matter, of a brownish-green colour, floating in the sea. They varied in size, from half to 3 or 4 inches square; and were quite irregular in figure. In an opaque vessel they could barely be distinguished, but in a glass one they were clearly visible. Under the microscope the flocculent matter was seen to consist of two kinds of confervæ, between which I am quite ignorant whether there exists any connexion. Minute cylindrical bodies, conical at each extremity, are involved in vast numbers, in a mass of fine threads. These threads have a diameter of about 2/3000 of an inch; they possess an internal lining, and are divided at irregular and very wide intervals by transverse septa. Their length is so great, that I could never with certainty ascertain the form of the uninjured extremity; they are all curvilinear, and resemble in mass a handful of hair, coiled up and squeezed together. In the midst of these threads, and probably connected by some viscid fluid, the other kind, or the cylindrical transparent bodies, float in great numbers. These have their two extremities terminated by cones, produced into the finest points: their diameter is tolerably constant between .006 and .008 of an inch; but their length varies considerably from .04 to .06, and even sometimes to .08. Near one extremity of the cylindrical part, a green septum, formed of granular matter, and thickest in the middle, may generally be seen. This, I believe, is the bottom of a most delicate, colourless sack, composed of a pulpy substance, which lines the exterior case, but does not extend within the extreme conical points. In some, small but perfect spheres of brownish granular matter supplied the place of the septa; and I observed the curious process by which they were produced. The pulpy matter of the internal coating suddenly grouped itself into lines, some of which assumed a form radiating from a common centre; it then continued, with an irregular and rapid movement, to contract itself, so that, in the course of a second, the whole was united into a perfect little sphere, which occupied the position of the septum at one end of the now quite hollow case. The appearance was as if an elastic membrane, for instance a thin Indian-rubber ball, had been distended with air, and then burst, in which case the edges would instantly shrink up and contract towards a point. The formation of the granular sphere was hastened by any accidental

injury. I may add, that frequently a pair of these bodies were attached to each other, as represented in the accompanying rude drawing, cone beside cone, at that end where the septum occurs. When floating uninjured in the sea, the formation of the spherical gemmules perhaps only takes place, when two of the plants (or rather animals, according to Bory St Vincent) thus become attached, and married to each other. Nevertheless, I certainly witnessed this curious process in several individuals, when separate, and where there was no apparent cause of disturbance. In any case it does not seem probable, from the fixed structure of the septum, that the whole of the granular matter is transferred from one to the other body, as with the true Conjugatæ.

I will here add a few other observations connected with the discoloration of the sea from organic causes. On the coast of Chile, a few leagues north of Concepción, the *Beagle* one day passed through great bands of muddy water; and again, a degree south of Valparaiso, the same appearance was still more extensive. Although we were nearly 50 miles from the coast, I at first attributed this circumstance to real streams of muddy water brought down by the river Maypo. Mr Sulivan, however, having drawn up some in a glass, thought he distinguished, by the aid of a lens, moving points. The water was slightly stained as if by red dust; and after leaving it for some time quiet, a cloud collected at the bottom. With a lens, of one-fourth of an inch focal distance, small hyaline points could be seen darting about with great rapidity, and frequently exploding. Examined with a much higher power, their shape was found to be oval, and contracted by a ring round the middle, from which line curved little setæ proceeded on all sides; and these were the organs of motion. One end of the body was narrower and more pointed than the other. According to the arrangement of Bory St Vincent, they are animalcula, belonging to the family of Trichodes: it was, however, very difficult to examine them with care, for almost the instant motion ceased, even while crossing the field of vision, their bodies burst. Sometimes both ends burst at once, sometimes only one, and a quantity of coarse brownish granular matter was ejected, which cohered very slightly. The ring with the setæ sometimes retained its irritability for a little while after the contents of the body had been emptied, and continued a riggling, uneven motion. The animal an instant before bursting expanded to half again its natural size; and the explosion took place about fifteen seconds after the rapid progressive motion had ceased: in a few cases it was preceded for a short interval by a rotatory movement on the longer axis. About two minutes after any number were isolated in a drop of water, they thus perished. The animals move with the narrow apex forwards, by the aid of their vibratory ciliæ, and generally by rapid starts. They are exceedingly minute, and quite invisible to the naked eye, only covering a space equal to the square of the thousandth of an inch. Their numbers were infinite; for the smallest drop of water which I could remove contained very many. In one day we passed through two spaces of water thus stained, one of which alone must have extended over several square miles. What incalculable numbers of these microscopical animals! The colour of the water, as seen at some distance, was like that of a river which has flowed through a red clay district; but under the shade of the vessel's side, it was quite as dark as chocolate. The line where the red and blue water joined was distinctly defined. The weather for some days previously had been calm, and the ocean abounded, to an unusual degree, with living creatures. In Ulloa's *Voyage* an account is given of crossing, in nearly the same latitude, some discoloured water, which was mistaken for a shoal: no soundings were obtained, and I have no doubt, from the description, that this little animalcule was the cause of the alarm.*

In the sea around Tierra del Fuego, and at no great distance from the land, I have seen narrow lines of water of a bright red colour, from the number of crustacea, which somewhat resemble in form large prawns. The sealers call them whale-food. Whether whales feed on them I do not know; but terns, cormorants, and immense herds of great unwieldly seals, on some parts of the coast, derive their chief sustenance from these swimming crabs. Seamen invariably attribute the discoloration of the water to spawn; but I found this to be the case only on one occasion. At the distance of several leagues from the Archipelago of the Galapagos, the ship sailed through three strips of a dark yellowish, or mud-like, water; these strips were some miles long, but only a few yards wide, and they were separated from the surrounding surface by a sinuous yet distinct margin. The colour was caused

---

* M. Lesson (Voyage de la Coquille, vol. i, p. 255) mentions red water off Lima, apparently produced by the same cause. Peron, the distinguished naturalist, in the 'Voyage Aux Terres Australes', gives no less than twelve references to voyagers who have alluded to the discoloured waters of the sea (vol. ii, p. 239). It was his intention to have written an essay on the subject. To the references given by Peron may be added, Humboldt's Pers. Narr., vol. vi., p. 804; Flinder's Voyage, vol. i, p. 92; Labillardiere, vol. i, p. 287; Ulloa's Voyage; Voyage of the Astrolabe and of the Coquille; Captain King's Survey of Australia, &c.

by little gelatinous balls, about the fifth of an inch in diameter, in which numerous minute spherical ovules were embedded: they were of two distinct kinds, one being of a reddish colour and of a different shape from the other. I cannot form a conjecture as to what two kinds of animals these belonged. Captain Colnett remarks, that this appearance is very common among the Galapagos Islands, and that the direction of the bands indicates that of the currents; in the described case, however, the line was caused by the wind. The only other appearance which I have to notice, is a thin oily coat on the surface which displays iridescent colours. I saw a considerable tract of the ocean thus covered on the coast of Brazil; the seamen attributed it to the putrefying carcass of some whale, which probably was floating at no great distance. I do not here mention the minute gelatinous particles which are frequently dispersed throughout the water, for they are not sufficiently abundant to create any change of colour.

There are two circumstances in the above accounts which appear very remarkable: first, how do the various bodies which form the bands with defined edges keep together? In the case of the prawn-like crabs, their movements were as co-instantaneous as in a regiment of soldiers; but this cannot happen from any thing like voluntary action with the ovules, or the confervæ, nor is it probable among the infusoria. Secondly, what causes the length and narrowness of the bands? The appearance so much resembles that which may be seen in every torrent, where the stream uncoils into long streaks, the froth collected in the eddies, that I must attribute the effect to a similar action either of the currents of the air, or sea. Under this supposition we must believe that the various organized bodies are produced in certain favourable places, and are thence removed by the set of either wind or water. I confess, however, there is a very great difficulty in imagining any one spot to be the birthplace of the millions of millions of animalcula and confervæ: for whence come the germs at such points? – the parent bodies having been distributed by the winds and waves over the immense ocean. But on no other hypothesis can I understand their linear grouping. I may add that Scoresby remarks, that green water abounding with pelagic animals, is invariably found in a certain part of the Arctic Sea.

CHAPTER II

Rio de Janeiro – Excursion north of Cape Frio – Great evaporation – Slavery – Botofogo Bay – Terrestrial Planariæ – Clouds on Corcovado – Heavy rain – Musical Hyla – Lampyris and its larvæ – Elater, springing powers of – Blue haze – Noise of butterfly – Entomology – Ants – Wasp-killing spider – Parasitical spider – Artifices of Epeira – Gregarious spider – Spider with imperfect web

RIO DE JANEIRO

APRIL 4TH to JULY 5TH, 1832 – A few days after our arrival I became acquainted with an Englishman who was going to visit his estate situated, rather more than 100 miles from the capital, to the northward of Cape Frio. As I was quite unused to travelling, I gladly accepted his kind offer of allowing me to accompany him.

APRIL 8TH – Our party amounted to seven. The first stage was very interesting. The day was powerfully hot, and as we passed through the woods, every thing was motionless, excepting the large and brilliant butterflies, which lazily fluttered about. The view seen when crossing the hills behind Praia Grande was most beautiful; the colours were intense, and the prevailing tint a dark blue; the sky and the calm waters of the bay vied with each other in splendour. After passing through some cultivated country, we entered a forest, which in the grandeur of all its parts could not be exceeded. We arrived by midday at Ithacaia; this small village is situated on a plain, and round the central house are the huts of the negroes. These, from their regular form and position, reminded me of the drawings of the Hottentot habitations in Southern Africa. As the moon rose early, we determined to start the same evening for our sleeping-place at the Lagoa Marica. As it was growing dark we passed under one of the massive, bare, and steep hills of granite which are so common in this country. This spot is notorious from having been, for a long time, the residence of some runaway slaves, who, by cultivating a little ground near the top, contrived to eke out a subsistence. At length they were discovered, and a party of soldiers being sent, the whole were seized with the exception of one old woman, who sooner than again be led into slavery, dashed herself to pieces from the summit of the mountain. In a Roman matron this would have been called the noble love of freedom: in a poor negress it is mere brutal obstinacy. We continued riding for some hours. For the few last miles the road was intricate, and it passed through a desert waste of marshes and lagoons. The scene by the dimmed light of the moon was most desolate. A few fireflies flitted by us; and the solitary snipe, as it rose, uttered its plaintive cry. The distant and sullen roar of the sea scarcely broke the stillness of the night.

APRIL 9TH — We left our miserable sleeping-place before sunrise. The road passed through a narrow sandy plain, lying between the sea and the interior salt lagoons. The number of beautiful fishing birds, such as egrets and cranes, and the succulent plants assuming most fantastical forms, gave to the scene an interest which it would not otherwise have possessed. The few stunted trees were loaded with parasitical plants, among which the beauty and delicious fragrance of some of the orchideæ were most to be admired. As the sun rose, the day became extremely hot, and the reflection of the light and heat from the white sand was very distressing. We dined at Mandetiba; the thermometer in the shade being 84°. The beautiful view of the distant wooded hills, reflected in the perfectly calm water of an extensive lagoon, quite refreshed us. As the vênda here was a very good one, and I have the pleasant, but rare remembrance, of an excellent dinner, I will be grateful and presently describe it, as the type of its class. These houses are often large, and are built of thick upright posts, with boughs interwoven, and afterwards plastered. They seldom have floors, and never glazed windows; but are generally pretty well roofed. Universally the front part is open, forming a kind of verandah, in which tables and benches are placed. The bed-rooms join on each side, and here the passenger may sleep as comfortably as he can, on a wooden platform, covered by a thin straw mat. The vênda stands in a courtyard, where the horses are fed. On first arriving, it was our custom to unsaddle the horses and give them their Indian corn; then, with a low bow, to ask the senhôr to do us the favour to give us something to eat. 'Any thing you choose, sir,' was his usual answer. For the few first times, vainly I thanked Providence for having guided us to so good a man. The conversation proceeding, the case universally became deplorable. 'Any fish can you do us the favour of giving?' – 'Oh! no, sir.' – 'Any soup?' – 'No, sir.' – 'Any bread?' – 'Oh! no, sir.' – 'Any dried meat?' – 'Oh! no, sir.' If we were lucky, by waiting a couple of hours, we obtained fowls, rice, and farinha. It not unfrequently happened, that we were obliged to kill, with stones, the poultry for our own supper. When thoroughly exhausted by fatigue and hunger, we timorously hinted that we should be glad of our meal, the pompous, and (though true) most unsatisfactory answer was, 'It will be ready when it is ready.' If we had dared to remonstrate any further, we should have been told to proceed on our journey, as being too impertinent. The hosts are most ungracious and disagreeable in their manners; their houses and their persons are often filthily dirty; the want of the accommodation of forks, knives, and spoons is common; and I am sure no cottage or hovel in England could be found in a state so utterly destitute of every comfort. At Campos Novos, however, we fared sumptuously; having rice and fowls, biscuit, wine, and spirits, for dinner; coffee in the evening, and fish with coffee for breakfast. All this, with good food for the horses, only cost 2s. 6d. per head. Yet the host of this vênda, being asked if he knew anything of a whip which one of the party had lost, gruffly answered, 'How should I know? why did you not take care of it? – I suppose the dogs have eat it.'

* * *

APRIL 13TH – After three days' travelling we arrived at Socêgo, the estate of Senhôr Manuel Figuireda, a relation of one of our party. The house was simple, and, though like a barn in form, was well suited to the climate. In the sitting-room gilded chairs and sofas were oddly contrasted with the whitewashed walls, thatched

roof, and windows without glass. The house, together with
the granaries, the stables, and workshops for the blacks, who had
been taught various trades, formed a rude kind of quadrangle; in
the centre of which a large pile of coffee was drying. These
buildings stand on a little hill, overlooking the cultivated ground,
and surrounded on every side by a wall of dark green luxuriant
forest. The chief produce of this part of the country is coffee. Each
tree is supposed to yield annually, on an average, two pounds; but
some give as much as eight. Mandioca or cassada is likewise
cultivated in great quantity. Every part of this plant is useful: the
leaves and stalks are eaten by the horses, and the roots are ground
into a pulp, which, when pressed dry and baked, forms the
farinha, the principal article of sustenance in the Brazils. It is a
curious, though well-known fact, that the expressed juice of this
most nutritious plant is highly poisonous. A few years ago a cow
died at this *fazênda*, in consequence of having drunk some of it.
Senhôr Figuireda told me that he had planted, the year before, one
bag of feijaô or beans, and three of rice; the former of which
produced 80, and the latter 320 fold. The pasturage supports a
fine stock of cattle, and the woods are so full of game, that a deer
had been killed on each of the three previous days. This profusion of food showed itself at dinner, where, if the tables did not
groan, the guests surely did: for each person is expected to eat of
every dish. One day, having, as I thought, nicely calculated so
that nothing should go away untasted, to my utter dismay a roast
turkey and a pig appeared in all their substantial reality. During
the meals, it was the employment of a man to drive out of the
room sundry old hounds, and dozens of little black children,
which crawled in together, at every opportunity. As long as the
idea of slavery could be banished, there was something exceedingly fascinating in this simple and patriarchal style of living:
it was such a perfect retirement and independence of the rest of the
world. As soon as any stranger is seen arriving, a large bell is set
tolling, and generally some small cannon are fired. The event is
thus announced to the rocks and woods, but to nothing else. One
morning I walked out an hour before daylight to admire the
solemn stillness of the scene; at last, the silence was broken by the
morning hymn, raised on high by the whole body of the blacks;

and in this manner, their daily work is generally begun. On such *fazêndas* as these, I have no doubt the slaves pass happy and contented lives. On Saturday and Sunday they work for themselves, and in this fertile climate the labour of two days is sufficient to support a man and his family for the whole week.

APRIL 14TH—Leaving Socêgo, we rode to another estate on the Rio Macâe, which was the last patch of cultivated ground in that direction. The estate was two and a half miles long, and the owner had forgotten how many broad. Only a very small piece had been cleared, yet almost every acre was capable of yielding all the various rich productions of a tropical land. Considering the enormous area of Brazil, the proportion of cultivated ground can scarcely be considered as any thing, compared to that which is left in the state of nature: at some future age, how vast a population it will support! During the second day's journey we found the road so shut up, that it was necessary that a man should go ahead with a sword to cut away the creepers. The forest abounded with beautiful objects; among which the tree ferns, though not large, were, from their bright green foliage, and the elegant curvature of their fronds, most worthy of admiration. In the evening it rained very heavily, and although the thermometer stood at 65°, I felt very cold. As soon as the rain ceased, it was curious to observe the extraordinary evaporation which commenced over the whole extent of the forest. At the height of 100 feet the hills were buried in a dense white vapour, which rose like columns of smoke from the most thickly wooded parts, and especially from the valleys. I observed this phenomenon on several occasions: I suppose it is owing to the large surface of foliage, previously heated by the sun's rays.

While staying at this estate, I was very nearly being an eyewitness to one of those atrocious acts, which can only take place in a slave country. Owing to a quarrel and a law-suit, the owner was on the point of taking all the women and children from the men, and selling them separately at the public auction at Rio. Interest, and not any feeling of compassion, prevented this act. Indeed, I do not believe the inhumanity of separating thirty families, who had lived together for many years, even occurred to the person. Yet I will pledge myself, that in humanity and good feeling, he was superior to the common run of men. It may be said there exists no limit to the blindness of interest and selfish habit. I may mention one very trifling anecdote, which at the time struck me more forcibly than any story of cruelty. I was crossing a ferry with a negro, who was uncommonly stupid. In endeavouring to make him understand, I talked loud, and made signs, in doing which I passed my hand near his face. He, I suppose, thought I was in a passion, and was going to strike him; for instantly, with a frightened look and half-shut eyes, he dropped his hands. I shall never forget my feelings of surprise, disgust, and shame, at seeing a great powerful man afraid even to ward off a blow, directed, as he thought, at his face. This man had been trained to a degradation lower than the slavery of the most helpless animal.

* * *

During the remainder of my stay at Rio, I resided in a cottage at Botofogo Bay. It was impossible to wish for any thing more delightful than thus to spend some weeks in so magnificent a country. In England any person fond of natural history enjoys in his walks a great advantage, by always having something to attract his attention; but in these fertile climates, teeming with life, the attractions are so numerous, that he is scarcely able to walk at all.

The few observations which I was enabled to make were almost exclusively confined to the invertebrate animals. The existence of a division of the genus Planaria, which inhabits the dry land, interested me much. These animals are of so simple a structure, that Cuvier has arranged them with the intestinal worms, though never found within the bodies of other animals. Numerous species inhabit both salt and fresh water; but those to which I allude were found beneath logs of rotten wood, even in the drier parts of the forest. In general form they resemble little slugs, but are very much narrower in proportion. I met with one specimen no less than 5 inches long. The lower surface, by which they crawl, is flat, the upper being convex: in this latter respect the terrestrial species all differ from the depressed forms of the aquatic. Their structure is very simple. Near the middle of the undersurface, there are two small transverse slits, from the anterior one of which a funnel-shaped organ, or cup, can be protruded. This seems to act as the mouth. It is soft, highly irritable, and capable of various movements; when drawn within the body it is generally folded up like the bud of a plant. From the central position of the orifice, the animal has its mouth in the middle of what would commonly be called its stomach! For some time after the rest of the animal has become dead from the effects of salt water, or other cause, this organ still retains its vitality. The body is soft and parenchymatous; in the central part a transparent space, with lateral ramifications, appears to act as a system of circulation. Minute, black, eye-like specks are scattered round the margin of the crawling surface, and more abundantly close to the anterior extremity, which is constantly used as a feeler. In a marine species, I extracted from the central parts of the body vast numbers of little spherical eggs; they were .006 of an inch in diameter, and contained a central opake mass or yolk.

The terrestrial Planariæ, of which I have found no less than eight species, occur from within the tropic to lat. 47° south, and are common to South America, New Zealand, Van Diemen's Land, and Mauritius. Some of the species are longitudinally striped with several bands of gay colours. At first sight there is a remarkable false analogy between these animals and snails, although so widely separated from each other in all essential points of organization. I suppose these Planariæ feed on rotten wood, for they are always found crawling on the under surface of old decayed trees; and some small specimens being kept with no other food, rapidly increased in size. Although gaily coloured little animals, they dislike, and are very sensitive to the light. Some specimens which I obtained at Van Diemen's Land, I kept alive for nearly two months. Having cut one of them transversely into two nearly equal parts, in the course of a fortnight both had the shape of perfect animals. I had, however, so divided the body, that one of the halves contained both the inferior orifices, and the other, in consequence, none. In the course of twenty-five days from the operation, the more perfect half could not have been distinguished from any other specimen. The other had increased much in size; and towards its posterior end, a clear space was formed in the parenchymatous mass, in which a rudimentary cup-shaped organ could clearly be distinguished; on the under surface, however, no corresponding slit was yet open. If the increased heat of the weather, as we approached the equator, had not destroyed all the individuals, there can be no doubt that this last step would have completed its structure. Although so well-known an experiment, it was interesting to watch the gradual production of every essential organ, out of the simple extremity of another animal. It is extremely difficult to preserve these Planariæ; immediately the cessation of life allows the ordinary laws of change to act, their entire bodies become soft and fluid, with a rapidity which I have never seen equalled. A method of preservation that I found answered pretty well, was to dry the whole animal rapidly on a thin plate of mica, for the body thus becomes transparent, and allows the internal structure to be seen.

I first visited the forest in which these Planariæ were found, in company with an old Portuguese priest who took me out to hunt with him. The sport consisted in turning into the cover a few dogs, and then patiently waiting to fire at any animal which might appear. We were accompanied by the son of a neighbouring farmer – a good specimen of a wild Brazilian youth. He was dressed in a tattered old shirt and trousers, and had his head uncovered: he carried an old-fashioned gun and a large knife. The habit of carrying the knife is universal; and in traversing a thick wood it is almost necessary, on account of the creeping plants. The frequent occurrence of murder may be partly attributed to this habit. The Brazilians are so dexterous with the knife, that they can throw it to some distance with precision, and with sufficient force to cause a fatal wound. I have seen a number of little boys practising this art as a game of play, and from their skill in hitting an upright stick, they promised well for more earnest attempts. My companion, the day before, had shot two large bearded monkeys. These animals have prehensile tails, the extremity of which, even after death, can support the whole weight of the body. One of them thus remained fast to a branch, and it was necessary to cut down a large tree to procure it. This was soon effected, and down came tree and monkey with an awful crash. Our day's sport, besides the monkey, was confined to sundry small green parrots and a few toucans. I profited, however, by my acquaintance with the Portuguese padre, for on another occasion he gave me a fine specimen of the Yaguarundi cat.

Every one has heard of the beauty of the scenery near Botofogo. The house in which I lived was seated close beneath the well-known mountain of the Corcovado. It has been remarked, with much truth, that abruptly conical hills are characteristic of the formation which Humboldt designates as gneiss-granite. Nothing can be more striking than the effect of these huge rounded masses of naked rock rising out of the most luxuriant vegetation.

I was often interested by watching the clouds, which, rolling in from seaward, formed a bank just beneath the highest point of the Corcovado. This mountain, like most others, when thus partly veiled, appeared to rise to a far prouder elevation than its real height of 2,300 feet. Mr Daniell has observed, in his meteorological essays, that a cloud sometimes appears fixed on a mountain summit, while the wind continues to blow over it. The same phenomenon here presented a slightly different appearance. In this case the cloud was clearly seen to curl over, and rapidly pass by the summit, and yet was neither diminished nor increased in size. The sun was setting, and a gentle southerly breeze, striking against the southern side of the rock, mingled its current with the colder air above; and the vapour was thus condensed: but as the light wreaths of cloud passed over the ridge, and came within the influence of the warmer atmosphere of the northern sloping bank, they were immediately redissolved.

The climate, during the months of May and June, or the beginning of winter, was delightful. The mean temperature, from observations taken at nine o'clock, both morning and evening, was only 72°. It often rained heavily, but the drying southerly winds soon again rendered the walks pleasant. One morning, in the course of six hours, 1.6 inches of rain fell. As this storm passed over the forests, which surround the Corcovado, the sound produced by the drops pattering on the countless multitude of leaves, was very remarkable; it could be heard at the distance of a quarter of a mile, and was like the rushing of a great body of water. After the hotter days, it was delicious to sit quietly in the garden and watch the evening pass into night. Nature, in these climes, chooses her vocalists from more humble performers than in Europe. A small frog, of the genus Hyla,* sits on a blade of glass about an inch above the surface of the water, and sends forth a pleasing chirp. When several were together they sung in harmony on different notes. Various cicadæ and crickets, at the same time, kept up a ceaseless shrill cry, but which, softened by the distance, was not unpleasant. Every evening after dark this great concert commenced; and often have I sat listening to it, until my attention has been drawn away by some curious passing insect.

At these times the fireflies are seen† flitting about from hedge to hedge. All that I caught belonged to the family of Lampyridæ, or glowworms, and the greater number were *Lampyris occidentalis*. I found that this insect emitted the most brilliant flashes when irritated: in the intervals the abdominal rings were obscured. The flash was almost co-instantaneous in the two rings, but it was first just perceptible in the anterior one. The shining matter was fluid and very adhesive: little spots, where the skin had been torn, continued bright with a slight scintillation, whilst the uninjured parts were obscured. When the insect was decapitated the rings remained uninterruptedly bright, but not so brilliant as before: local irritation with a needle always increased the vividness of the light. The rings in one instance retained their luminous property nearly twenty-four hours after the death of the insect. From these facts it would appear probable, that the animal has only the power of concealing or extinguishing the light for short intervals, and that at other times the display is involuntary. On the muddy and wet gravel-walks I found the larvæ of this lampyris in great numbers: they resembled in general form the female of the English glowworm. These larvæ possessed but feeble luminous

---

* I had some difficulty in catching a specimen of this frog. The genus Hyla has its toes terminated by small suckers; and I found this animal could crawl up a pane of glass, when placed absolutely perpendicular.

† On a dark night the light could be seen at about 200 paces distant. It is remarkable that in all the glowworms, shining elaters, and various marine animals, which I have observed (such as the crustacea, medusæ, nereidæ, a coralline of the genus Clytia, and Pyrosoma), the light has been of a well-marked green colour.

powers; very differently from their parents, on the slightest touch they feigned death, and ceased to shine; nor did irritation excite any fresh display. I kept several of them alive for some time: their tails are very singular organs, for they act, by a well-fitted contrivance, as suckers, or organs of attachment, and likewise as reservoirs for saliva, or some such fluid. I repeatedly fed them on raw meat; and I invariably observed, that every now and then the extremity of the tail was applied to the mouth, and a drop of fluid exuded on the meat, which was then in the act of being consumed. The tail, notwithstanding so much practice, does not seem to be able to find its way to the mouth; at least the neck was always touched first, and apparently as a guide.

* * *

During this day I was particularly struck with a remark of Humboldt's, who often alludes to 'the thin vapour which, without changing the transparency of the air, renders its tints more harmonious, softens its effects,' &c. This is an appearance which I have never observed in the temperate zones. The atmosphere, seen through a short space of half or three-quarters of a mile, was perfectly lucid, but at a greater distance all colours were blended into a most beautiful haze, of a pale French gray, mingled with a little blue. The condition of the atmosphere between the morning and about noon, when the effect was most evident, had undergone little change, excepting in its dryness. In the interval, the difference between the dew point and temperature had increased from 7.5° to 17°.

On another occasion I started early and walked to the Gavia, or topsail mountain. The air was delightfully cool and fragrant; and the drops of dew still glittered on the leaves of the large liliaceous plants, which shaded the streamlets of clear water. Sitting down on a block of granite, it was delightful to watch the various insects and birds as they flew past. The humming-birds seem particularly fond of such shady retired spots. Whenever I saw these little creatures buzzing round a flower, with their wings vibrating so rapidly as to be scarcely visible, I was reminded of the sphinx moths: their movements and habits are indeed, in many respects, very similar.

Following a pathway I entered a noble forest, and from a height of 500 or 600 feet, one of those splendid views was presented, which are so common on every side of Rio. At this elevation the landscape has attained its most brilliant tint; and every form, every shade, so completely surpasses in magnificence all that the European has ever beheld in his own country, that he knows not how to express his feelings. The general effect frequently recalled to my mind the gayest scenery of the Opera-house or the great theatres. I never returned from these excursions empty-handed. This day I found a specimen of a curious fungus, called Hymeno-phallus. Most people know the English Phallus, which in autumn taints the air with its odious smell: this, however, as the entomo-logist is aware, is to some of our beetles a delightful fragrance. So was it here; for a Strongylus, attracted by the odour, alighted on the fungus as I carried it in my hand. We here see in two distant countries a similar relation between plants and insects of the same families, though the species of both are different. When man is the agent in introducing into a country a new species, this relation is often broken: as one instance of this I may mention, that the leaves of the cabbages and lettuces, which in England afford food to such a multitude of slugs and caterpillars, in the gardens near Rio are untouched.

* * *

CHAPTER III

Monte Video – Maldonado – Excursion to R. Polanco – Lazo
and Bolas – Partridges – Geology – Absence of trees – Cervus
campestris – River hog – Tucutuco – Molothrus, cuckoo-like
habits – Tyrant-flycatcher – Mocking-bird – Carrion hawks –
Tubes formed by lightning – House struck

MALDONADO

JULY 5TH, 1832 – In the morning we got under way, and stood
out of the splendid harbour of Rio de Janeiro. In our passage to the
Plata, we saw nothing particular, excepting on one day a great
shoal of porpoises, many hundreds in number. The whole sea
was in places furrowed by them; and a most extraordinary
spectacle was presented, as hundreds, proceeding together by
jumps, in which their whole bodies were exposed, thus cut the
water. When the ship was running nine knots, these animals could
cross and recross the bows with the greatest ease, and then dash
away right ahead. As soon as we entered the estuary of the Plata,
the weather was very unsettled. One dark night we were surrounded by numerous seals and penguins, which made such
strange noises, that the officer on watch reported he could hear the
cattle bellowing on shore. On a second night we witnessed a
splendid scene of natural fireworks; the mast-head and yard-arm
ends shone with St Elmo's light; and the form of the vane could
almost be traced, as if it had been rubbed with phosphorus. The
sea was so highly luminous, that the tracks of the penguins were
marked by a fiery wake, and lastly, the darkness of the sky was
momentarily illuminated by the most vivid lightning.

When within the mouth of the river, I was interested by
observing how slowly the waters of the sea and river mixed. The

latter, muddy and discoloured, from its less specific gravity, floated on the surface of the salt water. This was curiously exhibited in the wake of the vessel, where a line of blue was seen mingling in little eddies, with the adjoining fluid.

JULY 26TH – We anchored at Monte Video. The *Beagle* was employed in surveying the extreme southern and eastern coasts of America, south of the Plata, during the two succeeding years. To prevent useless repetitions, I will extract those parts of my journal which refer to the same districts, without always attending to the order in which we visited them.

MALDONADO is situated on the northern bank of the Plata, and not very far from the mouth of the estuary. It is a most quiet, forlorn, little town; built, as is universally the case in these countries, with the streets running at right angles to each other, and having in the middle a large plaza or square, which, from its size, renders the scantiness of the population more evident and more unsociable. It possesses scarcely any trade; the exports being confined to a few hides and living cattle. The inhabitants are chiefly landowners, together with a few shopkeepers and the necessary tradesmen, such as blacksmiths and carpenters, who do nearly all the business for a circuit of 50 miles round. The town is separated from the river by a band of sand-hillocks, about a mile broad: it is surrounded on all other sides, by an open slightly undulating country, covered by one uniform layer of fine green turf, on which countless herds of cattle, sheep, and horses graze. There is very little land cultivated even close to the town. A few hedges, made of cacti and agave, mark out where some wheat or Indian corn has been planted. The features of the country are very similar along the whole northern bank of the Plata. The only difference is, that here the granitic hills are rather more boldly pronounced. The scenery is very uninteresting; there is scarcely a house, an enclosed piece of ground, or even a tree, to give it an air of cheerfulness. Yet, after being imprisoned for some time in a ship, there is a charm in the unconfined feeling of walking over boundless plains of turf. Moreover, if your view is limited to a small space, many objects possess beauty. Some of the smaller birds are brilliantly coloured; and the bright green sward, browsed short by the cattle, is ornamented by dwarf flowers, among which a plant, looking like the daisy, claimed the place of an old friend. What would a florist say to whole tracts so thickly covered by the *Verbena melindres*, as, even at a distance, to appear of the most gaudy scarlet?

I stayed ten weeks at Maldonado, in which time a nearly perfect collection of the animals, birds, and reptiles, was procured. Before making any observations respecting them, I will give an account of a little excursion I made as far as the river Polanco, which is about 70 miles distant, in a northerly direction. I may mention, as a proof how cheap every thing is in this country, that I paid only two dollars a day, or eight shillings, for two men, together with a troop of about a dozen riding-horses. My companions were well armed with pistols and sabres; a precaution which I thought rather unnecessary; but the first piece of news we heard was, that, the day before, a traveller from Monte Video had been found dead on the road, with his throat cut. This happened close to a cross, the record of a former murder.

On the first night we slept at a retired little country-house; and there I soon found out, that I possessed two or three articles, especially a pocket compass, which created unbounded astonishment. In every house I was asked to show the compass, and by its aid, together with a map, to point out the direction of various places. It excited the liveliest admiration that I, a perfect stranger, should know the road (for direction and road are synonymous in this open country) to places where I had never been. At one house a young woman, who was ill in bed, sent to entreat me to come and show her the compass. If their surprise was great, mine was greater, to find such ignorance among people who possessed their thousands of cattle, and *estancias* of great extent. It can only be accounted for by the circumstance that this retired part of the country is seldom visited by foreigners. I was asked whether the earth or sun moved; whether it was hotter or colder to the north; where Spain was, and many other such questions. The greater number of the inhabitants had an indistinct idea that England, London, and North America, were different names for the same place; but the better informed well knew that London and North America were separate countries close together, and that England was a large town in London! I carried with me some Promethean matches, which I ignited by biting; it was thought so wonderful that a man should strike fire with his teeth, that it was usual to collect the whole family to see it: I was once offered a dollar for a single one. Washing my face in the morning, caused much speculation at the village of Las Minas; a superior tradesman closely cross-questioned me about so singular a practice; and likewise why on board we wore our beards; for he had heard from my guide that we did so. He eyed me with much suspicion; perhaps he had heard of ablutions in the Mahomedan religion, and knowing me to be a heretic, probably he came to the conclusion that all heretics were Turks. It is the general custom in this country to ask for a night's lodging at the first convenient house. The astonishment at the compass, and my other feats in jugglery, was to a certain degree advantageous, as with that, and the long stories my guides told of my breaking stones, knowing venomous from harmless snakes, collecting insects, &c., I repaid them for their hospitality. I am writing as if I had been among the inhabitants of central Africa: Banda Oriental would not be flattered by the comparison; but such were my feelings at the time.

The next day we rode to the village of Las Minas. The country was rather more hilly, but otherwise continued the same; an inhabitant of the Pampas no doubt would have considered it as truly Alpine. The country is so thinly inhabited, that during the whole day we scarcely met a single person. Las Minas is much smaller even than Maldonado. It is seated on a little plain, and is surrounded by low rocky mountains. It is of the usual symmetrical form; and with its whitewashed church standing in the centre, had rather a pretty appearance. The outskirting houses rose out of the plain like isolated beings, without the accompaniment of gardens or courtyards. This is generally the case in the country, and all the houses have, in consequence, an uncomfortable aspect. At night we stopped at a pulperia, or drinking-shop. During the evening a great number of Gauchos came in to drink spirits and smoke cigars: their appearance is very striking; they are generally tall and handsome, but with a proud and dissolute expression of countenance. They frequently wear their moustaches, and long black hair curling down their backs. With their brightly coloured garments, great spurs clanking about their heels, and knives stuck as daggers (and often so used) at their waists, they look a very different race of men from what might be expected from their name of Gauchos, or simple countrymen. Their politeness is excessive: they never drink their spirits without expecting you to taste it; but whilst making their exceedingly graceful bow, they seem quite as ready, if occasion offered, to cut your throat.

On the third day we pursued rather an irregular course, as I was employed in examining some beds of marble. On the fine plains of turf we saw many ostriches (*Struthio rhea*). Some of the flocks contained as many as twenty or thirty birds. These, when standing on any little eminence, and seen against the clear sky, presented a very noble appearance. I never met with such tame ostriches in any other part of the country: it was easy to gallop up within a short distance of them; but then, expanding their wings, they made all sail right before the wind, and soon left the horse astern.

At night we came to the house of Don Juan Fuentes, a rich landed proprietor, but not personally known to either of my companions. On approaching the house of a stranger, it is usual to follow several little pieces of etiquette: riding up slowly to the door, the salutation of Ave Maria* is given, and until somebody comes out, and asks you to alight, it is not customary even to get off your horse. Having entered the house, some general conversation is kept up for a few minutes, till permission is asked to pass the night there. This is granted as a matter of course. The stranger then takes his meals with the family, and a room is assigned him, where with the horsecloths belonging to his *recado* (or saddle of the Pampas) he makes his bed. It is curious how similar circumstances produce such similar results in manners. At the Cape of Good Hope the same hospitality, and very nearly the same points of etiquette, are universally observed. The difference, however, between the character of the Spaniard and that of the Dutch boor is shown, by the former never asking his guest a single question beyond the strictest rule of politeness, whilst the honest Dutch-

---

* The formal answer of the owner of the place is, 'sin pecado concebida' – (conceived without sin).

man demands where he has been, where he is going, what is his business, and even how many brothers, sisters, or children he may happen to have.

Shortly after our arrival at Don Juan's, one of the large herds of cattle was driven in towards the house, and three beasts were picked out to be slaughtered for the supply of the establishment. These half-wild cattle are very active; and knowing full well the fatal lazo, they led the horses a long and laborious chase. After witnessing the rude wealth displayed in the number of cattle, men, and horses, Don Juan's miserable house was quite curious. The floor consisted of hardened mud, and the windows were without glass; the furniture of the sitting-room boasted only of a few of the roughest chairs and stools, with a couple of tables. The supper, although several strangers were present, consisted of two huge piles, one of roast beef, the other of boiled, with some pieces of pumpkin: besides this latter there was no other vegetable, and not even a morsel of bread. For drinking, a large earthenware jug of water served the whole party. Yet this man was the owner of several square miles of land, of which nearly every acre would produce corn, and, with a little trouble, all the common vegetables. The evening was spent in smoking, with a little impromptu singing, accompanied by the guitar. The signoritas all sat together in one corner of the room, and did not sup with the men.

So many works have been written about these countries, that it is almost superfluous to describe either the lazo or the bolas. The former consists of a very strong, but thin, well-plaited rope, made of raw hide. One end is attached to the broad surcingle, which fastens together the complicated gear of the *recado*, or saddle used in the Pampas; the other is terminated by a small ring of iron or brass, by which a noose can be formed. The Gaucho, when he is going to use the lazo, keeps a small coil in his bridle hand, and in the other holds the running noose, which is made very large, generally having a diameter of about 8 feet. This he whirls round his head, and by the dexterous movement of his wrist keeps the noose open; then, throwing it, he causes it to fall on any particular spot he chooses. The lazo, when not used, is tied up in a small coil to the after part of the *recado*. The bolas, or balls, are of two kinds:

the simplest, which is chiefly used for catching ostriches, consists of two round stones, covered with leather, and united by a thin plaited thong, about 8 feet long. The other kind differs only, in having three balls united by the thongs to a common centre. The Gaucho holds the smallest of the three in his hand, and whirls the other two round and round his head; then, taking aim, sends them like chain shot revolving through the air. The balls no sooner strike any object, than, winding round it, they cross each other, and become firmly hitched. The size and weight of the balls varies, according to the purpose for which they are made: when of stone, although not so large as a big apple, yet they are sent with such force as sometimes to break the leg even of a horse. I have seen the balls made of wood, and as large as a turnip, for the sake of catching these animals without injuring them. The balls are sometimes made of iron, and these can be hurled to the greatest distance. The main difficulty in using either lazo or bolas, is to ride so well, as to be able at full speed, and while suddenly turning about, to whirl them so steadily round the head, as to take aim: on foot any person would soon learn the art. One day, as I was amusing myself by galloping and whirling the balls round my head, by accident the free one struck a bush; and its revolving motion being thus destroyed, it immediately fell to the ground, and like magic caught one hind leg of my horse; the other ball was then jerked out of my hand, and the horse fairly secured. Luckily he was an old practised animal, and knew what it meant; otherwise he would probably have kicked till he had thrown himself down. The Gauchos roared with laughter; they cried they had seen every sort of animal caught, but had never before seen a man caught by himself.

* * *

During our stay at Maldonado I paid particular attention to the mammalia and birds. Of the latter I procured, within the distance of a morning's walk, no less than eighty species, of which many were exceedingly beautiful – I think even more so than those of Brazil. The other orders were not neglected. Reptiles were numerous, and nine different kinds of snakes were taken. Of the indigenous mammalia, the only one now left of any size, which is common, is the *Cervus Campestris*. This deer is exceedingly abundant throughout the countries bordering the Plata. It is found in Northern Patagonia as far south as the Rio Negro (lat. 41°); but further southward none were seen by the officers employed in surveying the coast. It appears to prefer a hilly country; I saw very many small herds, containing from five to seven animals each, near the Sierra Ventana, and among the hills north of Maldonado. If a person crawling close along the ground, slowly advances towards a herd, the deer frequently, out of curiosity, approach to reconnoitre him. I have by this means killed, from one spot, three out of the same herd. Although so tame and inquisitive, yet when approached on horseback, they are exceedingly wary. In this country nobody goes on foot, and the deer knows man as its enemy, only when he is mounted and armed with the bolas. At Bahia Blanca, a recent establishment in Northern Patagonia, I was surprised to find how little the deer cared for the noise of a gun: one day I fired ten times, from within 80 yards, at one animal; and it was much more startled at the ball cutting up the ground than at the report of the rifle. My powder being exhausted, I was obliged (to my shame as a sportsman be it spoken), to get up and halloo till the deer ran away.

The most curious fact with respect to this animal, is the overpoweringly strong and offensive odour which proceeds from the buck. It is quite indescribable: several times whilst skinning the specimen which is now mounted at the Zoological Museum, I was almost overcome by nausea. I tied up the skin in a silk pocket-handkerchief, and so carried it home: this handkerchief, after being well washed, I continually used, and it was, of course, as repeatedly washed; yet every time, for a space of one year and seven months, when first unfolded, I distinctly perceived the odour. This appears an astonishing instance of the permanence of some matter, which in its nature, nevertheless, must be most subtile and volatile. Frequently, when passing at the distance of half a mile to leeward of a herd, I have perceived the whole air tainted with the effluvium. I believe the smell from the buck is most powerful at the period when its horns are perfect, or free from the hairy skin. When in this state the meat is, of course, quite uneatable; but the Gauchos assert, that if buried for some time in fresh earth, the taint is removed. I have somewhere read that the islanders in the north of Scotland treat the rank carcasses of the fish-eating birds in the same manner.

The order Rodentia is here very numerous in species: of mice alone I obtained no less than eight kinds.* The largest gnawing animal in the world, the *Hydrocharus Capybara* (the water-hog), is here also common. One which I shot at Monte Video weighed 98 pounds: its length, from the end of the snout to the stump-like tail, was 3 feet 2 inches; and its girth, 3 feet 8. These great Rodents are generally called *carpinchos*: they occasionally frequent the islands in the mouth of the Plata, where the water is quite salt, but are far more abundant on the borders of fresh-water lakes and rivers. Near Maldonado three or four generally live together. In the daytime they either lie among the aquatic plants, or openly feed on the turf plain.† When viewed at a distance, from their manner of walking and colour, they resemble pigs: but when seated on their haunches, and attentively watching any object with one eye, they reassume the appearance of their congeners, the cavies. Both the front and side view of their head has quite a ludicrous aspect, from the great depth of their jaw. These animals, at Maldonado, were very tame; by cautiously walking, I approached within 3 yards of four old ones. This tameness may probably be accounted for, by the Jaguar having been banished for some years, and by the Gaucho not thinking it worth his while to hunt them. As I approached nearer and nearer they frequently made their peculiar noise, which is a low abrupt grunt; not having much actual sound, but rather arising from the sudden expulsion of air: the only noise I know at all like it, is the first hoarse bark of a

* These have been named and described by Mr Waterhouse at the meetings of the Zoological Society. I must be allowed to take this opportunity of returning my cordial thanks to Mr Waterhouse, and to the other gentlemen attached to that Society, for their kind and most liberal assistance on all occasions.

† In the stomach and duodenum of a *carpincho* which I opened, I found a very large quantity of a thin yellowish fluid, in which scarcely a fibre could be distinguished. Mr Owen informs me that a part of the œsophagus is so constructed, that nothing much larger than a crowquill can be passed down. Certainly the broad teeth and strong jaws of this animal are well fitted to grind into pulp the aquatic plants on which it feeds.

large dog. Having watched the four from almost within arm's length (and they me) for several minutes, they rushed into the water at full gallop, with the greatest impetuosity, and emitted, at the same time, their bark. After diving a short distance they came again to the surface, but only just showed the upper part of their heads. When the female is swimming in the water and has young ones, they are said to sit on her back. These animals are easily killed in numbers; but their skins are of trifling value, and the meat is very indifferent. I have never heard of the carpincho being found south of the Plata; but as I see in a map that there is a Laguna del Carpincho high up the Rio Salado, I suppose such must have occurred. On the islands in the Rio Parana they are exceedingly abundant, and afford the ordinary prey to the Jaguar.

The Tucutuco (*Ctenomys Braziliensis*) is a curious small animal, which may be briefly described as a Rodent, with the habits of a mole. It is extremely abundant in some parts of the country,* but is difficult to be procured, and still more difficult to be seen, when at liberty. It lives almost entirely under ground, and prefers a sandy soil with a gentle inclination. The burrows are said not to be deep, but of great length. They are seldom open; the earth being thrown up at the mouth into hillocks, not quite so large as those made by the mole. Considerable tracts of country are so completely undermined by these animals, that horses, in passing over, sink above their fetlocks. The tucutucos appear, to a certain degree, to be gregarious. The man who procured the specimens for me had caught six together, and he said this was a common occurrence. They are nocturnal in their habits; and their principal food is afforded by the roots of plants, which is the object of their extensive and superficial burrows. Azara says they are so difficult to be obtained, that he never saw more than one. He states that they lay up magazines of food within their burrows. This animal is universally known by a very peculiar noise, which it makes when beneath the ground. A person, the first time he hears it, is much surprised; for it is not easy to tell whence it comes, nor is it

---

* The wide plains north of the Rio Colorado are undermined by these animals; and near the Strait of Magellan, where Patagonia blends with Tierra del Fuego, the whole sandy country forms a great warren for the tucutuco.

possible to guess what kind of creature utters it. The noise consists in a short, but not rough, nasal grunt, which is repeated about four times in quick succession; the first grunt is not so loud, but a little longer, and more distinct than the three following: the musical time of the whole is constant, as often as it is uttered.* The name tucutuco is given in imitation of the sound. In all times of the day, where this animal is abundant, the noise may be heard, and sometimes directly beneath one's feet. When kept in a room, the tucutucos move both slowly and clumsily, which appears owing to the outward action of their hind legs; and they are likewise quite incapable of jumping even the smallest vertical height. Mr Reid, who dissected a specimen which I brought home in spirits, informs me that the socket of the thigh-bone is not attached by a ligamentum teres; and this explains, in a satisfactory manner, the awkward movements of their hinder extremities. When eating, they rest on their hind legs and hold the piece in their fore paws; they appeared also to wish to drag it into some corner. They are very stupid in making any attempt to escape; when angry or frightened, they uttered the tucu-tuco. Of those I kept alive, several, even the first day, became quite tame, not attempting to bite or to run away; others were a little wilder.

The man who caught them asserted that very many are invariably found blind. A specimen which I preserved in spirits was in this state; Mr Reid considers it to be the effect of inflammation in the nictitating membrane. When the animal was alive I placed my finger within half an inch of its head, and not the slightest notice was taken: it made its way, however, about the room nearly as well as the others. Considering the subterranean habits of the tucutuco, the blindness, though so frequent, cannot be a very serious evil; yet it appears strange that any animal should possess an organ constantly subject to injury. The mole, whose habits in nearly every respect, excepting in the kind of food, are so similar,

---

* At the R. Negro, in Northern Patagonia, there is an animal of the same habits, and probably a closely allied species, but which I never saw. Its noise is different from the Maldonado kind; it is repeated only twice instead of three or four times, and is more distinct and sonorous: when heard from a distance, it so closely resembles the sound made in cutting down a small tree with an axe, that I have sometimes remained in doubt concerning it.

has an extremely small and protected eye, which, although possessing a limited vision, at once seems adapted to its manner of life.

Birds of many kinds are extremely abundant on the undulating grassy plains around Maldonado. Several species, of the genus Cassicus, allied to our starlings in habits and structure, and of Tyrant-flycatchers, and a mocking-bird, from their numbers, give a character to the ornithology. Some of the Cassici are very beautiful, black and yellow being the prevailing colours; but *Oriolus ruber*, Gme., offers an exception, in having its head, shoulders, and thighs of the most splendid scarlet. This bird differs from its congeners in being solitary. It frequents marshes; and, seated on the summit of a low bush, with its mouth wide open, utters a plaintive agreeable cry, which can be heard at a long distance.

Another species,* of a purplish-black colour, with a metallic lustre, feeds on the plain in large flocks, mingled with other birds. Several may often be seen standing on the back of a cow or horse. While perched on a hedge, and pluming themselves in the sun, they sometimes attempt to sing, or rather to hiss: the noise is very peculiar; it resembles that of bubbles of air passing rapidly from a small orifice under water, so as to produce an acute sound. Azara states that this bird, like the cuckoo, deposits its eggs in other birds' nests. I was several times told by the country people, that there was some bird with this habit; and my assistant in collecting, who is a very accurate person, found a nest of the sparrow† of the country, with one egg in it larger than the others, and of a different colour and shape. Mr Swainson‡ has remarked that with the exception of the *Molothrus pecoris*, the cuckoos are the only birds which can be called truly parasitical; namely, such as 'fasten themselves, as it were, on another living animal, whose animal

---

* Le Troupiale commun of Azara (vol. iii, p. 169) – a second species of Molothrus.
† A Zonotrichia; – the *chingolo* of Azara. The egg is rather less than that of the missel-thrush; it is of a nearly globular form, but with one end rather smaller than the other. The ground colour is a pale pinkish-white, with irregular spots and blotches of a pinkish-brown, and others less distinct of a grayish hue. The egg is now in the museum of the Zoological Society.
‡ Magazine of Zoology and Botany, vol. i, p. 217.

heat brings their young into life, whose food they alone live upon, and whose death would cause theirs during the period of infancy'. The *Molothrus pecoris* is a North American bird, and is closely allied in general habits, even in such peculiarities as standing on the backs of cattle (as its name implies), and in appearance, with the species from the plains of La Plata; it only differs in being rather smaller and of a different colour, yet the two birds would be considered by every naturalist as distinct species. It is very interesting to see so close an agreement in structure, and in habits, between allied species coming from opposite parts of a great continent. It is also very remarkable, that the cuckoos and the molothri, although opposed to each other in almost every habit, should agree in the one strange one of their parasitical propagation. The molothrus, like our starling, is eminently sociable, and lives on the open plains without art or disguise:* the cuckoo, as every one knows, is a singularly shy bird; it frequents the most retired thickets, and feeds on fruit and caterpillars. In structure these birds are likewise widely removed from each other.

* * * *

* See Azara, vol. iii, p. 170.

CHAPTER IV

Rio Negro – Estancias attacked by Indians – Salt lakes, geological position of – Flamingoes – R. Negro to Colorado – Sacred tree – Patagonian hare – Indian families – General Rosas – Proceed to Bahia Blanca – Sand dunes – Negro lieutenant – Bahia Blanca – Ground incrusted with Glauber salt – Punta Alta – Zorillo

RIO NEGRO TO BAHIA BLANCA

JULY 24TH 1833 – The *Beagle* sailed from Maldonado, and on August the 3rd she arrived off the mouth of the Rio Negro. This is the principal river on the whole line of coast between the Straits of Magellan and the Plata. It enters the sea about 300 miles south of the estuary of the latter. About fifty years since, under the old Spanish government, a small colony was established here; and it is still the most southern position (lat. 41°) on this eastern coast of America which is inhabited by civilized man.

* * *

To the northward of the Rio Negro, between it and the inhabited country near Buenos Ayres, the Spaniards have only one small settlement, recently established at Bahia Blanca. The distance in a straight line to the capital is very nearly 500 British miles. The wandering tribes of horse Indians, which have always occupied the greater part of this country, having of late much harassed the outlying *estancias*, the government at Buenos Ayres equipped some time since an army under the command of General Rosas for the purpose of exterminating them. The troops were now encamped on the banks of the Colorado; a river lying about 80 miles to the northward of the Rio Negro. When General Rosas left Buenos Ayres, he struck in a direct line across the unexplored plains: and as the country was thus pretty well cleared of Indians, he left behind him, at wide intervals, a small party of soldiers, with a troop of horses (a *posta*), so as to be enabled to keep up a communication with the capital. As the *Beagle* intended to call at Bahia Blanca, I determined to proceed there by land; and ultimately I extended my plan so as to travel the whole way by the *postas* to Buenos Ayres.

AUGUST 11TH – Mr Harris, an Englishman residing at Patagones, a guide, and five Gauchos, who were proceeding to the army on business, were my companions on the journey. The Colorado, as I have already said, is nearly 80 miles distant: and as we travelled slowly, we were two days and a half on the road. The whole line of country deserves scarcely a better name than that of a desert. Water is found only in two small wells: it is called fresh; but even at this time of the year, during the rainy season, it was quite brackish. In the summer this must be a distressing passage; for now it was sufficiently desolate. The valley of the Rio Negro, broad as it is, has merely been excavated out of the sandstone plain; for immediately above the bank on which the town stands, a level country commences, which is interrupted only by a few trifling valleys and depressions. Everywhere the landscape wears the same sterile aspect; a dry gravelly soil supports tufts of brown withered grass, and low scattered bushes, armed with thorns.

Shortly after passing the first spring we came in sight of a famous tree, which the Indians reverence as the altar of Walleechu. It is situated on a high part of the plain, and hence is a landmark visible at a great distance. As soon as a tribe of Indians come in sight of it, they offer their adorations by loud shouts. The tree itself is low, much branched, and thorny. Just above the root it has a diameter of about 3 feet. It stands by itself without any neighbour, and was indeed the first tree we saw; afterwards we met with a few others of the same kind, but they were far from common. Being winter the tree had no leaves, but in their place numberless threads, by which the various offerings, such as cigars, bread, meat, pieces of cloth, &c., had been suspended. Poor people not having any thing better, only pulled a thread out of their ponchos, and fastened it to the tree. The Indians, more-over, were accustomed to pour spirits and maté into a certain hole, and likewise to smoke upwards, thinking thus to afford all possible gratification to Walleechu. To complete the scene, the tree was surrounded by the bleached bones of the horses which had been slaughtered as sacrifices. All Indians of every age and sex, made their offerings; they then thought that their horses would not tire, and that they themselves should be prosperous. The Gaucho who told me this, said that in the time of peace he had witnessed this scene, and that he and others used to wait till the Indians had passed by, for the sake of stealing their offerings from Walleechu.

The Gauchos think that the Indians consider the tree as the god itself; but it seems far more probable that they regard it as the altar. The only cause which I can imagine for this choice, is its being a landmark in a dangerous passage. The Sierra de la Ventana is visible at an immense distance; and a Gaucho told me that he was once riding with an Indian a few miles to the north of the Rio Colorado, when the latter commenced making the same loud noise, which is usual at the first sight of the distant tree; putting his hand to his head, and then pointing it in the direction of the Sierra. Upon being asked the reason of this, the Indian said in broken Spanish, 'First see the Sierra.' This likewise would render it probable that the utility of a distant landmark is the first cause of its adoration. About two leagues beyond this curious tree we halted for the night: at this instant an unfortunate cow was spied by the lynx-eyed Gauchos. Off they set in chase, and in a few minutes she was dragged in by the lazo, and slaughtered. We here had the four necessaries of life 'en el campo', – pasture for the horses, water (only a muddy puddle), meat, and firewood. The Gauchos were in high spirits at finding all these luxuries; and we soon set to work at the poor cow. This was the first night which I had ever passed under the open sky, with the gear of the *recado* for my bed. There is high enjoyment in the independence of the Gaucho life – to be able at any moment to pull up your horse, and say, 'Here we will pass the night.' The deathlike stillness of the plain, the dogs keeping watch, the gipsy-group of Gauchos making their beds round the fire, have left in my mind a strongly marked picture of this first night, which will not soon be forgotten.

The next day the country continued similar to that above described. It is inhabited by few birds or animals. Occasionally a deer, or a Guanaco (wild Llama) may be seen; but the Agouti (*Cavia Patagonica*) is the commonest quadruped. This animal here represents our hares. It differs, however, from that genus in many essential respects; for instance, it has only three toes behind. It is also nearly twice the size, weighing from 20 to 25 pounds. The Agouti is a true friend to the desert; it is a common feature in the landscape to see two or three hopping quickly one after the other in a straight line across these wild plains. On the eastern side of America their northern limit is formed by the Sierra Tapalguen (lat. 37° 30′), where the plains rather suddenly become greener and more humid. The limit certainly depends on this change, for near Mendoza (lat. 33° 30′), which is much further north, but where the country is very sterile, I again met the Agouti. It is not evident by what circumstances their southern limit is governed; it occurs between Port Desire and St Julian (about 48° 30′), where there is no change in the kind of land, and only a trifling and gradual one of temperature. It is a singular fact, that although the Agouti is not now found so far south as Port St Julian, yet that Captain Wood, in his voyage in 1670, talks of them as being numerous there. What cause can have altered, in a wide, uninhabited, and rarely visited country, the range of an animal like this? It appears, also, from the number shot in one day at Port Desire, that they must have been considerably more abundant there formerly than at present. Azara states that the Agouti never excavates its own burrow, but uses that of the Bizcacha. Wherever this animal is present, without doubt this is true; but on the sandy plains of Bahia Blanca, where the Bizcacha is not found, the Gauchos maintain that the Agouti is its own workman. The same thing occurs with the little owls of the Pampas (*Noctua cunicularia*), which have so often been described as standing like sentinels at the mouths of the burrows; for in Banda Oriental, owing to the absence of the Bizcacha, they are obliged to hollow out their own habitations. Azara also says that the Agouti, except when pressed by danger, does not enter its burrow: on this point I must again differ from that high authority. At Bahia Blanca I have repeatedly seen two or three of these animals sitting on their haunches by the mouths of their holes, which, as I passed by at a distance, they quietly entered. Daily in the neighbourhood of these spots the Agouti were abundant: but differently from most burrowing animals, it wanders, commonly two or three together, to miles or leagues from its home; nor do I know whether it returns at night. The Agouti feeds and roams about by day; is shy and watchful; does not squat, or so rarely that I never saw an instance of this; it cannot run very fast; and, therefore, is frequently caught by a couple of dogs, even of mixed breed. Its manner of running more resembles that of a rabbit than of a hare. The Agouti generally produces two young ones at a birth, which are brought forth within the burrow. The flesh, when cooked, is very white; it is, however, rather tasteless and dry.

The next morning, as we approached the Rio Colorado, the appearance of the country changed; we soon came on a plain covered with turf, which, from its flowers, tall clover, and little owls, resembled the Pampas. We passed also a muddy swamp of considerable extent, which in summer dries, and becomes incrusted with various salts; and hence is called a salitral. It was covered by low succulent plants, of the same kind with those growing on the sea-shore. The Colorado, at the pass where we crossed it, is only about 60 yards wide; generally it must be nearly double that width. Its course is very tortuous, being marked by willow-trees and beds of reeds: in a direct line the distance to the mouth of the river is said to be 9 leagues, but by water 25. We were delayed crossing in the canoe by some immense troops of mares, which were swimming the river in order to follow a division of troops into the interior. A more ludicrous spectacle I never beheld, than the hundreds of heads, all directed one way, with pointed ears and distended nostrils, appearing just above the water like a great shoal of some amphibious animals. Mare's flesh is the only food which the soldiers have when on an expedition. This gives them a very great facility of movement; for the distance to which horses can be driven over these plains is quite surprising: I have been assured that an unloaded horse can travel a hundred miles a day for many days successively.

The encampment of General Rosas was close to the river. It consisted of a square formed by waggons, artillery, straw huts, &c. The soldiers were nearly all cavalry; and I should think such a villanous, banditti-like army, was never before collected together. The greater number of men were of a mixed breed, between Negro, Indian, and Spaniard. I know not the reason, but men of such origin seldom have a good expression of countenance. I called on the secretary to show my passport. He began to cross-question me in the most dignified and mysterious manner. By good luck I had a letter of recommendation from the government of Buenos Ayres* to the commandant of Patagones. This was taken to General Rosas, who sent me a very obliging message; and the secretary returned all smiles and graciousness. We took up our residence in the *rancho*, or hovel, of a curious old Spaniard, who had served wth Napoleon in the expedition against Russia.

* * *

General Rosas intimated a wish to see me; a circumstance which I was afterwards very glad of. He is a man of an extraordinary character, and has a most predominant influence in the country, which it seems probable he will use to its prosperity and advancement. He is said to be the owner of 74 square leagues of land, and to have about 300,000 head of cattle. His estates are admirably managed, and are far more productive of corn than any others. He first gained his celebrity by his laws for his own *estancias*, and by disciplining several hundred men, so as to resist with success the attacks of the Indians. There are many stories current about the rigid manner in which his laws were enforced. One of these was, that no man, on penalty of being put into the stocks, should carry his knife on a Sunday: this day being the principal one for gambling and drinking, many quarrels arose, which from the general manner of fighting with the knife often proved fatal. One

---

* I am bound to express, in the strongest terms, my obligation to the government of Buenos Ayres for the obliging manner in which passports to all parts of the country were given me, as naturalist of the *Beagle*.

Sunday the Governor came in great form to pay the *Estancia* a visit, and General Rosas, in his hurry, walked out to receive him with his knife, as usual, stuck in his belt. The steward touched his arm, and reminded him of the law; upon which turning to the Governor, he said he was extremely sorry, but that he must go into the stocks, and that till let out, he possessed no power even in his own house. After a little time the steward was persuaded to open the stocks, and to let him out, but no sooner was this done, than he turned to the steward and said, 'You now have broken the laws, so you must take my place in the stocks.' Such actions as these delighted the Gauchos, who all possess high notions of their own equality and dignity.

General Rosas is also a perfect horseman – an accomplishment of no small consequence in a country where an assembled army elected its general by the following trial: a troop of unbroken horses being driven into a corral, were let out through a gateway, above which was a cross-bar: it was agreed, whoever should drop from the bar on one of these wild animals, as it rushed out, and should be able without saddle or bridle, not only to ride it, but also to bring it back to the door of the corral, should be their general. The person who succeeded was accordingly elected; and doubtless made a fit general for such an army. This extraordinary feat has also been performed by Rosas.

By these means, and by conforming to the dress and habits of the Gauchos, he has obtained an unbounded popularity in the country, and in consequence a despotic power. I was assured by an English merchant, that a man who had murdered another, when arrested and questioned concerning his motive, answered, 'He spoke disrespectfully of General Rosas, so I killed him.' At the end of a week the murderer was at liberty. This doubtless was the act of the general's party, and not of the general himself.

In conversation he is enthusiastic, sensible, and very grave. His gravity is carried to a high pitch: I heard one of his mad buffoons (for he keeps two, like the barons of old) relate the following anecdote: 'I wanted very much to hear a certain piece of music, so I went to the general two or three times to ask him; he said to me, "Go about your business, for I am engaged." I went a second time; he said, "If you come again I will punish you." A third time

I asked, and he laughed. I rushed out of the tent, but it was too late; he ordered two soldiers to catch and stake me. I begged by all the Saints in heaven, he would let me off; but it would not do; when the general laughs he spares neither mad man or sound.' The poor flighty gentleman looked quite dolorous, at the very recollection of the staking. This is a very severe punishment; four posts are driven into the ground, and the man is extended by his arms and legs horizontally, and there left to stretch for several hours. The idea is evidently taken from the usual method of drying hides. My interview passed away without a smile, and I obtained a passport and order for the government post-horses, and this he gave me in the most obliging and ready manner.

In the morning we started for Bahia Blanca, which we reached in two days. Leaving the regular encampment, we passed by the *toldos* of the Indians. These are round like ovens, and covered with hides; by the mouth of each a tapering *chuzo* was stuck in the ground. The *toldos* were divided into separate groups, which belonged to the different caciques' tribes, and the groups were again divided into smaller ones, according to the relationship of the owners. For several miles we travelled along the valley of the Colorado. The alluvial plains on the side appeared fertile, and it is supposed that they are well adapted to the growth of corn. Turning northward from the river, we soon entered on a country, differing from those plains that extend south of the river. The land still continued dry and sterile; but it supported many different kinds of plants, and the grass, though brown and withered, was more abundant, as the thorny bushes were less so. These latter in a short space entirely disappeared, and the plains were left without a thicket to cover their nakedness. This change in the vegetation marks the commencement of the grand calcareo-argillaceous deposit, which I have already noticed as forming the wide extent of the Pampas, and as covering the granitic rocks of Banda Oriental. From the Strait of Magellan to the Colorado, a distance of about 800 miles, the face of the country is every where composed of shingle: the pebbles are chiefly of porphyry, and probably owe their origin to the rocks of the Cordillera. North of the Colorado the bed thins out, and the pebbles become exceedingly small, and here the characteristic vegetation of Patagonia ceases.

* * *

We found the *Beagle* had not arrived, and consequently set out on our return, but the horses soon tiring, we were obliged to bivouac on the plain. In the morning we had caught an armadillo, which, although a most excellent dish when roasted in its shell, did not make a very substantial breakfast and dinner for two hungry men. The ground at the place where we stopped for the night, was incrusted with a layer of Glauber salt, and hence, of course, was without water. Yet many of the smaller rodents managed to exist even here, and the tucutuco was making its odd little grunt beneath my head, during half the night. Our horses were very poor ones, and in the morning they were soon exhausted from not having had any thing to drink, so that we were obliged to walk. About noon the dogs killed a kid, which we roasted. I eat some of it, but it made me intolerably thirsty. This was the more distressing as the road, from some recent rain, was full of little puddles of clear water, yet not a drop was drinkable. I had scarcely been twenty hours without water, and only part of the time under a hot sun, yet the thirst rendered me very weak. How people survive two or three days under such circumstances, I cannot imagine: at the same time, I must confess that my guide did not suffer at all, and was astonished that one day's deprivation should be so troublesome to me.

I have several times alluded to the surface of the ground being incrusted with salt. This phenomenon is quite different from that of the salinas, and much more extraordinary. In many parts of South America, wherever the climate is moderately dry, these incrustations occur; but I have nowhere seen them so abundant as near Bahia Blanca. The salt here consists of a large proportion of sulphate of soda mixed with a very little of the common muriate. As long as the ground remains moist in these *salitrales* (as the Spaniards improperly call them, mistaking this substance for saltpetre), nothing is to be seen but an extensive plain composed of a black, muddy soil, supporting scattered tufts of succulent plants. I was therefore much surprised, after a week's hot weather, when I first saw square miles of country, that I had previously ridden over in the former condition, white, as if from a slight fall of snow which the wind had heaped up into partial drifts. This latter appearance is chiefly due to the tendency which the salt has to crystallize, like hoar-frost, round the blades of grass, stumps of wood, or on the top of the broken ground, in lieu of the bottoms of the puddles of water. The salinas, as a general rule, occur in depressions on the more elevated plains; the *salitrales*, either on level tracts elevated a few feet above the level of the sea, and appearing as if lately inundated, or on alluvial land bordering rivers. In this latter case, although I am not absolutely certain, I have strong reasons for believing that the salt is often removed by the waters of the river, and is again reproduced. Several circumstances incline me to think that the black, muddy soil, generates the sulphate of soda. The whole phenomenon is well worth the attention of naturalists: what can be more singular than thus to see square miles of country thinly crusted over with Glauber salt? It may be asked whether plants do not decompose the muriate of soda? but whence comes the sulphuric acid? In Peru, nitrate of soda occurs in beds far thicker than these of the sulphate. Both cases are equally mysterious. I suspect that, as a general rule, the salts of soda are infinitely more common in South America than those of potash.

Two days afterwards, I again rode to the harbour, but to a nearer part of it. When not far from our destination, my companion, the same man as before, spied three people hunting on horseback. He immediately dismounted, and watching them intently, said, 'They don't ride like Christians, and nobody can leave the fort.' The three hunters joined company, and likewise dismounted from their horses. At last one mounted again and rode over the hill out of sight. My companion said, 'We must now get on our horses: load your pistol', and he looked to his own sword. I asked, 'Are they Indians?' – 'Quien sabe? (who knows?) If there are no more than three, it does not signify.' It then struck me, that the one man had gone over the hill to fetch the rest of his tribe. I suggested this; but all the answer I could extort was, 'Quien sabe?' His head and eye never for a minute ceased scanning slowly the distant horizon. I thought his uncommon coolness too good a joke, and asked him why he did not return home. I was startled when he answered, 'We are returning, but in a line so as to pass near a swamp, into which we can gallop the horses as far as they can go, and then trust to our own legs; so that there is no danger.' I did not feel quite so confident of this, and wanted to increase our pace. He said, 'No, not until they do.' When any little inequality concealed us, we galloped; but when in sight continued walking. At last we reached a valley, and turning to the left, galloped quickly to the foot of a hill, he gave me his horse to hold, made the dogs lie down, and then crawled on his hands and knees to reconnoitre. He remained in this position for some time, and at last, bursting out in laughter, exclaimed, 'Mugeres!' (women!) He knew them to be the wife and sister-in-law of the major's son, hunting for ostrich's eggs. I have described this man's conduct, because he acted under the full impression that they were Indians. As soon, however, as the absurd mistake was found out, he gave me a hundred reasons, why they could not have been Indians; but all these were forgotten at the time. We then rode on in peace and quietness to a low point called Punta Alta, whence we could see nearly the whole of the great harbour of Bahia Blanca.

The wide expanse of water is choked up by numerous great mud-banks, which the inhabitants call *Cangrejales*, or crabberies, from the number of small crabs. The mud is so soft, that it is impossible to walk over them, even for the shortest distance. Many of the banks have their surfaces covered with long rushes, the tops of which alone are visible at high water. On one occasion, when in a boat, we were so entangled by these shallows, that we could hardly find our way. Nothing was visible, but the flat beds of mud: the day was not very clear, and there was much refraction, or as the sailors expressed it, 'things loomed high'. The only object within our view, which was not level, was the horizon; rushes looked like bushes unsupported in the air, and water like mud-banks, and mud-banks like water.

We passed the night in Punta Alta, and I employed myself in searching for fossil bones; this point, being a perfect catacomb for monsters of extinct races. The evening was perfectly calm and clear; the extreme monotony of the view gave it an interest, even in the midst of mud-banks and gulls, sand-hillocks, and solitary vultures. In riding back in the morning, we came across a very fresh track of a Puma, but did not succeed in finding him. We saw also a couple of Zorillos, or skunks – odious animals, which are far from uncommon. In general appearance the Zorillo resembles a polecat, but it is rather larger, and much thicker in proportion. Conscious of its power, it roams by day about the open plain, and fears neither dog nor man. If a dog is urged to the attack, its courage is instantly checked by a few drops of the fetid oil, which brings on violent sickness and running at the nose. Whatever is once polluted by it, is for ever useless. Azara says the smell can be perceived at a league distant; more than once, when entering the harbour of Monte Video, the wind being off shore, we have perceived the odour on board the *Beagle*. Certain it is, that every animal most willingly makes room for the Zorillo.

CHAPTER V

Bahia Blanca – Geology – Extinct quadrupeds, four Edentata,
horse, ctenomys – Recent extinction – Longevity of species
– Large animals do not require luxuriant vegetation –
Southern Africa – Siberian fossils – Catalogue of extinct quad-
rupeds of South America – Two species of Ostrich, habits of –
Tinochorus – Oven-bird – Armadilloes – Venomous snake,
toad, lizard – Hybernation of animals – Habits of sea-pen –
Indian wars and massacres – Arrow-head, antiquarian relic

BAHIA BLANCA

THE *BEAGLE* arrived on the 24th of August, and a week afterwards sailed for the Plata. With Captain Fitzroy's consent I was
left behind, to travel by land to Buenos Ayres. I will here add
some observations, which were made during this visit, and on a
previous occasion, when the *Beagle* was employed in surveying
the harbour. Not much can be made out respecting the geology.
At the distance of some miles inland, an escarpment of a great
argillaceo–calcareous formation of rock extends. The space near
the coast consists of plains of hardened mud, and broad bands of
sand-dunes, which present appearances, that can easily be
accounted for by a rise of the land; and of this phenomenon,*
although to a trifling amount, we have other proofs.

At Punta Alta, a low cliff, about 20 feet high, exposes a mass of
partly consolidated shingle, irregularly interstratified with a reddish muddy clay, and containing numerous recent shells. We may
believe a similar accumulation would now take place, on any

---

* A few leagues further south, near the Bay of San Blas, M. D'Orbigny found
  great beds of recent shells elevated between 25 and 30 feet above the level of the
  sea – vol. ii, p. 43.

point, where tides and waves were opposed. In the gravel a considerable number of bones were embedded. Mr Owen, who has undertaken the description of these remains, has not yet examined them with care; but the following list may give some idea of their nature: first, a tolerably perfect head of a megatherium, and a fragment and teeth of two others; second, an animal of the order Edentata, as large as a pony, and with great scratching claws; third and fourth, two great Edentata related to the megatherium, and both fully as large as an ox or horse; fifth, another equally large animal, closely allied or perhaps identical with the Toxodon (hereafter to be described), which had very flat grinding teeth, somewhat resembling those of a rodent; sixth, a large piece of the tessellated covering like that of the armadillo, but of gigantic size; seventh, a tusk which in its prismatic form, and in the disposition of the enamel, closely resembles that of the African boar; it is probable that it belonged to the same animal with the singular flat grinders. Lastly, a tooth in the same state of decay with the others: its broken condition does not allow Mr Owen, without further comparison, to come to any definite conclusion; but the part that is perfect, resembles in every respect the tooth of the common horse.* All these remains were found embedded in a beach which is covered at spring tides; and the space in which they were collected could not have exceeded 150 yards square. It is a remarkable circumstance that so many different species should be found together; and it proves how numerous in kind the ancient inhabitants of this country must have been.

At the distance of about 30 miles, in another cliff of red earth, I found several fragments of bones. Among them were the teeth of a rodent, much narrower, but even larger than those of the *Hydrochœrus capybara*; the animal which has been mentioned as exceeding in dimensions every existing member of its order. There was also part of the head of a Ctenomys; the species

---

* With respect to the remains of the last animal, as some doubt may be entertained by others, respecting its origin; it must be remarked, that it was fairly embedded in the gravel with the other bones; and that its state of decay was equal. To this circumstance it may be added, that the surrounding country is without fresh water, and is uninhabited, and that the settlement, itself only of five years standing, is 25 miles distant.

being different from the Tucutuco, but with a close general resemblance.

The remains at Punta Alta were associated, as before remarked, with shells of existing species. These have not as yet been examined with scrupulous care, but it may be safely asserted, that they are most closely similar to the species now living in the same bay: it is also very remarkable, that not only the species, but the proportional numbers of each kind, are nearly the same with those now cast up on the pebble beaches. There are eleven marine species (some in an imperfect state), and one terrestrial. If I had not collected living specimens from the same bay, some of the fossils would have been thought extinct; for Mr Sowerby, who was kind enough to look at my collection, had not previously seen them. We may feel certain that the bones have not been washed out of an older formation, and embedded in a more recent one, because the remains of one of the Edentata were lying in their proper relative position (and partly so in a second case); which could not have happened, without the carcass had been washed to the spot where the skeleton is now entombed.

We here have a strong confirmation of the remarkable law so often insisted on by Mr Lyell, namely, that the 'longevity of the species in the mammalia, is upon the whole inferior to that of the testacea'.* When we proceed to the southern part of Patagonia, I shall have occasion to describe the case of an extinct camel, from which the same result may be deduced.

From the shells being littoral species (including one terrestrial), and from the character of the deposit, we may feel absolutely certain that the remains were embedded in a shallow sea, not far from the coast. From the position of the skeleton being undisturbed, and likewise from the fact that full-grown serpulæ were attached to some of the bones, we know that the mass could not have been accumulated on the beach itself. At the present time, part of the bed is daily washed by the tide, while another part has been raised a few feet above the level of the sea. Hence we may infer, that the elevation has here been trifling, since the period when the mammalia, now extinct, were living. This

---

* Principles of Geology, vol. iv, p. 40.

conclusion is in harmony with several other considerations (such as the recent character of the beds underlying the Pampas deposit), but which I have not space in this work to enter on.

From the general structure of the coast of this part of South America, we are compelled to believe, that the changes of level have all (at least of late) been in one direction, and that they have been very gradual. If, then, we look back to the period when these quadrupeds lived, the land probably stood at a level, less elevated only by a few fathoms than at present. Therefore, its general configuration since that epoch cannot have been greatly modified; a conclusion which certainly would be drawn from the close similarity in every respect, between the shells now living in the bay (as well as in the case of the one terrestrial species) with those which formerly lived there.

The surrounding country, as may have been gathered from this journal, is of a very desert character. Trees nowhere occur, and only a few bushes, which are chiefly confined to depressions among the sand-hillocks, or to the borders of the saline marshes. Here, then, is an apparent difficulty: we have the strongest evidence that there has occurred no great physical change to modify the features of the country, yet in former days, numerous large animals were supported on the plains now covered by a thin and scanty vegetation.

That large animals require a luxuriant vegetation, has been a general assumption, which has passed from one work to another. I do not hesitate, however, to say that it is completely false; and that it has vitiated the reasoning of geologists, on some points of great interest in the ancient history of the world. The prejudice has probably been derived from India, and the Indian islands, where troops of elephants, noble forests, and impenetrable jungles, are associated together in every account. If, on the other hand, we refer to any work of travels through the southern parts of Africa, we shall find allusions in almost every page either to the desert character of the country, or to the numbers of large animals inhabiting it. The same thing is rendered evident by the many sketches which have been published of various parts of the interior. When the *Beagle* was at Cape Town, I rode a few leagues into the country, which at least was sufficient to render that which I had read more fully intelligible.

Dr Andrew Smith, who, at the head of his adventurous party, has so lately succeeded in passing the Tropic of Capricorn, informs me that, taking into consideration the whole of the southern part of Africa, there can be no doubt of its being sterile country. On the southern and south-eastern coasts there are some fine forests, but with these exceptions, the traveller may pass, for days together, through open plains, covered by a poor and scanty vegetation. It is difficult to convey any accurate idea of degrees of comparative fertility; but it may be safely said, that the amount of vegetation supported at any one time* by Great Britain, exceeds, perhaps even tenfold, the quantity on an equal area, in the interior parts of Southern Africa. The fact that bullock-waggons can travel in any direction, excepting near the coast, without more than occasionally half an hour's delay, gives, perhaps, a more definite notion of its scantiness. Now if we look to the animals inhabiting these wide plains, we shall find their numbers extraordinarily great, and their bulk immense. We must enumerate the elephant, three species of rhinoceros, and as Dr Smith is convinced two others also, the hippopotamus, giraffe, the bos caffer— as large as a full-grown bull, and the elan – but little less, two zebras, and the quaccha, two gnus, and several antelopes even larger than these latter animals. It may be supposed that although the species are numerous, the individuals of each kind are few. By the kindness of Dr Smith, I am enabled to show that the case is very different. He informs me, that in lat. 24°, in one day's march with the bullock-waggons, he saw, without wandering to any great distance on either side, between 100 and 150 rhinoceroses, which belonged to three species. That the same day he saw several herds of giraffes, amounting together to nearly 100; and that, although no elephant was observed, yet they are found in this district. At the distance of a little more than one hour's march from their place of encampment on the previous night, his party actually killed at one spot eight hippopotamuses, and saw many

---

* I mean by this to exclude the total amount, which may have been successively produced and consumed during a given period.

more. In this same river there were likewise crocodiles. Of course it was a case quite extraordinary, to see so many great animals crowded together, but it evidently proves that they must exist in great numbers. Dr Smith describes the country passed through that day, as 'being thinly covered with grass, and bushes about four feet high, and still more thinly with mimosa-trees'. The waggons were not prevented travelling in a nearly direct line.

Besides these large animals, every one the least acquainted with the natural history of the Cape, has read of the herds of antelopes, which can be compared only to flocks of migratory birds. The numbers indeed of the lion,* panther, and hyæna, and the multitude of birds of prey, plainly tell of the abundance of the smaller quadrupeds. As Dr Smith remarked to me, the carnage each day in Southern Africa must indeed be terrific! I confess it is truly surprising, how such a number of animals can find support in a country producing so little food. The larger quadrupeds no doubt roam over wide extents in search of it; and their food chiefly consists of underwood, which probably contains much nutriment in a small bulk. Dr Smith also informs me that the vegetation has a rapid growth; no sooner is a part consumed, than its place is supplied by a fresh stock. I apprehend, however, that our ideas respecting the quantity necessary for the support of large quadrupeds are exaggerated. It should have been remembered that the camel, an animal of no mean bulk, has always been considered as the emblem of the desert.

The belief that where large quadrupeds exist, the vegetation must necessarily be luxuriant, is the more remarkable, because the converse is far from true. Mr Burchell observed to me that when entering Brazil, nothing struck him more forcibly than the splendour of the South American vegetation, contrasted with that of South Africa, together with the absence of all large quadrupeds. In his Travels,† he has suggested that the comparison of the respective weights (if there were sufficient data) of an equal number of the largest herbivorous quadrupeds of each country would be extremely curious. If we take on the one side the

---

* Dr Smith mentioned to me, that one evening seven lions were counted at one time walking on the plain, round the encampment.
† Travels in the interior of South Africa, vol. ii, p. 207.

elephant,* hippopotamus, giraffe, bos caffer, elan, certainly three species of rhinoceros, and probably five; and on the other side, two tapirs, the guanaco, three deer, the vicuna, peccari, capybara (after which we must choose from the monkeys to complete the number), and then place these two groups alongside each other, it is not easy to conceive ranks more disproportionate. After the above facts, we are compelled to conclude, against anterior probability,† that among the mammalia there exists no close relation between the *bulk* of the species, and the *quantity* of the vegetation, in the countries they inhabit.

With regard to the number of large quadrupeds, there certainly exists no quarter of the globe which will bear comparison with Southern Africa. After the different statements which have been given, the extremely desert character of that region will not be disputed. In the European division of the world, we must look back to the tertiary epochs, to find a condition of things among the mammalia resembling that which is now found at the Cape of Good Hope. That tertiary epoch, which we are apt to consider as abounding to an astonishing degree with large animals, because we find the remains of many ages accumulated at certain spots, could boast of but few more of the large quadrupeds, than Southern Africa does at present. If we speculate on the condition

---

* The elephant which was killed at Exeter Change, was estimated (being partly weighed) at five tons and a half. The elephant actress, as I was informed, weighed one ton less; so that we may take five as the average of a full-grown elephant. I was told at the Surrey Gardens, that a hippopotamus which was sent to England cut up into pieces, was estimated at three tons and a half; we will call it three. From these premises we may give three tons and a half to each of the five rhinoceroses; perhaps a ton to the giraffe, and half to the bos caffer as well as the elan, (a large ox weighs from 1,200 to 1,500 pounds). This will give an average (from the above conjectures) of 2.7 of a ton for the ten largest herbivorous animals of Southern Africa. In South America, allowing 1,200 pounds for the two tapirs together, 550 for the guanaco and vicuna, 500 for three deer, 300 for capybara, peccari, and a monkey, we shall have an average of 250 pounds, which I believe is overstating the result. The ratio will therefore be, as 6,048 to 250, for the ten largest animals from the two countries.

† If we suppose the case of the discovery of a skeleton of a Greenland whale in a fossil state, not a single cetaceous animal being known to exist, what naturalist would even conjecture on the possibility of a carcass so gigantic, being supported on the minute crustacea and mollusca, living in the frozen seas of the extreme North?

of the vegetation during that epoch, we are at least bound so far to consider existing analogies, as not to urge as absolutely necessary a luxuriant vegetation, when we see a state of things so totally different in the region to which we refer.

We know* that the extreme regions of North America, many degrees beyond the limit where the ground at the depth of a few feet remains perpetually congealed, are covered by forests of large and tall trees. In a like manner, in Siberia, we have woods of birch, fir, aspen, and larch, growing in a latitude† (64°), where the mean temperature of the air falls below the freezing-point, and where the earth is so completely frozen, that the carcass of an animal embedded in it is perfectly preserved. With these facts we must grant, as far as *quantity alone* of vegetation is concerned, that the great quadrupeds of the later tertiary epochs might, in most parts of Northern Europe and Asia, have lived on the spots where their remains are now found. I do not here speak of the *kind* of vegetation necessary for their support; because, as there is evidence of physical changes, and as the animals have become extinct, so may we suppose that the species of plants have likewise been changed.

These remarks directly bear on the case of the Siberian animals preserved in ice. The firm conviction of the necessity of a vegetation, possessing a character of tropical luxuriance, to support such large animals, and the impossibility of reconciling this with the proximity of perpetual congelation, was one chief cause of the several theories of sudden revolutions of climate, and of overwhelming catastrophes, which were invented to account for their entombment. I am far from supposing that the climate has not changed since the period when those animals lived, which now lie buried in the ice. At present I only wish to show, that as

* See Zoological Remarks to Capt. Back's Expedition, by Dr Richardson. He says, 'The subsoil north of latitude 56° is perpetually frozen, the thaw on the coast not penetrating above three feet, and at Bear Lake, in latitude 64°, not more than twenty inches. The frozen substratum does not of itself destroy vegetation, for forests flourish on the surface, at a distance from the coast.'

† See Humboldt Fragmens Asiatiques, p. 386, Barton's Geography of Plants and Malte Brun. In the latter work it is said, that the limit of the growth of trees in Siberia may be drawn under the parallel of 70°.

far as *quantity* of food *alone* is concerned, the ancient rhinoceroses might have roamed over the *steppes* of central Siberia (the northern parts probably being under water) even in their present condition, as well as the living rhinoceroses and elephants over the *Karros* of Southern Africa.

After our long digression, if we return to the case of the fossil animals at Bahia Blanca, there is a difficulty from our not knowing on what food the great Edentata probably lived. If on insects and larvæ, like their nearest representatives the armadilloes and anteaters, there is an end to all conjecture. But as vegetation is the first source of life in every part of the world, I think we may safely conclude that the country around Bahia Blanca, with a very little increase of fertility, would support large animals. The plains of the Rio Negro, thickly scattered over with thorny bushes, I do not doubt would supply sufficient food equally well with the *Karros* of Africa. As there is evidence of a physical change to a small amount, so may we allow it to be probable that the productiveness of the soil has decreased in an equally small degree. With this concession I apprehend every difficulty is removed. On the other hand, if we imagine a luxuriant vegetation to be necessary for the support of these animals, we become involved in a series of contradictions and improbabilities.

As the notices of the remains of several quadrupeds, which I discovered in South America, are scattered in different parts of this volume, I will here give a catalogue of them. After having enlarged on the diminutive size of the present races, it may be of interest to see that formerly a very different order of things prevailed. First, the megatherium, and the four or five other large edentata, already alluded to; sixth, an immense mastodon, which must have abounded over the whole country; seventh, the horse (I do not now refer to the broken tooth at Bahia Blanca, but to more certain evidence); eighth, the toxodon, an extraordinary animal as large as a hippopotamus; ninth, a fragment of the head of an animal larger than a horse, and of a very singular character; tenth, eleventh, and twelfth, parts of rodents – one of considerable size; lastly, a llama or guanaco, fully as large as the camel. All these animals coexisted during an epoch which, geologically speaking is so recent, that it may be considered as only just gone by. These remains have been presented to the College of Surgeons, where they are now in the hands of those best qualified to appreciate whatever value they may possess.

I will now give an account of the habits of some of the more interesting birds, which are common on these wild plains; and first of the *Struthio Rhea*, or South American ostrich. This bird is well known to abound over the plains of Northern Patagonia, and the united provinces of La Plata. It has not crossed the Cordillera; but I have seen it within the first range of mountains on the Uspallata plain, elevated between 6,000 and 7,000 feet. The ordinary habits of the ostrich are familiar to every one. They feed on vegetable matter; such as roots and grass; but at Bahia Blanca, I have repeatedly seen three or four come down at low water to the extensive mud-banks which are then dry, for the sake, as the Gauchos say, of catching small fish. Although the ostrich in its habits is so shy, wary, and solitary, and although so fleet in its pace, it falls a prey, without much difficulty, to the Indian or Gaucho armed with the bolas. When several horsemen appear in a semicircle, it becomes confounded, and does not know which way to escape. They generally prefer running against the wind; yet at the first start they expand their wings, and like a vessel make all sail. On one fine hot day I saw several ostriches enter a bed of tall rushes, where they squatted concealed, till quite closely approached. It is not generally known that ostriches readily take to the water. Mr King informs me that at the Bay of San Blas, and at Port Valdes in Patagonia, he saw these birds swimming several times from island to island. They ran into the water both when driven down to a point, and likewise of their own accord when not frightened: the distance crossed was about 200 yards. When swimming, very little of their bodies appear above water, and their necks are extended a little forward: their progress is slow. On two occasions, I saw some ostriches swimming across the Santa Cruz river, where its course was about 400 yards wide, and the stream rapid. Captain Sturt, when descending the Murrumbidgee, in Australia, saw two emus in the act of swimming.

The inhabitants who live in the country readily distinguish, even at a distance, the cock bird from the hen. The former is larger and darker coloured,* and has a bigger head. The ostrich, I believe the cock, emits a singular deep-toned, hissing note. When first I heard it, standing in the midst of some sand-hillocks, I thought it was made by some wild beast, for it is a sound that one cannot tell whence it comes, or from how far distant. When we were at Bahia Blanca in the months of September and October, the eggs, in extraordinary numbers, were found all over the country. They either lie scattered single, in which case they are never hatched, and are called by the Spaniards, *huachos*; or they are collected together into a shallow excavation, which forms the nest. Out of the four nests which I saw, three contained twenty-two eggs each, and the fourth twenty-seven. In one day's hunting on horseback sixty-four eggs were found; forty-four of these were in two nests, and the remaining twenty scattered *huachos*. The Gauchos unanimously affirm, and there is no reason to doubt their statement, that the male bird alone hatches the eggs, and for some time afterwards accompanies the young. The cock when on the nest lies very close; I have myself almost ridden over one. It is asserted that at such times they are occasionally fierce, and even dangerous, and that they have been known to attack a man on horseback, trying to kick and leap on him. My informer pointed out to me an old man, whom he had seen much terrified by one chasing him. I observe in Burchell's travels in South Africa, that he remarks, 'having killed a male ostrich, and the feathers being dirty, it was said by the Hottentots to be a nest bird.' I understand that the male emu, in the Zoological Garden, takes charge of the nest: this habit, therefore, is common to the family.

The Gauchos unanimously affirm that several females lay in one nest. I have been positively told, that four or five hen birds have been seen to go, in the middle of the day, one after the other, to the same nest. I may add, also, that it is believed in Africa, that two females lay in one nest.† Although this habit at first appears very strange, I think the cause may be explained in a simple manner. The number of eggs in the nest varies from twenty to forty, and even to fifty; and according to Azara to seventy or

* A Gaucho assured me, that he had once seen a snow-white, or Albino variety, and that it was a most beautiful bird.
† Burchell's Travels, vol. i, p. 280.

eighty. Now although it is most probable, from the number of eggs found in one district being so extraordinarily great, in proportion to that of the parent birds, and likewise from the state of the ovarium of the hen, that she may in the course of the season lay a large number, yet the time required must be very long. Azara states,* that a female in a state of domestication laid seventeen eggs, each at the interval of three days one from another. If the hen were obliged to hatch her own eggs, before the last was laid the first probably would be addled; but if each laid a few eggs at successive periods, in different nests, and several hens, as is stated to be the case, combined together, then the eggs in one collection would be nearly of the same age. If the number of eggs in one of these nests is, as I believe, not greater on an average than the number laid by one female in the season, then there must be as many nests as females, and each cock bird will have its fair share of the labour of incubation; and that during a period when the females could not sit, on account of not having finished laying. I have before mentioned the great numbers of *huachos*, or scattered eggs; so that in one day's hunting the third part were found in this state. It appears odd that so many should be wasted. Does it not arise from the difficulty of several females associating together, and persuading an old cock to undertake the office of incubation? It is evident that there must at first be some degree of association, between at least two females; otherwise the eggs would remain scattered over the wide plains, at distances far too great to allow of the male collecting them into one nest. Some have believed that the scattered eggs were deposited for the young birds to feed on. This can hardly be the case in America, because the *huachos*, although oftentimes found addled and putrid, are generally whole.

When at the Rio Negro, in Northern Patagonia, I repeatedly heard the Gauchos talking of a very rare bird which they called Avestruz Petise. They described it as being less than the common ostrich (which is there abundant), but with a very close general resemblance. They said its colour was dark and mottled, and that its legs were shorter, and feathered lower down than those of the

* Azara, vol. iv, p. 173.

common ostrich. It is more easily caught by the bolas than the other species. The few inhabitants who had seen both kinds, affirmed they could distinguish them apart from a long distance. The eggs of the small species appeared, however, more generally known; and it was remarked, with surprise, that they were very little less than those of the Rhea, but of a slightly different form, and with a tinge of pale blue. Some eggs, picked up on the plains of Patagonia, agree pretty well with this description, and I do not doubt are those of the Petise. This species occurs most rarely on the plains bordering the Rio Negro; but about a degree and a half further south they are tolerably abundant. One Gaucho, however, said he distinctly recollected having seen one, many years before, near the mouth of the Rio Colorado, which is to the north of the Rio Negro. They are said to prefer the plains near the sea. When at Port Desire, in Patagonia (lat. 48°), Mr Martens shot an ostrich; and I looked at it, forgetting at the moment, in the most unaccountable manner, the whole subject of the Petises, and thought it was a two-third grown one of the common sort. The bird was cooked and eaten before my memory returned. Fortunately the head, neck, legs, wings, many of the larger feathers, and a large part of the skin, had been preserved. From these a very nearly perfect specimen has been put together, and is now exhibited in the museum of the Zoological Society. Mr Gould, who in describing this new species did me the honour of calling it after my name, states, that besides the smaller size and different colour of the plumage, the beak is of considerably less proportional dimensions than in the common Rhea; that the tarsi are covered with differently-shaped scales, and that they are feathered six inches beneath the knee. In this latter respect, and in the broader feathers of the wing, this bird perhaps shows more affinity to the gallinaceous family than any other of the Struthionidæ.

Among the Patagonian Indians in the Strait of Magellan, we found a half-Indian, who had lived some years with the tribe, but had been born in the northern provinces. I asked him if he had ever heard of the Avestruz Petise? He answered by saying, 'Why there are none others in these southern countries.' He informed me that the number of eggs in the nest of the petise is considerably less than with the other kind, namely, not more than fifteen on an average; but he asserted that more than one female deposited them. At Santa Cruz we saw several of these birds. They were excessively wary: I think they could see a person approaching when he was so far off as not to distinguish the ostrich. In ascending the river few were seen; but in our quiet and rapid descent, many, in pairs and by fours or fives, were observed. It was remarked, and I think with truth, that this bird did not expand its wings, when first starting at full speed, after the manner of the northern kind. The fact of these ostriches swimming across the river has been mentioned. In conclusion, I may repeat that the *Struthio Rhea* inhabits the country of La Plata as far as a little south of the Rio Negro, in lat. 41°, and that the Petise takes its place in Southern Patagonia; the part about the Rio Negro being neutral territory. Wallis saw ostriches at Batchelor's river (lat. 53° 54'), in the Strait of Magellan, which must be the extreme southern possible range of the Petise. M. D'Orbigny, when at the Rio Negro, made great exertions to procure this bird, but never had the good fortune to succeed. He mentions it in his Travels,* and proposes (in case, I presume, of a specimen being obtained) to call it *Rhea pennata*: a fuller notice was given long before in Dobrizhoffer's Account of the Abipones† (A.D. 1749). He says, 'You must know, moreover, that Emus differ in size and habits in different tracts of land; for those that inhabit the plains of Buenos Ayres and Tucuman are larger, and have black, white, and gray feathers; those near to the Strait of Magellan are smaller and more beautiful, for their white feathers are tipped with black at the extremity, and their black ones in like manner terminate in white.'

* * *

Amongst the Batrachian reptiles, I found only one little toad, which was most singular from its colour. If we imagine, first, that

---

* Vol. ii, p. 76 – When at the Rio Negro, we heard much of the indefatigable labours of this naturalist. M. D'Alcide D'Orbigny, during the years 1826 to 1833, traversed several large portions of South America, and has made a collection, and is now publishing the results on a scale of magnificence, which at once places him in the list of American travellers second only to Humboldt.

† Vol. i (English translation), p. 314.

it had been steeped in the blackest ink, and then when dry, allowed to crawl over a board, freshly painted with the brightest vermilion, so as to colour the soles of its feet and parts of its stomach, a good idea of its appearance will be gained. If it is an unnamed species, surely it ought to be called *diabolicus*, for it is a fit toad to preach in the ear of Eve. Instead of being nocturnal in its habits, as other toads are, and living in damp obscure recesses, it crawls during the heat of the day about the dry sand–hillocks and arid plains, where not a single drop of water can be found. It must necessarily depend on the dew for its moisture; and this probably is absorbed by the skin, for it is known, that these reptiles possess great powers of cutaneous absorption. At Maldonado, I found one in a situation nearly as dry as at Bahia Blanca, and thinking to give it a great treat, carried it to a pool of water; not only was the little animal unable to swim, but, I think, without help would soon have been drowned.

* * *

During my stay at Bahia Blanca, while waiting for the *Beagle*, the place was in a constant state of excitement, from rumours of wars and victories, between the troops of Rosas and the wild Indians. One day an account came, that a small party forming one of the *postas* on the line to Buenos Ayres, had been found all murdered. The next day, three hundred men arrived from the Colorado, under the command of Commandant Miranda. A large portion of these men were Indians (*mansos*, or tame), belonging to the tribe of the Cacique Bernantio. They passed the night here; and it was impossible to conceive any thing more wild and savage than the scene of their bivouac. Some drank till they were intoxicated; others swallowed the steaming blood of the cattle slaughtered for their suppers, and then, being sick from drunkenness, they cast it up again, and were besmeared with filth and gore.

> Nam simul expletus dapibus, vinoque sepultus
> Cervicem inflexam posuit, jacuitque per antrum
> Immensus, saniem eructans, ac frusta cruenta
> Per somnum commixta mero.

In the morning they started for the scene of the murder, with orders to follow the *rastro*, or track, even if it led them to Chile. We subsequently heard that the wild Indians had escaped into the great Pampas, and from some cause the track had been missed. One glance at the *rastro* tells these people a whole history. Supposing they examine the track of 1,000 horses, they will soon guess by seeing how many have cantered the number of men; by the depth of the other impressions, whether any horses were loaded with cargoes; by the irregularity of the footsteps, how far tired; by the manner in which the food has been cooked, whether the pursued travelled in haste; by the general appearance, how long it has been since they passed. They consider a *rastro* of ten days or a fortnight, quite recent enough to be hunted out. We also heard that Miranda struck from the west end of the Sierra Ventana, in a direct line to the island of Cholechel, situated seventy leagues up the Rio Negro. This is a distance of between 200 and 300 miles, through a country completely unknown. What other troops in the world are so independent? With the sun for their guide, mares' flesh for food, their saddle-cloths for beds, – as long as there is a little water, these men would penetrate to the land's end.

A few days afterwards I saw another troop of these banditti-like soldiers start on an expedition against a tribe of Indians at the small salinas, who had been betrayed by a prisoner cacique. The Spaniard who brought the orders for this expedition was a very intelligent man. He gave me an account of the last engagement at which he was present. Some Indians, who had been taken prisoners, gave information of a tribe living north of the Colorado. Two hundred soldiers were sent; and they first discovered the Indians by a cloud of dust from their horses' feet, as they chanced to be travelling. The country was mountainous and wild, and it must have been far in the interior, for the Cordillera was in sight. The Indians, men, women, and children, were about 110 in number, and they were nearly all taken or killed, for the soldiers sabre every man. The Indians are now so terrified, that they offer no resistance in a body, but each flies, neglecting even his wife and children; but when overtaken, like wild animals, they fight against any number to the last moment. One dying Indian seized with his teeth the thumb of his adversary, and allowed his own eye to be forced out, sooner than relinquish his hold. Another, who was wounded, feigned death, keeping a knife ready to strike one more fatal blow. My informer said, when he was pursuing an Indian, the man cried out for mercy, at the same time that he was covertly loosing the bolas from his waist, meaning to whirl it round his head and so strike his pursuer. 'I however struck him with my sabre to the ground, and then got off my horse, and cut his throat with my knife.' This is a dark picture; but how much more shocking is the unquestionable fact, that all the women who appear above twenty years old, are massacred in cold blood. When I exclaimed that this appeared rather inhuman, he answered, 'Why, what can be done? they breed so!'

Every one here is fully convinced that this is the most just war because it is against barbarians. Who would believe in this age, in a Christian civilized country, that such atrocities were committed? The children of the Indians are saved, to be sold or given away as servants, or rather slaves, for as long a time as the owners can deceive them; but I believe in this respect there is little to complain of.

In the battle four men ran away together. They were pursued, and one was killed, but the other three were taken alive. They turned out to be messengers or ambassadors from a large body of Indians, united in the common cause of defence, near the Cordillera. The tribe to which they had been sent was on the point of holding a grand council; the feast of mare's flesh was ready, and the dance prepared: in the morning the ambassadors were to have returned to the Cordillera. They were remarkably fine men, very fair, above 6 feet high, and all under thirty years of age. The three survivors of course possessed very valuable information; and to extort this they were placed in a line. The two first being questioned, answered, 'No se' (I do not know), and were one after the other shot. The third also said, 'No se', adding, 'Fire, I am a man, and can die!' Not one syllable would they breathe to injure the united cause of their country! The conduct of the cacique was very different: he saved his life by betraying the intended plan of warfare, and the point of union in the Andes. It was believed that there were already 600 or 700 Indians together, and that in summer their numbers would be doubled. Ambassadors were to have been sent to the Indians at the small salinas, near Bahia Blanca, whom I mentioned that a cacique, this same man, had betrayed. The communication, therefore, extends from the Cordillera to the east coast.

General Rosas's plan is to kill all stragglers, and having driven the remainder to a common point, in the summer, with the assistance of the Chilenos, to attack them in a body. This operation is to be repeated for three successive years. I imagine the summer is chosen as the time for the main attack, because the plains are then without water, and the Indians can only travel in particular directions. The escape of the Indians to the south of the Rio Negro, where in such a vast unknown country they would be safe, is prevented by a treaty with the Tehuelches to this effect; — that Rosas pays them so much to slaughter every Indian who passes to the south of the river, but if they fail in so doing, they themselves are to be exterminated. The war is waged chiefly against the Indians near the Cordillera; for many of the tribes on this eastern side are fighting with Rosas. The general, however, like Lord Chesterfield, thinking that his friends may in a future day become his enemies, always places them in the front ranks, so that their numbers may be thinned. Since leaving South America we have heard that this war of extermination completely failed.

Among the captive girls taken in the same engagement, there were two very pretty Spanish ones, who had been carried away by the Indians when young, and could now only speak the Indian tongue. From their account, they must have come from Salta, a distance in a straight line of nearly 1,000 miles. This gives one a grand idea of the immense territory over which the Indians roam: yet, great as it is, I think there will not, in another half-century, be a wild Indian northward of the Rio Negro. The warfare is too bloody to last; the Christians killing every Indian, and the Indians doing the same by the Christians. It is melancholy to trace how the Indians have given way before the Spanish invaders. Schirdel* says, that in 1535, when Buenos Ayres was founded, there were

* Purchas's Collection of Voyages.

villages containing two and three thousand inhabitants. Even in Falconer's time (1750) the Indians made inroads as far as Lucan, Areco, and Arrecife, but now they are driven beyond the Salado. Not only have whole tribes been wholly exterminated, but the remaining Indians have become more barbarous: instead of living in large villages, and being employed in the arts of fishing, as well as of the chase, they now wander about the open plains, without home or fixed occupation.

I heard also some account of an engagement which took place, a few weeks previously to the one mentioned, at Cholechel. This is a very important station, on account of being a pass for horses; and it was, in consequence, for some time the head-quarters of a division of the army. When the troops first arrived there, they found a tribe of Indians, of whom they killed twenty or thirty. The cacique escaped in a manner which astonished every one. The chief Indians always have one or two picked horses, which they keep ready for any urgent occasion. On one of these, an old white horse, the cacique sprung, taking with him his little son. The horse had neither saddle nor bridle. To avoid the shots, the Indian rode in the peculiar method of his nation; namely, with an arm round the horse's neck, and one leg only on its back. Thus hanging on one side, he was seen patting the horse's head, and talking to him. The pursuers urged every effort in the chase; the Commandant three times changed his horse, but all in vain. The old Indian father and his son escaped, and were free. What a fine picture one can form in one's mind, – the naked bronze-like figure of the old man with his little boy, riding like a Mazeppa on the white horse, thus leaving far behind him the host of his pursuers!

I saw one day a soldier striking fire with a piece of flint, which I immediately recognized as having been a part of the head of an arrow. He told me it was found near the island of Cholechel, and that they are frequently picked up there. It was between two and three inches long, and therefore twice as large as those now used in Tierra del Fuego: it was made of opake cream-coloured flint, but the point and barbs had been intentionally broken off. It is well known that no Pampas Indians now use bows and arrows. I believe a small tribe in Banda Oriental must be excepted; but they are widely separated from the Pampas Indians, and border close on those tribes that inhabit the forest, and live on foot. It appears, therefore, that these arrow-heads are antiquarian* relics of the Indians, before the great change in habits consequent on the introduction of the horse into South America.

* Azara has even doubted whether the Pampas Indians ever used bows.

CHAPTER VI

BAHIA BLANCA TO BUENOS AYRES

SEPTEMBER 8TH – Having with some difficulty hired a Gaucho
to accompany me, on my ride to Buenos Ayres, we started early
in the morning. The distance is about 400 miles, and nearly the
whole way through an uninhabited country. Ascending a few
hundred feet from the basin of green turf on which Bahia Blanca
stands, we entered on a wide desolate plain. It consists of a
crumbling argillaceo-calcareous rock, which, from the dry nature
of the climate, supports only scattered tufts of withered grass,
without a single bush or tree to break the monotonous uniformity. The weather was fine, but the atmosphere remarkably
hazy; I thought the appearance foreboded a gale, but the Gaucho
said it was owing to the plain, at some great distance in the
interior, being on fire. After a long gallop, having changed horses
twice, we reached the Rio Sauce. It is a deep, rapid, little stream,
but not above twenty-five feet wide. The second *posta* on the road
to Buenos Ayres stands on its banks; a little above there is a pass
for horses, where the water does not reach to the horse's belly; but
from that point, in its course to the sea, it is quite impassable, and
hence makes a most useful barrier against the Indians.

Insignificant as this stream is, the Jesuit Falconer, whose information is generally so very correct, figures it as a considerable river, rising at the foot of the Cordillera. With respect to its source, I do not doubt this is the case; for the Gauchos assured me, that in the middle of the dry summer, this stream, at the same time with the Colorado, has periodical floods; which can only originate in the snow melting on the Andes. It is extremely improbable that a stream, so small as the Sauce then was, should traverse the entire width of the continent; and indeed, if it were the residue of a large river, its waters, as in other ascertained cases, would be saline. During the winter we must look to the springs round the Sierra Ventana as the source of its pure and limpid stream. I suspect the plains of Patagonia, like those of Australia, are traversed by many water-courses, which only perform their proper parts at certain periods. Probably this is the case with the water which flows into the head of Port Desire, and likewise with the Rio Chupat, on the banks of which masses of highly cellular scoriæ were found by the officers employed in the survey.

As it was early in the afternoon when we arrived, we took fresh horses, and a soldier for a guide, and started for the Sierra de la Ventana. This mountain is visible from the anchorage at Bahia Blanca; and Capt. FitzRoy calculates its height to be 3,500 feet; – an altitude very remarkable on this eastern side of the continent. I am not aware that any foreigner, previous to my visit, had ascended this mountain; and indeed very few of the soldiers at Bahia Blanca knew any thing about it. Hence we heard of beds of coal, of gold and silver, of caves, and of forests, all of which inflamed my curiosity, only to disappoint it. The distance from the *posta* was about six leagues, over a level plain of the same character as before. The ride was, however, interesting, as the mountain began to show its true form. When we reached the foot of the main ridge, we had much difficulty in finding any water, and we thought we should have been obliged to have passed the night without any. At last we discovered some, by looking close to the mountain, for at the distance even of a few hundred yards, the streamlets were buried and entirely lost in the friable calcareous stone, and loose detritus. I do not think nature ever made a more solitary, desolate pile of rock; it well deserves its name of *Hurtado*, or separated. The mountain is steep, extremely rugged, and broken, and so entirely destitute of trees and even bushes, that we actually could not find a skewer to stretch out our meat over the fire made of thistle* stalks. The strange aspect of this mountain is contrasted by the sea-like plain, which not only abuts against its steep sides, but likewise separates the parallel ranges. The uniformity of the colouring gives, also, an extreme quietness to the view; the whitish gray of the quartz rock, and the light brown of the withered grass of the plain, being unrelieved by any brighter tint. From custom, one expects to see in the neighbourhood of a lofty and bold mountain a broken country, strewed over with huge fragments. Here nature shows, that the last movement before the bed of the sea is changed into dry land, may sometimes be one of tranquillity. Under these circumstances, I was curious to observe how far from the parent rock any pebbles could be found. On the shores of Bahia Blanca, and near the settlement, there were some of quartz, which certainly must have come from this source: the distance is forty-five miles.

The dew, which in the early part of the night wetted the saddle-cloths, under which we slept, was in the morning frozen. From the sharpness of the cold, I supposed we were already at a considerable elevation, although, to the eye, the plain had appeared horizontal. In the morning (9th September) the guide told me to ascend the nearest ridge, which he thought would lead me to the four peaks that crown the summit. The climbing up such rough rocks was very fatiguing, the sides were so indented, that, what was gained in one five minutes, was often lost in the next. At last, when I reached the ridge, my disappointment was extreme in finding a precipitous valley as deep as the plain, which cut the chain transversely in two, and separated me from the four points. This valley is very narrow, but flatbottomed, and it forms a fine horse-pass for the Indians, as it connects the plains on the northern and southern sides of the range. Having descended, and while crossing it, I saw two horses grazing: I immediately hid myself in the long grass, and began to reconnoitre; but as I could see no signs of Indians, I proceeded cautiously on my second ascent. It was late in the day, and this part of the mountain, like the other, was steep and rugged. I was on the top of the second peak

* I call these thistle stalks for the want of a more correct name. I believe it is a species of Eryngium.

by two o'clock, but got there with extreme difficulty; every twenty yards I had the cramp in the upper part of both thighs, so that I was afraid I should not have been able to have descended. It was also necessary to return by another road, as it was out of the question to pass over the saddle-back. I was therefore obliged to give up the two higher peaks. Their altitude was but little greater, and every purpose of geology had been answered; so that the attempt was not worth the hazard of any further exertion. I presume the cause of the cramp was the great change in the kind of muscular action, from that of hard riding to that of still harder climbing. It is a lesson worth remembering, as in some cases it might cause much difficulty.

I have already said the mountain is composed of white quartz rock, and with it a little glossy clay-slate is associated. At the height of a few hundred feet above the plain, patches of conglomerate adhered in several places to the solid rock. They resembled in hardness, and in the nature of the cement, the masses which may be seen daily forming on some coasts. I do not doubt these pebbles were, in a similar manner, aggregated, at a period when the great calcareous formation was depositing beneath the surrounding sea. We may believe that the jagged and battered forms of the hard quartz yet show the effects of the waves of an open ocean.

I was, on the whole, disappointed with this ascent. Even the view was insignificant; a plain like the sea, but without its beautiful colour and defined outline. The scene, however, was novel, and a little danger, like salt to meat, gives it a relish. That the danger was very little was certain, for my two companions made a good fire – a thing which is never done when it is suspected that Indians are near. I reached the place of our bivouac by sunset, and drinking much maté, and smoking several cigaritos, soon made up my bed for the night. The wind was very strong and cold, but I never slept more comfortably.

* * *

SEPTEMBER 19TH – Passed the Guardia del Monte. This is a nice scattered little town, with many gardens, full of peach- and quince-trees. The plain here looked like that around Buenos Ayres; the turf being short and bright green, with beds of clover and thistles, and with bizcacha holes. I was very much struck with the marked change in the aspect of the country after having crossed the Salado. From a coarse herbage we passed on to a carpet of fine green verdure. I at first attributed this to some change in the nature of the soil, but the inhabitants assured me that in this part, as well as in Banda Oriental, where there was as great a difference between the country around Monte Video and the thinly inhabited savannahs of Colonia, that the whole was to be attributed to the manuring and grazing of the cattle. I am not botanist enough to say, whether the change is owing to the introduction of new species, to the altered growth of the same, or to a difference in their proportional numbers. Azara has also observed with astonishment this change: he is likewise much perplexed by the immediate appearance of plants not occurring in the neighbourhood, on the borders of any track that leads to a newly constructed hovel. In another part he says,* 'Ces chevaux (sauvages) ont la manie de préférer les chemins, et le bord des routes pour déposer leurs excrémens, dont on trouve des monceaux dans ces endroits.' Does this not partly explain the circumstance? We thus have lines of richly manured land serving as channels of communication across wide districts.

Near the Guardia we find the southern limit of two European plants, now become excessively common. The fennel in great profusion covers the ditch banks in the neighbourhood of Buenos Ayres, Monte Video, and other towns. But the cardoon (*Cynara cardunculus*)† has a far wider range: it occurs in these latitudes on

---

* Azara's Voyage, vol. i, p. 373.

† D'Orbigny (vol. i, p. 474), says that the cardoon and artichoke are both found wild. Dr Hooker (Botanical Magazine, vol. lv, p. 2862), has described a variety of the Cynara from this part of South America under the name of *inermis*. He states that botanists are now generally agreed that the cardoon and the artichoke are varieties of one plant. I may add, that an intelligent farmer assured me, he had observed in a deserted garden, some artichokes changing into the common cardoon. Dr Hooker believes that Head's vivid description of the thistle of the Pampas applies to the cardoon; but this is a mistake. Captain Head referred to the plant, which I have mentioned a few lines lower down, under the title of giant thistle. Whether it is a true thistle I do not know; but it is quite different from the cardoon, and more like a thistle properly so called.

both sides of the Cordillera, across the continent. I saw it in unfrequented spots in Chile, Entre Rios, and Banda Oriental. In the latter country alone, very many (probably several hundred) square miles are covered by one mass of these prickly plants, and are impenetrable by man or beast. Over the undulating plains, where these great beds occur, nothing else can live. Before their introduction, however, I apprehend the surface supported as in other parts a rank herbage. I doubt whether any case is on record, of an invasion on so grand a scale of one plant over the aborigines. As I have already said, I nowhere saw the cardoon south of the Salado; but it is probable that in proportion as that country becomes inhabited, the cardoon will extend its limits. The case is different with the giant thistle (with variegated leaves) of the Pampas, for I met with it in the valley of the Sauce. According to the principles so well laid down by Mr Lyell, few countries have undergone more remarkable changes, since the year 1535, when the first colonist of La Plata landed with seventy-two horses. The countless herds of horses, cattle, and sheep, not only have altered the whole aspect of the vegetation, but they have almost banished the guanaco, deer, and ostrich. Numberless other changes must likewise have taken place; the wild pig in some parts probably replaces the peccari; packs of wild dogs may be heard howling on the wooded banks of the less frequented streams; and the common cat, altered into a large and fierce animal, inhabits rocky hills. I have alluded to the invasion of the cardoon: in a like manner, the islands near the mouth of the Parana, are thickly clothed with peaches and orange-trees, springing from seeds carried there by the waters of the river.

While changing horses at the Guardia, several people questioned us much about the army – I never saw any thing like the enthusiasm for Rosas, and for the success of the 'most just of all wars, because against barbarians'. This expression it must be confessed is very natural, for till lately, neither man, woman, nor horse, was safe from the attacks of the Indians. We had a long day's ride over the same rich green plain abounding with various flocks, and with here and there a solitary *estancia*, and its one *ombu* tree. In the evening it rained heavily: on arriving at a post-house, we were told by the owner that if we had not a regular passport we must pass on, for there were so many robbers he would trust no one. When he read, however, my passport, which began with 'El Naturalista Don Carlos, &c.' his respect and civility were as unbounded, as his suspicions had been before. What a naturalist may be, neither he nor his countrymen, I suspect, had any idea; but probably my title lost nothing of its value from that cause.

SEPTEMBER 20TH – We arrived by the middle of the day at Buenos Ayres. The outskirts of the city looked quite pretty, with the agave hedges, and groves of olive, peach, and willow trees, all just throwing out their fresh green leaves. I rode to the house of Mr Lumb, an English merchant, to whose kindness and hospitality, during my stay in the country, I was greatly indebted.

The city of Buenos Ayres is large*; and I should think one of the most regular in the world. Every street is at right angles to the one it crosses, and the parallel ones being equidistant, the houses are collected into solid squares of equal dimensions, which are called quadras. On the other hand, the houses themselves are hollow squares; all the rooms opening into a neat little courtyard. They are generally only one story high, with flat roofs, which are fitted with seats, and are much frequented by the inhabitants in summer. In the centre of the town is the Plaza, where the public offices, fortress, cathedral, &c., stand. Here also, the old viceroys, before the revolution, had their palaces. The general assemblage of buildings possesses considerable architectural beauty, although none individually can boast of any.

The great *corral* where the animals are kept for slaughter to supply food to this beef-eating population, is one of the spectacles best worth seeing. The strength of the horse as compared to that of the bullock is quite astonishing: a man on horseback having thrown his lazo round the horns of a beast, can drag it any where he chooses. The animal having ploughed up the ground with outstretched legs, in vain efforts to resist the force, generally dashes at full speed to one side; but the horse immediately turning to receive the shock, stands so firmly, that the bullock is almost thrown down, and one would think, would certainly have its

---

* Is said to contain 60,000 inhabitants. Monte Video, the second town of importance on the banks of the Plata, has 15,000.

neck dislocated. The struggle is not, however, one of fair strength; the horse's girth being matched against the bullock's extended neck. In a similar manner a man can hold the wildest horse, if caught with the lazo, just behind the ears. When the bullock has been dragged to the spot where it is to be slaughtered, the *matador* with great caution cuts the hamstrings. Then is given the death bellow; a noise more expressive of fierce agony than any I know: I have often distinguished it from a long distance, and have always known that the struggle was then drawing to a close. The whole sight is horrible and revolting, the ground is almost made of bones; and the horses, and riders are drenched with gore.

CHAPTER VII

BUENOS AYRES TO SANTA FE

SEPTEMBER 27TH – In the evening I set out on an excursion to St Fe, which is situated nearly 300 English miles from Buenos Ayres, on the banks of the Parana. The roads in the neighbourhood of the city, after the rainy weather were extraordinarily bad. I should never have thought it possible for a bullock-waggon to have crawled along: as it was, they scarcely went at the rate of a mile an hour, and a man was kept ahead, to survey the best line for making the attempt. The bullocks were terribly jaded: it is a great mistake to suppose that with improved roads, and an accelerated velocity of travelling, the sufferings of the animals increase in the same proportion. We passed a train of waggons and a troop of beasts on their road to Mendoza. The distance is about 580 geographical miles, and the journey is generally performed in fifty days. These waggons are very long, narrow, and thatched with reeds; they have only two wheels, the diameter of which in some cases is even 10 feet. Each is drawn by six bullocks which are urged on by a goad at least 20 feet long: this is suspended from within the roof; for the wheel bullocks a smaller one is kept; and for the intermediate pair, a point projects at right angles from the middle of the long one. The whole apparatus looked like some implement of war.

*   *   *

OCTOBER 1ST – We started by moonlight and arrived at the Rio Tercero by sunrise. This river is also called the Saladillo, and it deserves the name, for the water is brackish.

I stayed here the greater part of the day, searching for fossil bones. Falconer mentions having seen, in the bed of this river, great bones, and the case of a giant armadillo. By good fortune, I discovered a tooth embedded in a layer of rock marl, which was afterwards found exactly to fit the socket in the head of a strange animal, the Toxodon, which will presently be mentioned. Hearing also of the remains of one of the old giants which a man told me he had seen on the banks of the Parana, I procured a canoe, and proceeded to the place. Two groups of immense bones projected in bold relief from the perpendicular cliff. They were, however, so completely decayed, that I could only bring away small fragments of one of the great molar-teeth; but these were sufficient to show that the remains belonged to a species of Mastodon. The men who took me in the canoe, said they had long known of them, and had often wondered how they had got there: the necessity of a theory being felt, they came to the conclusion, that, like the bizcacha, the mastodon formerly was a burrowing animal! In the evening we rode another stage, and crossed the Monge, another brackish stream, bearing the dregs of the washings of the Pampas.

OCTOBER 2ND – We passed through Corunda, which, from the luxuriance of its gardens, was one of the prettiest villages I saw. From this point to St Fe the road is not very safe. The western side of the Parana further northward, ceases to be inhabited; and hence the Indians sometimes come down, and waylay travellers. The nature of the country also favours this, for instead of a grassy plain, there is an open woodland, composed of low prickly mimosas. We passed some houses that had been ransacked and since deserted; we saw also a spectacle, which my guides viewed with high satisfaction; it was the skeleton of an Indian with the dried skin hanging on the bones, suspended to the branch of a tree.

In the morning we arrived at St Fe. I was surprised to observe how great a change of climate a difference of only three degrees of latitude between this place and Buenos Ayres had caused. This was evident from the dress and complexion of the men – from the increased size of the ombu-trees – the number of new cacti and other plants – and especially from the birds. In the course of an hour I remarked half-a-dozen of the latter, which I had never seen at Buenos Ayres. Considering that there is no natural boundary between the two places, and that the character of the country is nearly similar, the difference was much greater than I should have expected.

OCTOBER 3RD AND 4TH – I was confined to my bed by a headache for these two days. A goodnatured old woman, who attended me, wished me to try many odd remedies. A common practice is, to bind an orange-leaf, or a bit of black plaster, to each temple; and a still more general plan is, to split a bean into halves, moisten them, and place one on each temple, where they will easily adhere. It is not thought proper ever to remove the beans or plaster, but to allow them to drop off; and sometimes, if a man, with patches on his head, is asked, what is the matter? he will answer, 'I had a headache the day before yesterday.'

St Fe is a quiet little town, and is kept clean, and in good order. The governor, Lopez, was a common soldier at the time of the revolution; but has now been seventeen years in power. This stability of government is owing to his tyrannical habits; for tyranny seems as yet better adapted to these countries, than republicanism. The governor's favourite occupation is hunting Indians: a short time since he slaughtered forty-eight, and sold the children at the rate of £3 or £4 apiece.

OCTOBER 5TH – We crossed the Parana to St Fe Bajada, a town on the opposite shore. The passage took some hours, as the river here consisted of a labyrinth of small streams, separated by low wooded islands. I had a letter of introduction to an old Catalonian Spaniard, who treated me with the most uncommon hospitality. The Bajada is the capital of Entre Rios. In 1825 the town contained 6,000 inhabitants, and the province 30,000; yet, few as they are, none have suffered more from bloody and desperate revolutions. They boast here of representatives, ministers, a standing army, and governors: so it is no wonder that they have their revolutions. At some future day this must be one of the richest countries of La Plata. The soil is varied and productive, and its almost insular form gives it two grand lines of communication by the rivers Parana and Uruguay.

I was delayed here five days, and employed myself in examining the geology of the surrounding country, which was very interesting. We here see beds of sand, clay, and limestone, containing sea-shells and sharks' teeth, passing above into an indurated marl, and from that into the red clayey earth of the Pampas, with its calcareous concretions and the bones of terrestrial animals. This vertical section clearly tells us, of a large bay of pure salt-water, gradually encroached on, and at last becoming the bed of a muddy estuary, into which floating carcasses were swept. I found near the Bajada a large piece, nearly 4 feet across, of the giant armadillo-like case; also a molar tooth of a mastodon, and fragments of very many bones, the greater number of which were rotten, and as soft as clay.

A tooth which I discovered by one point projecting from the side of a bank, interested me much, for I at once perceived that it had belonged to a horse. Feeling much surprise at this, I carefully examined its geological position, and was compelled to come to the conclusion,* that a horse, which cannot from a comparison of the tooth alone, be distinguished from the existing species,† lived as a contemporary with the various great monsters that formerly inhabited South America. Mr Owen and myself, at the College of Surgeons, compared this tooth with a fragment of another, probably belonging to the Toxodon, which was embedded at the

---

* The broken tooth mentioned at Bahia Blanca must not be forgotten.

† As this horse existed at the same time with animals now extinct, it is not probable, that it is the same species with the recent kind, although from the similarity of the teeth it must have been closely allied. Cuvier, talking of the remains of the horse, found fossil under similar conditions in Europe, remarks, 'It is not possible to say whether it was one of the species now existing or not, because the skeletons of these species are so like each other, that they cannot be distinguished by the mere comparison of isolated fragments.' *Theory of the Earth*, English translation, p. 285.

distance only of a few yards in the same earthy mass. No sensible difference in their state of decay could be perceived; they were both tender, and partially stained red. If the horse did not coexist with the Toxodon, the tooth must by some accident, not very easily understood, have been embedded within the last three centuries (the period of the introduction of the horse), with the remains of those animals, which ages since perished, when the Pampas was covered by the waters of the sea. Now, I may ask, will any one credit that two teeth of nearly equal size, buried in the same substance close together, after a period of so vast an inequality, could exist in the same condition of decay? We must conclude otherwise. Certainly it is a marvellous event in the history of animals, that a native kind should have disappeared* to be succeeded in after ages by the countless herds introduced with the Spanish colonist! But our surprise should be modified when it is already known, that the remains of the *Mastodon angustidens* (the tooth formerly alluded to as embedded near that of the horse, probably belonged to this species) have been found both in South America, and in the southern parts of Europe.

With regard to North America, Cuvier says† the *Elephas primigenus* 'has left thousands of its carcasses from Spain to the shores of Siberia, and it has been found in the whole of North America'. The fossil ox, in a like manner, he writes,‡ is buried 'dans toute la partie boréale des deux continens, puisque on en a d'Allemagne, d'Italie, de Prussie, de la Sibérie occidentale et orientale, et de l'Amérique'. I may here add that horses' bones, mingled with those of the mastodon, have several times been transmitted for sale from North America to England; but it has always been imagined, from the simple fact of their being horses' bones, that they had been accidentally mingled with the fossils. Among the remains brought home by Captain Beechey from the west coast of the same continent, in the frozen region of 66° north, Dr Buckland§ has described the astragalus metacarpus, and

---

* I need not here state, that there is no kind of evidence to support the belief, that a horse existed in America previously to the age of Columbus.
† Theory of the Earth, p. 281 (English translation).
‡ Ossemens Fossiles, vol. iv, p. 147.
§ See the admirable Appendix to Beechey's Voyage, p. 592 (quarto edition).

metatarsus of the horse, which were associated with the remains of the *Elephas primigenus*, and of the fossil ox. Thus we have an elephant, an ox, and a horse (the species of the latter is only presumed to be identical), common to Europe and to North America.

Very few species of living quadrupeds,* which are altogether terrestrial in their habits, are common to the two continents, and these few are chiefly confined to the extreme frozen regions of the north. The separation, therefore, of the Asiatic and American zoological provinces appears formerly to have been less perfect than at present. The remains of the elephant and of the ox have been found on the banks of the Anadir (long. 175° E), on the extreme part of Siberia, nearest the American coast: and the former remains, according to Chamisso, are common in the peninsula of Kamtschatka. On the opposite shores, likewise, of the narrow strait which divides these two great continents, we know, from the discoveries of Kotzebue and Beechey, that the remains of both animals occur abundantly: and as Dr Buckland has shown they are associated with the bones of the horse, the teeth of which animal in Europe, according to Cuvier, accompany by thousands the remains of the pachydermata of the later periods. With these facts, we may safely look at this quarter, as the line of communication (now interrupted by the steady progress of geological change) by which the elephant, the ox, and the horse, entered America, and peopled its wide extent.†

The occurrence of the fossil horse and of *Mastodon angustidens* in South America, is a much more remarkable circumstance than that of the animals mentioned above in the northern half of the continent; for if we divide America, not by the Isthmus of

---

* See Dr Richardson's very interesting Report on North American Zoology for the British Association of 1836.

† I do not here mean to fix on the more northern parts of the old world as the parent country of these two animals. I only want to point out the channel of communication – not the course of the stream – whether from west to east, or the reverse. Perhaps, when we recollect how extraordinarily the Pachydermata abounded during the Tertiary epochs in the Old World, and that the representatives of these animals now only exist in that quarter, it may seem most probable that the migration took place from Asia to America.

Panama, but by the southern part of Mexico,* in lat. 20°, where the great table-land presents an obstacle to the migration of species, by affecting the climate, and by forming, with the exception of some valleys and of a fringe of low land on the coast, a broad barrier; we shall then have two zoological provinces strongly contrasted with each other. Some few species alone have passed the barrier, and may be considered as wanderers, such as the puma, opossum, kinkajou, and peccari. The mammalogy of South America is characterized by possessing several species of the genera of llama, cavy (and the allied animals), tapir, peccari, opossum, anteater, sloth, and armadillo. If North America had possessed species of these genera proper to it, the distinction of the two provinces could not have been drawn; but the presence of a few wanderers scarcely affects the case. North America, on the other hand, is characterized by its numerous rodents,† and by four genera of solid horned ruminants,‡ of which section the southern half does not possess a single species.

This distinction of the two zoological provinces does not appear always to have existed. At the present day the order of Edentata is much more strongly developed in South America, than in any other part of the world: and concluding from the fossil remains, which were discovered at Bahia Blanca, such must have been the case during a former epoch. In America, north of Mexico, not one of this order is now found: yet, as is well known, the gigantic megalonyx, considered by Cuvier as a species of Megatherium, has been found only in that country; and as it appears from recent observations,§ the Megatherium Cuvierii itself likewise occurs there. Mr Owen showed me the tibia of

* This is the division followed by Lichtenstein, Swainson, and Richardson. The section from Vera Cruz to Acapulco, given by Humboldt in the Atlas to Polit. Essay on Kingdom of N Spain, will show how immense a barrier the Mexican table-land forms.

† Dr Richardson (Report to Brit. Assoc., p. 157), talking of the identification of a Mexican animal with the *Synetheres prehensilis*, says, 'We do not know with what propriety, but, if correct, it is, if not a solitary instance, at least very nearly so, of a rodent animal being common to North and South America.'

‡ *Dicranocerus furcifer, Capra Americana, Ovis montana, Bos Americana*, and *Moschatus* – Report to Brit. Assoc., p. 159.

§ Ed. New Phil. Journal. July, 1828, p. 327. From a paper by Mr Cooper in the Lyceum of Natural History of New York.

some large animal, which Sir Philip Egerton had purchased out of a collection of the remains of the mastodon brought from North America. Mr Owen says it certainly belongs to one of the Edentata, and it so closely resembles a bone which I found embedded, together with fragments of the great armadillo-like covering, in Banda Oriental, that it probably forms a species of the same genus. Lastly, among the fossils brought home by Captain Beechey from the NW coast, there was a cervical vertebra, which, when compared by Mr Pentland* with the skeletons at Paris, was found to resemble that of the sloth and anteater more than that of any other animal, although having some points of essential difference.

Of the Pachydermata four or five species are now found in America; but, as in the case of the Edentata, none are peculiar to the continent north of Mexico; and one alone seems to exist there as a wanderer. Yet the account of the multitude of bones of the mastodon and elephant, which have been discovered in the salt-licks of North America, is familiar to every one. The remains of the *Mastodon giganteum* have been found nowhere else; but those of the *Elephas primigenus* are common to a large part of the terrestrial globe.† This elephant must have existed in Mexico; and Cuvier,‡ judging from a fragment of a tusk, thinks it even extended to the neighbourhood of Quinto in South America. In the latter country three species of Mastodon have been discovered. One of these, *M. angustidens*, is common to Europe. It is singular that its remains, as yet, have never been brought from North America; nevertheless, considering that it was a contemporary of the extinct animals above mentioned, it seems highly probable that it arrived by the same line of communication on the NW coast. As its remains have frequently been found at a great

* See Dr Buckland. Appendix to Beechey's Voyage, p. 597.
† I may observe that at the present day both species of elephants have wide ranges. The African one is found from the Senegal to the Cape of Good Hope, a distance of about 3,000 miles. The Asiatic kind formerly had an equal range, namely, from the banks of the Indus to the East Indian Isles. The hippopotamus is believed to have reached from the Cape to Egypt.
‡ Ossemens Fossiles, vol. i, p. 158. Cuvier says he cannot decide positively, not having seen a molar tooth.

elevation in the Cordillera, perhaps its habits led it to follow that chain of mountains from north to south.

After these facts, it is only in conformity with what we might almost have expected, that the horse, belonging to the same order of Pachydermata, should formerly have inhabited both North and South America. It is interesting thus to discover an epoch anterior to the division, as far at least as two important orders among the mammalia are concerned, of the continent into two separate zoological provinces. The geologist who believes in considerable oscillations of level in the crust of the globe within recent periods, will not fear to speculate either on the elevation of the Mexican platform, as a cause of the distinction, or on the submergence of land in the West Indian seas, – a circumstance which is perhaps indicated by the zoology of those islands. *

The number of bones embedded in the grand estuary deposit of the Pampas must be very great; I myself heard of, and saw many groups. The names of such places as 'the stream of the animal', 'the hill of the giant', tell the same story. At other times I heard of the marvellous property of certain rivers, which had the power of changing small bones into large; or as some maintained, the bones themselves grew. As far as I am aware, not one of these animals, as was formerly supposed, perished in the marshes, or muddy river-beds of the present land, but their bones have been exposed by the streams intersecting the deposit in which their remains were formerly buried. We may therefore conclude that the whole area of the Pampas is one wide sepulchre for these extinct quadrupeds.

While travelling through the country, I received several vivid descriptions of the effect of a great drought; and the account of this may throw some light on the cases, where vast numbers of animals of all kinds, have been embedded together. The period

* Dr Richardson (Report for 1836, to Brit. Assoc., p. 157) says, 'the spotted cavy (*cælogenys*), and perhaps a species of *cavia*, and one *dasyprocta*, extend from South America to the West Indies and Mexico.' Cuvier says the Kinkajou is found in the larger Antilles, but others affirm that this is an error: according to M. Gervais, the *Didelphis crancrivora* inhabits the Antilles. A tooth of the Mastodon has been brought from Bahama (Ed. New Phil. Journal, July 1826, p. 395). We cannot, however, from this conclude, that the Mastodon formerly inhabited those islands, for the carcass might have been floated there. Some mammalia certainly are peculiar to the Archipelago.

included between the years 1827 and 1830 is called the 'gran seco'
or the great drought. During this time, so little rain fell, that the
vegetation, even to the thistles, failed; the brooks were dried up,
and the whole country assumed the appearance of a dusty high-
road. This was especially the case in the northern part of the
province of Buenos Ayres, and the southern part of St Fe. Very
great numbers of birds, wild animals, cattle, and horses, perished
from the want of food and water. A man told me, that the deer*
used to come into his courtyard to the well, which he had been
obliged to dig to supply his own family with water; and that the
partridges had hardly strength to fly away when pursued. The
lowest estimation of the loss of cattle in the province of Buenos
Ayres alone, was taken at one million head. A proprietor at San
Pedro had previously to these years 20,000 cattle; at the end not
one remained. San Pedro is situated in the middle of the finest
country; and even now again abounds with animals; yet, during
the latter part of the 'gran seco', live cattle were brought in vessels
for the consumption of the inhabitants. The animals roamed
from their *estancias*, and wandering far to the southward, were
mingled together in such multitudes, that a government com-
mission was sent from Buenos Ayres to settle the disputes of the
owners. Sir Woodbine Parish informed me of another and very
curious source of dispute; the ground being so long dry, such
quantities of dust were blown about, that in this open country the
landmarks became obliterated, and people could not tell the limits
of their estates.

I was informed by an eyewitness, that the cattle in herds of
thousands rushed into the Parana,† and being exhausted by

---

* In Capt. Owen's Surveying Voyage (vol. ii, p. 274) there is a curious account of
the effects of a drought on the elephants, at Benguela (west coast of Africa). 'A
number of these animals had sometime since entered the town, in a body, to
possess themselves of the wells, not being able to procure any water in the
country. The inhabitants mustered, when a desperate conflict ensued, which
terminated in the ultimate discomfiture of the invaders, but not until they had
killed one man, and wounded several others.' The town is said to have a
population of nearly 3,000!

† Azara talks of the fury of the wild horses rushing into the marshes during a dry
season: 'et les premiers arrivés sont foulés, et écrasés par ceux, qui les suivent. Il
m'est arrivé plus d'une fois de trouver plus de mille cadavres de chevaux
sauvages morts de cette façon.' – Vol. i, p. 374.

hunger they were unable to crawl up the muddy banks, and thus were drowned. The arm which runs by San Pedro was so full of putrid carcasses, that the master of a vessel told me, that the smell rendered it quite impossible to pass that way. Without doubt several hundred thousand animals thus perished in the river. Their bodies when putrid floated down the stream, and many in all probability were deposited in the estuary of the Plata. All the small rivers became highly saline, and this caused the death of vast numbers in particular spots; for when an animal drinks of such water it does not recover. I noticed, but probably it was the effect of a gradual increase, rather than of any one period, that the smaller streams in the Pampas were paved with a breccia of bones.* Subsequently to this unusual drought a very rainy season commenced, which caused great floods. Hence it is almost certain, that some thousands of these skeletons were buried by the deposits of the very next year. What would be the opinion of a geologist, viewing such an enormous collection of bones, of all kinds of animals and of all ages, thus embedded in one thick earthy mass? Would he not attribute it to a flood having swept over the surface of the land, rather than to the common order of things?

These droughts to a certain degree seem to be periodical; I was told the dates of several others, and the intervals were about fifteen years. A tendency to periodical droughts is, I believe, common in most dry climates:† such certainly is the case in Australia. Captain Sturt says they return after every ten and twelve years, and are then followed by excessive rains, which gradually become less and less, till another drought is the

---

* In the neighbourhood of the great towns on the shores of the Plata, the number of bones strewed over the ground is truly astonishing. Since our return I have been informed, that ships have been freighted to this country with a cargo of bones. That cattle should be fattened on turnips manured with the bones of animals that lived in the southern hemisphere, is a curious fact in the commerce of the world. In the East Indies the luxurious drink wine cooled with North American ice, which in its journey has twice crossed the equator!

† Perhaps in every country, but the effect is more marked where the mean annual quantity of rain is small. I have seen the trunk of an old tree in England, in which the successive rings showed a tendency to periodical increase and diminution of size; about every tenth ring being small. See Mr Babbage's Ninth Bridgewater Treatise. Note M.

consequence. The year 1826 and the two following were singularly dry in Australia, and the latter were the first of the 'gran seco'. I mention this, because General Beatson in his account of the island of St Helena, has remarked that variations in climate sometimes appear to be the effect of the operation of some very general cause. He says (page 43), 'The severe drought felt here in 1791 and 1792, was far more calamitous in India. Doctor Anderson states, in a letter to Colonel Kyd, dated the 9th of August, 1792, that, owing to a failure of rain, during the above two years, one half of the inhabitants in the northern provinces had perished by famine; and the remainder were so feeble and weak, that on the report of rice coming from the Malabar coast, 5,000 poor people left Rajamundy, and very few of them reached the sea-side, although the distance is only 50 miles. It appears by Mr Bryan Edwards's History of the West Indies, that the seasons 1791–2 were unusually dry at the island of Montserrat.' Barrow* in the latter part of 1792, when at the Cape de Verd islands says, 'In fact a drought of three years' continuance, and consequent famine for almost the same period, had nearly desolated the island.'

OCTOBER 12TH – I had intended to have pushed my excursion further, but not being quite well, I was compelled to return by a *balandra*, or one-masted vessel of about a hundred tons burden, which was bound to Buenos Ayres. As the weather was not fair, we moored early in the day to a branch of a tree on one of the islands. The Parana is full of islands, which undergo a constant round of decay and renovation. In the memory of the master several large ones had disappeared, and others again had been formed and protected by vegetation. They are composed of muddy sand, without even the smallest pebble, and were then about four feet above the level of the river; but during the periodical floods they are inundated. They all present one character; numerous willows and a few other trees are bound together by a great variety of creeping plants, thus forming a thick jungle. These thickets afford a retreat for carpinchos and jaguars. The fear of the latter animal, quite destroyed all pleasure in scrambling through the woods. This evening I had not proceeded a hundred

* Voyage to Cochin China, p. 67.

yards, before finding indubitable signs of the recent presence of the tiger, I was obliged to come back. On every island there are tracks; and as on the former excursion *el rastro de los Indios* had been the subject of conversation, so in this was *el rastro del tigre*.

* * *

18TH AND 19TH – We continued slowly to sail down the noble stream: the current helped us but little. Azara has estimated that even near the sources between latitudes 16° 24' and 22° 57', the river has only a fall of one foot for each mile of latitude; lower down, this must be much diminished. It is stated that a rise of seven feet at Buenos Ayres can be perceived sixty leagues up the course of the Parana. We met, during our descent, very few vessels. One of the best gifts of nature seems here wilfully thrown away, in so grand a channel of communication being left unoccupied. A river in which ships might navigate from a temperate country, as surprisingly abundant in certain productions as destitute of others, to another possessing a tropical climate, and a soil which, according to the best of judges, M. Bonpland, is perhaps unequalled in fertility, in any part of the world. How different would have been the aspect of this river, if English colonists had by good fortune first sailed up the Plata! What noble towns would now have occupied its shores! Till the death of Francia, the Dictator of Paraguay, these two countries must remain distinct, as if placed on opposite sides of the globe. And when the old, bloody-minded tyrant is gone to his long account, Paraguay will be torn by revolutions, violent in proportion to the previous unnatural calm. That country will have to learn, like every other South American state, that a republic cannot succeed, till it contains a certain body of men imbued with the principles of justice and honour.

OCTOBER 20TH – Being arrived at the mouth of the Parana, and as I was very anxious to reach Buenos Ayres, I went on shore at Las Conchas, with the intention of riding there. Upon landing, I found to my great surprise, that I was to a certain degree a prisoner. A violent revolution having broken out, all the ports were laid under an embargo. I could not return to my vessel, and as for going by land to the city, it was out of the question. After a long conversation with the Commandant, I obtained permission to go, the next day, to General Rolor, who commanded a division of the rebels, on this side of the capital. In the morning I rode to the encampment. The general, officers, and soldiers, all appeared, and I believe really were, great villains. The general the very evening before he left the city, voluntarily went to the governor, and with his hand to his heart, pledged his word of honour, that he would remain faithful to the last. The general told me, that the city was in a state of close blockade, and that all he could do was to give me a passport to the commander-in-chief of the rebels at Quilmes. We had, therefore, to take a great sweep round the city, and it was with much difficulty that we procured horses. My reception at the encampment was quite civil, but I was told it was impossible that I could be allowed to enter the city. I was very anxious about this, as I anticipated the *Beagle*'s departure from the Rio Plata, earlier than it took place. Having mentioned, however, General Rosas's obliging kindness to me when at the Colorado, magic itself could not have altered circumstances quicker than did this conversation. I was instantly told that though they could not give me a passport, if I chose to leave my guide and horses, I might pass their sentinels. I was too glad to accept of this, and an officer was sent with me to give directions, that I might not be stopped at the bridge. The road for the space of a league was quite deserted. I met one party of soldiers, who were satisfied by gravely looking at an old passport: and at length I was not a little pleased, to find myself within the city.

This revolution was supported by scarcely any pretext of grievances. But in a state which, in the course of nine months (from February to October, 1820), underwent fifteen changes in its government – each governor, according to the constitution, being elected for three years – it would be the height of illiberality, to ask for pretexts. In this case a party of men, who being attached to Rosas, were disgusted with the governor Balcarce, to the number of seventy left the city, and with the cry of Rosas, the whole country took arms. The city was then blockaded, no provisions, cattle, or horses, were allowed to enter; besides this, there was only a little skirmishing, and a few men daily killed.

The outside party well knew, that by stopping the supply of meat, they would certainly be victorious. General Rosas could not have known of this rising; but it appears to me quite consonant with the plans of his party. A year ago he was elected governor, but he refused it, without the Sala would also confer on him extraordinary powers. This was refused, and since then his party have shown, that no other governor can keep his place. The warfare on both sides was avowedly protracted, till it was possible to hear from Rosas. A note arrived a few days after I left Buenos Ayres, which stated that the General disapproved of peace having been broken, but that he thought the outside party had justice on their side. On the bare reception of this, the Governor, ministers, and part of the military, to the number of some hundreds, fled from the city. The rebels entered, elected a new governor, and were paid for their services to the number of 5,500 men. From these proceedings, it was clear that Rosas ultimately would become the dictator: to the term king, the people in this, as in other republics, have a particular dislike. Since leaving South America, we have heard that Rosas has been elected, with powers, and for a time altogether opposed to the constitutional principles of the republic.

CHAPTER VIII

BANDA ORIENTAL

HAVING BEEN delayed for nearly a fortnight in the city, I was glad to escape on board a packet bound for Monte Video. A town in a state of blockade must always be a disagreeable place of residence; in this case moreover there were constant apprehensions from robbers within. The sentinels were the worst of all; for, from their office and from having arms in their hands, they robbed with a degree of authority, which other men could not imitate.

Our passage was a very long and tedious one. The Plata looks like a noble estuary on the map; but it is in truth a poor affair. A wide expanse of muddy water, has neither grandeur nor beauty. At one time of the day the two shores, both of which are extremely low, could just be distinguished from the deck. On arriving at Monte Video I found the *Beagle* would not sail for some time, so I prepared for a short excursion in this part of Banda Oriental. Every thing which I said about the country near Maldonado is applicable to this; the land, however, with the one exception of the Green Mount, 450 feet high, from which it takes its name, is far more level. Very little of the undulating grassy plain is enclosed; but near the town there are a few hedge banks, covered with agaves, cacti, and fennel.

NOVEMBER 14TH – We left Monte Video in the afternoon. I intended to proceed to Colonia del Sacramiento, situated on the northern bank of the Plata and opposite to Buenos Ayres, and thence following up the Uruguay, to the village of Mercedes on the Rio Negro (one of the many rivers of this name in South America), and from this point to return direct to Monte Video. We slept at the house of my guide at Canelones. In the morning we rose early in the hopes of being able to ride a good distance; but it was a vain attempt, for all the rivers were flooded. We passed in boats the streams of Canelones, St Lucia, and San José, and thus lost much time. On a former excursion I crossed the Lucia, near its mouth, and I was surprised to observe how easily our horses, although not used to swim, passed over a width of at least 600 yards. On mentioning this at Monte Video I was told that a vessel containing some mountebanks and their horses, being wrecked in the Plata, one horse swam seven miles to the shore. In the course of the day I was amused by the dexterity with which a Gaucho forced a restive horse to swim a river. He stripped off his clothes, and jumping on its back, rode into the water till it was out of its depth; then slipping off over the crupper, he caught hold of the tail, and as often as the horse turned round, the man frightened it back, by splashing water in its face. As soon as the horse touched the bottom on the other side, the man pulled himself on, and was firmly seated, bridle in hand, before the horse gained the bank. A naked man, on a naked horse, is a fine spectacle; I had no idea how well the two animals suited each other. The tail of a horse is a very useful appendage; I have passed a river in a boat with four people in it, which was ferried across in the same way as the Gaucho. If a man and horse have to cross a broad river, the best plan is for the man to catch hold of the pummel or mane, and help himself with the other arm.

We slept, and stayed the following day at the post of Cufre. In the evening the postman or letter-carrier arrived. He was a day after his time, owing to the Rio Rozario being flooded. It would not, however, be of much consequence; for, although he had passed through some of the principal towns in Banda Oriental, his luggage consisted of two letters. The view from the house was pleasing; an undulating green surface, with distant glimpses of the

Plata. I find I look at this province with very different eyes, from what I did upon my first arrival. I recollect I then thought it singularly level; but now, after galloping over the Pampas, my only surprise is, what could have induced me ever to have called it level. The country is a series of undulations, in themselves perhaps not absolutely great, but as compared to the plains of St Fe, real mountains. From these inequalities, there is an abundance of small rivulets, and the turf is green and luxuriant.

* * *

NOVEMBER 19TH — Passing the village of Las Vacas, we slept at a house of a North American, who worked a lime-kiln on the Arroyo de las Vivoras. In the morning we rode to a projecting headland on the banks of the river, called Punta Gorda. On the way we tried to find a jaguar. There were plenty of fresh tracks, and we visited the trees, on which they are said to sharpen their claws; but we did not succeed in disturbing one. From this point the Rio Uruguay presented to our view a noble volume of water. From the clearness and rapidity of the stream, its appearance was far superior to that of its neighbour the Parana. On the opposite coast, several branches from the latter river entered the Uruguay. As the sun was shining, the two colours of the waters could be seen quite distinct. The geological section presented by the cliffs was interesting. At St Fe, a stratum with marine remains was seen gradually passing into an estuary deposit. Here we have an alternation of action; a circumstance no ways improbable in a great bay. A formation of red earthy clay, with nodules of marl, and in every respect identical with that of the Pampas, is covered by a white limestone, containing large extinct oysters, and other marine shells; and over this again, is placed the reddish earthy matter, as in the rest of Banda Oriental.

In the evening we proceeded on our road towards Mercedes on the Rio Negro. At night we asked permission to sleep at an *estancia*, at which we happened to arrive. It was a very large estate, being ten leagues square, and the owner is one of the greatest landowners in the country. His nephew had charge of it, and with him there was a captain in the army, who the other day ran away from Buenos Ayres. Considering their station, their conversation was rather amusing. They expressed, as was usual, unbounded astonishment at the globe being round, and could scarcely credit that a hole would, if deep enough, come out on the other side. They had, however, heard of a country where there were six months light and six of darkness, and where the inhabitants were very tall and thin! They were curious about the price and condition of horses and cattle in England. Upon finding out we did not catch our animals with the lazo, they cried out, 'Ah then, you use nothing but the bolas': the idea of an enclosed country was quite novel to them. The captain at last said, he had one question to ask me, which he should be very much obliged if I would answer with all truth. I trembled to think how deeply scientific it would be: it was, 'Whether the ladies of Buenos Ayres were not the handsomest in the world.' I replied, 'Charmingly so.' He added, 'I have one other question. Do ladies in any other part of the world wear such large combs?' I solemnly assured him they did not. They were absolutely delighted. The captain exclaimed, 'Look there! a man who has seen half the world says it is the case; we always thought so, but now we know it.' My excellent judgment in beauty procured me a most hospitable reception; the captain forced me to take his bed, and he would sleep on his *recado*.

* * *

NOVEMBER 26TH – I set out on my return in a direct line for Monte Video. Having heard of some giant's bones at a neighbouring farm-house on the Sarandis, a small stream entering the Rio Negro, I rode there accompanied by my host, and purchased for the value of 18 pence, the head of an animal equalling in size that of the hippopotamus. Mr Owen in a paper read before the Geological Society,* has called this very extraordinary animal, Toxodon, from the curvature of its teeth. The following notice is

---

* Read, April 19th, 1837. A detailed account will appear in the first part of the zoology of the voyage of the *Beagle*.

taken from the proceedings of that society: Mr Owen says, judging from the portion of the skeleton preserved, the Toxodon, as far as dental characters have weight, must be referred to the rodent order. But from that order it deviates in the relative position of its supernumerary incisors, in the number and direction of the curvature of its molars, and in some other respects. It again deviates, in several parts of its structure which Mr Owen enumerated, both from the *Rodentia*, and the existing *Pachydermata*, and it manifests an affinity to the *Dinotherium* and the *Cetaceous* order. Mr Owen, however, observed, that 'the development of the nasal cavity and the presence of frontal sinuses, renders it extremely improbable that the habits of the *Toxodon* were so exclusively aquatic as would result from the total absence of hinder extremities; and concludes, therefore, that it was a quadruped, and not a Cetacean; and that it manifested an additional step in the gradation of mammiferous forms leading from the *Rodentia*, through the *Pachydermata* to the *Cetacea*; a gradation of which the water-hog of South America (*Hydrochærus capybara*) already indicates the commencement amongst existing *Rodentia*, of which order it is interesting to observe this species is the largest, while at the same time it is peculiar to the continent in which the remains of the gigantic *Toxodon* were discovered.'

The people at the farm-house told me that the remains were exposed, by a flood having washed down part of a bank of earth. When found, the head was quite perfect; but the boys knocked the teeth out with stones, and then set up the head as a mark to throw at. By a most fortunate chance, I found a perfect tooth, which exactly fits one of the sockets in this skull, embedded by itself on the banks of the Rio Tercero, at the distance of about 180 miles from this place. Near the Toxodon I found the fragments of the head of an animal, rather larger than the horse, which has some points of resemblance with the Toxodon, and others perhaps with the Edentata. The head of this animal, as well as that of the Toxodon, and especially the former, appear so fresh, that it is difficult to believe they have lain buried for ages under ground. The bone contains so much animal matter, that when heated in the flame of a spirit-lamp, it not only exhales a very strong animal odour, but likewise burns with a slight flame.*

At the distance of a few leagues I visited a place where the remains of another great animal, associated with large pieces of armadillo-like covering, had been found. Similar pieces were likewise lying in the bed of the stream, close to the spot where the skeleton of the Toxodon had been exposed. These portions are dissimilar from those mentioned at Bahia Blanca. It is a most interesting fact thus to discover, that more than one gigantic animal in former ages was protected by a coat of mail,† very similar to the kind now found on the numerous species of armadillo, and exclusively confined to that South American genus.

By the middle of the day on the 28th we arrived at Monte Video, having been two days and a half on the road. The country for the whole way was of a very uniform character, some parts being rather more rocky and hilly than near the Plata. Not far from Monte Video we passed through the village of Las Pietras, so named from some large rounded masses of syenite. Its appearance was rather pretty. In this country, a few fig-trees around the houses, and a site elevated a hundred feet above the general level, ought always to be called picturesque.

During the last six months, I have had an opportunity of seeing a little of the character of the inhabitants of these provinces. The Gauchos, or countrymen, are very superior to those who reside in the towns. The Gaucho is invariably most obliging, polite, and hospitable. I did not meet even with one instance of rudeness or inhospitality. He is modest, both respecting himself and country, at the same time being a spirited, bold fellow. On the other hand,

* I must express my obligation to Mr Keane, at whose house I was staying on the Berquelo, and to Mr Lumb at Buenos Ayres, for without their assistance, these valuable remains would never have reached England.

† I may here just mention, that I saw in the possession of a clergyman near Monte Video, the terminal portion of a tail, which precisely resembled, but on a gigantic scale, that of the common armadillo. The fragment was 17 inches long, eleven and a half in circumference at the upper end, and eight and a half at the *extreme* point. As we do not know what proportion the tail bore to the body of the animal, we cannot compare it with that of any living species. But at the same time we may conjecture that, in all probability, this extinct monster was from 6 to 10 feet long.

there is much blood shed, and many robberies committed. The constant presence of the knife is the chief cause of the former. It is lamentable to hear how many lives are lost in trifling quarrels. In fighting, each party tries to mark the face of his adversary, by slashing his nose or eyes; as is often attested by deep and horrid-looking scars. Robberies are a natural consequence of universal gambling, much drinking, and extreme indolence. At Mercedes, I asked two men why they did not work. One gravely said the days were too long, the other that he was too poor. The number of horses, and the profusion of food, are the destruction of all industry. Moreover, there are so many feast-days; and then nothing can succeed without it is begun when the moon is on the increase; so that half the month is lost from these two causes.

Police and justice are quite inefficient. If a man who is poor, commits murder, and is taken, he will be imprisoned, and perhaps even shot; but if he is rich and has friends, he may rely on it, no very severe consequence will ensue. It is curious that the most respectable people in the country invariably assist a murderer to escape. They seem to think the individual sins against the governing powers and not against the state. A traveller has no protection besides his fire-arms: and the constant habit of carrying them, is the main check to a more frequent occurrence of robbery.

The character of the higher and more educated classes, who reside in the towns, partakes, but perhaps in a lesser degree, of the good parts of the Gaucho, but is I fear stained by many vices of which he is free. Sensuality, mockery of all religion, and the grossest corruption, are far from uncommon. Nearly every public officer can be bribed. The head man in the post-office sold forged government franks. The governor and prime minister openly combined to plunder the state. Justice, where gold came into play, was hardly expected by any one. I knew an Englishman, who went to the chief justice (he told me that not understanding the ways of the place, he trembled as he entered the room), and said, 'Sir, I have come to offer you 200 dollars (value about £5 sterling) if you will arrest before a certain time a man who has cheated me. I know it is against the law, but my lawyer (naming him) recommended me to take this step.' The chief justice smiled acquiescence, thanked him, and the man before night was safe in prison. With this entire want of principle in many of the leading men, with the country full of ill-paid turbulent officers, the people yet hope that a democratic form of government can succeed!

On first entering society in these countries, two or three features strike one as particularly remarkable. The polite and dignified manners pervading every grade of life; the excellent taste displayed by the women in their dresses; and the equality amongst all ranks. At the Rio Colorado some men who kept the humblest shops, used to dine with General Rosas. A son of a major at Bahia Blanca gained his livelihood by making paper cigars, and he wished to accompany me, as guide or servant, to Buenos Ayres, but his father objected on the score of the danger alone. Many officers in the army can neither read nor write, yet all meet in society as equals. In Entre Rios, the Sala consisted of only six representatives. One of them kept a common shop, and evidently was not degraded by the office. All this is what would be expected in a new country; nevertheless, the absence of gentlemen by profession appears to an Englishman something strange.

When speaking of these countries, the manner in which they have been brought up by their unnatural parent, Spain, should always be borne in mind. On the whole, perhaps, more credit is due for what has been done, than blame for that which may be deficient. It is impossible to doubt but that the extreme liberalism of these countries, must ultimately lead to good results. The very general toleration of foreign religions, the regard paid to the means of education, the freedom of the press, the facilities offered to all foreigners, and especially, as I am bound to add, to every one professing the humblest pretensions to science, should be recollected with gratitude, by those who have visited Spanish South America.*

---

* I cannot conclude without adding my testimony to the spirit and accuracy of Head's Rough Notes. I do not think the picture is at all more exaggerated than every good one must be – that is, by taking the strong examples, and neglecting those of lesser interest.

CHAPTER IX

Rio Plata – Flocks of butterflies – Beetles alive in the sea – Aeronaut spiders – Pelagic animals – Phosphorescence of sea – Port Desire – Spanish settlements – Zoology – Guanaco – Excursion to head of harbour – Indian grave – Port St Julian – Geology of Patagonia, successive terraces, transport of pebbles – Fossil gigantic llama – Types of organization constant – Change in zoology of America – Causes of extinction

PATAGONIA

DECEMBER 6TH, 1833 – The *Beagle* sailed from the Rio Plata, never again to enter its muddy stream. Our course was directed to Port Desire, on the coast of Patagonia. Before proceeding any further, I will here put together a few observations made at sea.

Several times when the ship has been some miles off the mouth of the Plata, and at other times when off the shores of Northern Patagonia, we have been surrounded by insects. One evening, when we were about ten miles from the Bay of San Blas, vast numbers of butterflies, in bands or flocks of countless myriads, extended as far as the eye could range. Even by the aid of a glass it was not possible to see a space free from butterflies. The seamen cried out 'it was snowing butterflies', and such in fact was the appearance. More species than one were present, but the main part belonged to a kind very similar to, but not identical with, the common English *Colias edusa*.* Some moths and hymenoptera accompanied the butterflies; and a fine Calosoma flew on board. Other instances are known of this beetle having been caught far out at sea; and this is the more remarkable, as the greater number

---

* I am indebted to Mr Waterhouse for naming these and other insects.

of the Carabidæ seldom or never take wing. The day had been fine and calm, and the one previous to it equally so, with light and variable airs. Hence we cannot suppose that the insects were blown off the land, but we must conclude that they voluntarily took flight. The great bands of the Colias seem at first to afford an instance like those on record of the migrations of *Vanessa cardui*;* but the presence of other insects makes the case distinct, and not so easily intelligible. Before sunset, a strong breeze sprung up from the north, and this must have been the cause of tens of thousands of the butterflies and other insects having perished.

On another occasion, when 17 miles off Cape Corrientes, I had a net overboard to catch pelagic animals. Upon drawing it up, to my surprise I found a considerable number of beetles in it, and although in the open sea, they did not appear much injured by the salt water. I lost some of the specimens, but those which I preserved, belonged to the genera, colymbetes, hydroporus, hydrobius (two species), notaphus, cynucus, adimonia, and scarabæus. At first, I thought that these insects had been blown from the shore; but upon reflecting that out of the eight species, four were aquatic, and two others partly so in their habits, it appeared to me most probable that they were floated into the sea, by a small stream which drains a lake near Cape Corrientes. On any supposition, it is an interesting circumstance to find insects, quite alive, swimming in the open ocean, 17 miles from the nearest point of land. There are several accounts of insects having been blown off the Patagonian shore. Captain Cook observed it, as did more lately Captain King in the *Adventure*. The cause probably is due to the want of shelter, both of trees and hills, so that an insect on the wing with an off-shore breeze, would be very apt to be blown out to sea. The most remarkable instance I ever knew of an insect being caught far from the land, was that of a large grasshopper (*Acrydium*), which flew on board, when the *Beagle* was to windward of the Cape de Verd Islands, and when the nearest point of land, not directly opposed to the trade-wind, was Cape Blanco on the coast of Africa, 370 miles distant.†

* Lyell's Geology, vol. iii, p. 63.
† The flies which frequently accompany a ship for some days on its passage from harbour to harbour, wandering from the vessel, are soon lost, and all disappear.

On several occasions, when the vessel has been within the mouth of the Plata, the rigging has been coated with the web of the Gossamer Spider. One day (November 1st, 1832) I paid particular attention to the phenomenon. The weather had been fine and clear, and in the morning the air was full of patches of the flocculent web, as on an autumnal day in England. The ship was sixty miles distant from the land, in the direction of a steady though light breeze. Vast numbers of a small spider, about one-tenth of an inch in length, and of a dusky red colour were attached to the webs. There must have been, I should suppose, some thousands on the ship. The little spider when first coming in contact with the rigging, was always seated on a single thread, and not on the flocculent mass. This latter seems merely to be produced by the entanglement of the single threads. The spiders were all of one species, but of both sexes, together with young ones. These latter were distinguished by their smaller size, and more dusky colour. I will not give the description of this spider, but merely state that it does not appear to me to be included in any of Latreille's genera. The little aeronaut as soon as it arrived on board, was very active, running about; sometimes letting itself fall, and then reascending the same thread; sometimes employing itself in making a small and very irregular mesh in the corners between the ropes. It could run with facility on the surface of water. When disturbed it lifted up its front legs, in the attitude of attention. On its first arrival it appeared very thirsty, and with exserted maxillæ drank eagerly of the fluid; this same circumstance has been observed by Strack: may it not be in consequence of the little insect having passed through a dry and rarefied atmosphere? Its stock of web seemed inexhaustible. While watching some that were suspended by a single thread, I several times observed that the slightest breath of air bore them away out of sight, in a horizontal line. On another occasion (25th) under similar circumstances, I repeatedly observed the same kind of small spider, either when placed, or having crawled, on some little eminence, elevate its abdomen, send forth a thread, and then sail away in a lateral course, but with a rapidity which was quite unaccountable. I thought I could perceive that the spider before performing the above preparatory steps, connected its legs together with the most delicate threads, but I am not sure, whether this observation is correct.

One day, at St Fe, I had a better opportunity of observing some similar facts. A spider which was about three-tenths of an inch in length, and which in its general appearance resembled a Citigrade (therefore quite different from the gossamer), while standing on the summit of a post, darted forth four or five threads from its spinners. These glittering in the sunshine, might be compared to rays of light; they were not, however, straight, but in undulations like a film of silk blown by the wind. They were more than a yard in length, and diverged in an ascending direction from the orifices. The spider then suddenly let go its hold, and was quickly borne out of sight. The day was hot and apparently quite calm; yet under such circumstances the atmosphere can never be so tranquil, as not to affect a vane so delicate as the thread of a spider's web. If during a warm day we look either at the shadow of any object cast on a bank, or over a level plain at a distant landmark, the effect of an ascending current of heated air will almost always be evident. And this probably would be sufficient to carry with it so light an object as the little spider on its thread. The circumstance of spiders of the same species but of different sexes and ages, being found on several occasions at the distance of many leagues from the land, attached in vast numbers to the lines, proves that they are the manufacturers of the mesh, and that the habit of sailing through the air, is probably as characteristic of some tribe, as that of diving is of the Argyroneta. We may then reject Latreille's supposition, that the gossamer owes its origin to the webs of the young of several genera, as Epeira or Thomisa: although, as we have seen that the young of other spiders do possess the power of performing aerial voyages.*

During our different passages south of the Plata, I often towed

---

* I was not at the time aware of M. Virey's very curious observations, (Bulletin des Sciences Natur., tom. xix, p. 130) which seem to prove that small spiders, in an atmosphere perfectly tranquil, and without the aid of any web, have the power of darting through the air. M. Virey, believes that by means of a rapid vibration of their feet, they *walk the air*. Although in his case, the conclusion seems almost inevitable, yet in the one I have described, we must suppose that the several threads which were sent forth, served as sails for the atmospheric currents to act on. After reading M. Virey's account, it appears to me far from

footnote continued overleaf

astern a net made of bunting, and thus caught many curious animals. The structure of the Beroe (a kind of jelly fish) is most extraordinary, with its rows of vibratory ciliæ, and complicated though irregular system of circulation. Of Crustacea, there were many strange and undescribed genera. One, which in some respects is allied to the Notopods (or those crabs which have their posterior legs placed almost on their backs, for the purpose of adhering to the under side of ledges), is very remarkable from the structure of its hind pair of legs. The penultimate joint, instead of being terminated by a simple claw, ends in three bristle-like appendages of dissimilar lengths, the longest equalling that of the entire leg. These claws are very thin, and are serrated with teeth of an excessive fineness, which are directed towards the base. The curved extremities are flattened, and on this part five most minute cups are placed, which seem to act in the same manner as the suckers on the arms of the cuttle-fish. As the animal lives in the open sea, and probably wants a place of rest, I suppose this beautiful structure is adapted to take hold of the globular bodies of the Medusæ, and other floating marine animals.

In deep water, far from the land, the number of living creatures is extremely small: south of the latitude 35°, I never succeeded in catching any thing besides some beroe, and a few species of minute crustacea belonging to the Entomostraca. In shoaler water, at the distance of a few miles from the coast, very many kinds of crustacea and some other animals were numerous, but only during the night. Between latitudes 56° and 57° south of Cape Horn the net was put astern several times; it never, however, brought up any thing besides a few of two extremely minute species of Entomostraca. Yet whales and seals, petrels and albatross, are exceedingly abundant throughout this part of the ocean. It has always been a source of mystery to me, on what the latter, which live far from the shore, can subsist. I presume the albatross, like the condor, is able to fast long; and that one good feast on the

---

footnote continued

improbable, that the little aeronaut actually did unite, as was suspected, its feet together by some fine lines; thus forming artificial wings. I regret I did not determine this point with accuracy; for it would be a curious fact, that a spider should thus be able to take flight by the aid of temporary wings.

carcass of a putrid whale lasts for a long siege of hunger. It does not lessen the difficulty to say, they feed on fish; for on what can the fish feed? It often occurred to me, when observing how the waters of the central and intertropical parts of the Atlantic,* swarmed with Pteropoda, Crustacea, and Radiata, and with their devourers the flying-fish, and again with *their* devourers the bonitos and albicores, that the lowest of these pelagic animals perhaps possess the power of decomposing carbonic acid gas, like the members of the vegetable kingdom.

While sailing in these latitudes on one very dark night, the sea presented a wonderful and most beautiful spectacle. There was a fresh breeze, and every part of the surface, which during the day is seen as foam, now glowed with a pale light. The vessel drove before her bows two billows of liquid phosphorus, and in her wake she was followed by a milky train. As far as the eye reached, the crest of every wave was bright, and the sky above the horizon, from the reflected glare of these livid flames, was not so utterly obscure, as over the rest of the heavens.

As we proceed further southward, the sea is seldom phosphorescent; and off Cape Horn, I do not recollect more than once having seen it so, and then it was far from being brilliant. This circumstance probably has a close connexion with the scarcity of organic beings in that part of the ocean. After the elaborate paper† by Ehrenberg, on the phosphorescence of the sea, it is almost superfluous on my part to make any observations on the subject. I may however add, that the same torn and irregular particles of gelatinous matter, described by Ehrenberg, seem in the southern as well as in the northern hemisphere, to be the common cause of this phenomenon. The particles were so minute as easily to pass through fine gauze; yet many were distinctly visible by the naked eye. The water when placed in a tumbler and agitated gave out sparks, but a small portion in a watch-glass, scarcely ever was luminous. Ehrenberg states, that these particles all retain a certain degree of irritability. My observations, some of which were made

---

* From my experience, which has been but little, I should say that the Atlantic was far more prolific than the Pacific, at least, than in that immense open area, between the west coast of America and the extreme eastern isles of Polynesia.

† An abstract is given in No. IV of the Magazine of Zoology and Botany.

directly after taking up the water, would give a different result. I may also mention, that having used the net during one night I allowed it to become partially dry, and having occasion twelve hours afterwards, to employ it again, I found the whole surface sparkled as brightly as when first taken out of the water. It does not appear probable in this case, that the particles could have remained so long alive. I remark also in my notes, that having kept a Medusa of the genus Dianæa, till it was dead, the water in which it was placed became luminous. When the waves scintillate with bright green sparks, I believe it is generally owing to minute crustacea. But there can be no doubt that very many other pelagic animals, when alive, are phosphorescent.

On two occasions I have observed the sea luminous at considerable depths beneath the surface. Near the mouth of the Plata some circular and oval patches, from 2 to 4 yards in diameter, and with defined outlines, shone with a steady, but pale light; while the surrounding water only gave out a few sparks. The appearance resembled the reflection of the moon, or some luminous body; for the edges were sinuous from the undulation of the surface. The ship, which drew thirteen feet water, passed over, without disturbing, these patches. Therefore we must suppose that some animals were congregated together at a greater depth than the bottom of the vessel.

Near Fernando Noronha the sea gave out light in flashes. The appearance was very similar to that which might be expected from a large fish moving rapidly through a luminous fluid. To this cause the sailors attributed it; at the time, however, I entertained some doubts, on account of the frequency and rapidity of the flashes. With respect to any general observations, I have already stated that the display is very much more common in warm than in cold countries. I have sometimes imagined that a disturbed electrical condition of the atmosphere was most favourable to its production. Certainly I think the sea is most luminous after a few days of more calm weather than ordinary, during which time it has swarmed with various animals. Observing that the water charged with gelatinous particles is in an impure state, and that the luminous appearance in all common cases is produced by the agitation of the fluid in contact with the atmosphere, I have always been inclined to consider that the phosphorescence was the result of the decomposition of the organic particles, by which process (one is tempted almost to call it a kind of respiration) the ocean becomes purified.

DECEMBER 23RD – We arrived at Port Desire, situated in lat. 47°, on the coast of Patagonia. The creek runs for about twenty miles inland, with an irregular width. The *Beagle* anchored a few miles within the entrance in front of the ruins of an old Spanish settlement.

The same evening I went on shore. The first landing in any new country is very interesting, and especially when, as in this case, the whole aspect bears the stamp of a marked and individual character. At the height of between 200 and 300 feet, above some masses of porphyry, a wide plain extends, which is truly characteristic of Patagonia. The surface is quite level, and is composed of well-rounded shingle mixed with a whitish earth. Here and there scattered tufts of brown wiry grass are supported, and still more rarely some low thorny bushes. The weather is dry and pleasant, for the fine blue sky is but seldom obscured. When standing in the middle of one of these desert plains, the view on one side is generally bounded by the escarpment of another plain, rather higher, but equally level and desolate; and on the other side it becomes indistinct from the trembling mirage which seems to rise from the heated surface.

The plains are traversed by many broad, flat-bottomed valleys, and in these the bushes grow rather more abundantly. The present drainage of the country is quite insufficient to excavate such large channels. In some of the valleys ancient stunted trees, growing in the very centre of the dry watercourse, seem as if placed to prove how long a time had elapsed, since any flood had passed that way. We have evidence, from shells lying on the surface, that the plains of gravel have been elevated within a recent epoch above the level of the sea; and we must look to that period for the excavation of the valleys by the slowly retiring waters. From the dryness of the climate, a man may walk for days together over these plains without finding a single drop of water. Even at the base of the porphyry hills, there are only a few small wells containing but little water, and that rather saline and half putrid.

In such a country the fate of the Spanish settlement was soon decided; the dryness of the climate during the greater part of the year, and the occasional hostile attacks of the wandering Indians, compelled the colonists to desert their half-finished buildings. The style, however, in which they were commenced, showed the strong and liberal hand of Spain in the old time. The end of all the attempts to colonize this side of America south of 41°, have been miserable. At Port Famine, the name expresses the lingering and extreme sufferings of several hundred wretched people, of whom one alone survived to relate their misfortunes. At St Joseph's bay, on the coast of Patagonia, a small settlement was made; but during one Sunday the Indians made an attack and massacred the whole party, excepting two men, who were led captive many years among the wandering tribes. At the Rio Negro I conversed with one of these men, now in extreme old age.

The zoology of Patagonia is as limited as its Flora.* On the arid plains a few black beetles (Heteromera) might be seen slowly crawling about, and occasionally a lizard darting from side to side. Of birds we have three carrion hawks, and in the valleys a few finches and insect feeders. The *Ibis malanops* (a species said to be found in central Africa) is not uncommon on the most desert parts. In the stomachs of these birds I found grasshoppers, cicadæ, small lizards, and even scorpions.† At one time of the year they go in flocks, at another in pairs: their cry is very loud and singular, and resembles the neighing of the guanaco.

I will here give an account of this latter animal, which is very common, and is the characteristic quadruped of the plains of Patagonia. The Guanaco, which by some naturalists is considered as the same animal with the Llama, but in its wild state, is the South American representative of the camel of the East. In size it may be compared to an ass, mounted on taller legs, and with a very long neck. The guanaco abounds over the whole of the

---

* I found here a species of cactus, described by Professor Henslow under the name of *Opuntia darwinii* (Magazine of Zoology and Botany, vol. i, p. 466), which was remarkable by the irritability manifested by the stamens, when I inserted in the flower either a piece of stick, or the end of my finger. The segments of the periant also closed on the pistil, but more slowly than the stamens.

† These insects were not uncommon beneath stones. I found one cannibal scorpion quietly devouring another.

temperate parts of South America, from the wooded islands of Tierra del Fuego, through Patagonia, the hilly parts of La Plata, Chile, even to the Cordillera of Peru. Although preferring an elevated site, it yields in this respect to its near relative the Vicuna. On the plains of Southern Patagonia, we saw them in greater numbers than in any other part. Generally they go in small herds, from half a dozen to thirty together; but on the banks of the St Cruz we saw one herd which must have contained at least 500. On the northern shores of the Strait of Magellan they are also very numerous.

Generally the guanacoes are wild and extremely wary. Mr Stokes told me, that he one day saw through a glass a herd of these beasts, which evidently had been frightened, running away at full speed, although their distance was so great that they could not be distinguished by the naked eye. The sportsman frequently receives the first intimation of their presence, by hearing, from a long distance, the peculiar shrill neighing note of alarm. If he then looks attentively, he will perhaps see the herd standing in a line on the side of some distant hill. On approaching them, a few more squeals are given, and then off they set at an apparently slow, but really quick canter, along some narrow beaten track to a neighbouring hill. If, however, by chance he should abruptly meet a single animal, or several together, they will generally stand motionless, and intently gaze at him; then perhaps move on a few yards, turn round, and look again. What is the cause of this difference in their shyness? Do they mistake a man in the distance for their chief enemy the puma? Or does curiosity overcome their timidity? That they are curious is certain; for if a person lies on the ground, and plays strange antics, such as throwing up his feet in the air, they will amost always approach by degrees to reconnoitre him. It was an artifice that was repeatedly practised by our sportsmen with success, and it had moreover the advantage of allowing several shots to be fired, which were all taken as parts of the performance. On the mountains of Tierra del Fuego, and in other places, I have more than once seen a guanaco, on being approached, not only neigh and squeal, but prance and leap about in the most ridiculous manner, apparently in defiance as a challenge. These animals are very easily domesticated, and I have seen some thus kept near the houses, although at large on their native plains. They are in this state very bold, and readily attack a man, by striking him from behind with both knees. It is asserted, that the motive for these attacks is jealousy on account of their females. The wild guanacoes, however, have no idea of defence; even a single dog will secure one of these large animals, till the huntsman can come up. In many of their habits they are like sheep in a flock. Thus when they see men approaching in several directions on horseback, they soon became bewildered and know not which way to run. This greatly facilitates the Indian method of hunting, for they are thus easily driven to a central point, and are encompassed.

The guanacoes readily take to the water: several times at Port Valdes they were seen swimming from island to island. Byron, in his voyage, says he saw them drinking salt water. Some of our officers likewise saw a herd apparently drinking the briny fluid from a salina near Cape Blanco. I imagine in several parts of the country, if they do not drink salt water, they drink none at all. In the middle of the day, they frequently roll in the dust, in saucer-shaped hollows. The males fight together; two one day passed quite close to me, squealing and trying to bite each other; and several were shot with their hides deeply scored. Herds sometimes appear to set out on exploring-parties: at Bahia Blanca, where, within 30 miles of the coast, these animals are extremely unfrequent, I one day saw the tracks of thirty or forty, which had come in a direct line to a muddy salt-water creek. They then must have perceived that they were approaching the sea, for they had wheeled with the regularity of cavalry, and had returned back in as straight a line as they had advanced. The guanacoes have one singular habit, which is to me quite inexplicable; namely, that on successive days they drop their dung in the same defined heap. I saw one of these heaps which was eight feet in diameter, and necessarily was composed of a large quantity. Frezier remarks on this habit as common to the guanaco as well as to the llama;* he says it is very useful to the Indians, who use the dung for fuel, and are thus saved the trouble of collecting it.

---

* D'Orbigny says (vol. ii, p. 69) that all the species of the genus have this habit.

The guanacoes appear to have favourite spots for dying in. On the banks of the St Cruz, the ground was actually white with bones, in certain circumscribed spaces, which were generally bushy and all near the river. On one such spot I counted between ten and twenty heads. I particularly examined the bones; they did not appear, as some scattered ones which I had seen, gnawed or broken, as if dragged together by beasts of prey. The animals in most cases, must have crawled, before dying, beneath and amongst the bushes. Mr Bynoe informs me that during the last voyage, he observed the same circumstance on the banks of the Rio Gallegos. I do not at all understand the reason of this, but I may observe, that the wounded guanacoes at the St Cruz, invariably walked towards the river. At St Jago in the Cape de Verd islands I remember having seen in a retired ravine a corner under a cliff, where numerous goats' bones were collected: we at the time exclaimed, that it was the burial-ground of all the goats in the island. I mention these trifling circumstances, because in certain cases they might explain the occurrence of a number of uninjured bones in a cave, or buried under alluvial accumulations; and likewise the cause, why certain mammalia are more commonly embedded than others in sedimentary deposits. Any great flood of the St Cruz, would wash down many bones of the guanaco, but probably not a single one of the puma, ostrich, or fox. I may also observe, that almost every kind of waterfowl when wounded takes to the shore to die; so that the remains of birds, from this cause alone and independently of other reasons, would but rarely be preserved in a fossil state.

* * *

JANUARY 9TH, 1834 – Before it was dark the *Beagle* anchored in the fine spacious harbour of Port St Julian, situated about 110 miles to the south of Port Desire. We remained here eight days. The country is nearly similar to that of Port Desire, but, perhaps, rather more sterile. One day a party accompanied Captain FitzRoy on a long walk round the head of the harbour. We were eleven hours without tasting any water, and some of the party were quite exhausted. From the summit of a hill (since well-named

Thirsty Hill) a fine lake was spied, and two of the party proceeded with concerted signals to show whether it was fresh water. What was our disappointment to find a snow-white expanse of salt, crystallized in great cubes! We attributed our extreme thirst to the dryness of the atmosphere; but whatever the cause might be, we were exceedingly glad late in the evening to get back to the boats. Although we could nowhere find, during our whole visit, a single drop of fresh water, yet some must exist; for by an odd chance I found on the surface of the salt water, near the head of the bay, a Colymbetes not quite dead, which in all probability had lived in some not far distant pool. Three other kinds of insects – a Cincindela, like *hybrida*, Cymindis, and a Harpalus, which all live on muddy flats occasionally overflowed by the sea, and one other beetle found dead on the plain – completes the list of coleoptera. A good-sized fly (Tabanus) was extremely numerous, and tormented us by its painful bite. The common horsefly, which is so troublesome in the shady lanes of England, belongs to this genus. We here have the puzzle, that so frequently occurs in the case of musquitoes; on the blood of what animals do these insects commonly feed? The guanaco is nearly the only warmblooded quadruped, and they are present in numbers quite inconsiderable, compared to the multitude of flies.

* * *

At first I could only understand the grand covering of gravel, by the supposition of some epoch of extreme violence, and the successive lines of cliff, by as many great elevations, the precise action of which I could not however follow out. Guided by the *Principles of Geology*, and having under my view the vast changes going on in this continent, which at the present day seems the great workshop of nature, I came to another, and I hope more satisfactory conclusion. The importance of any view which may explain the agency by which such vast beds of shingle have been transported over the surface of the successive plains, cannot be doubted. Whatever the cause may have been, it has determined the condition of this desert country, with respect to its form, nature, and capabilities of supporting life.

There are proofs, that the whole coast has been elevated to a considerable height within the recent period; and on the shores of the Pacific, where successive terraces likewise occur, we know that these changes have latterly been very gradual. There is indeed reason for believing, that the uplifting of the ground during the earthquakes in Chile, although only to the height of 2 or 3 feet, has been a disturbance which may be considered as a great one, in comparison to the series of lesser and scarcely sensible movements which are likewise in progress. Let us then imagine the consequence of the shoaling bed of an ocean, elevated at a perfectly equable rate, so that the same number of feet should be converted into dry land in each succeeding century. Every part of the surface would then have been exposed for an equal length of time to the action of the beach-line, and the whole in consequence equally modified. The shoaling bed of the ocean would thus be changed into a sloping land, with no marked line on it. If, however, there should occur a long period of repose in the elevations, and the currents of the sea should tend to wear away the land (as happens along this whole coast), then there would be formed a line of cliff. Accordingly as the repose was long, so would be the quantity of land consumed, and the consequent height of such cliffs. Let the elevations recommence, and another sloping bank (of shingle, or sand, or mud, according to the nature of the successive beach-lines) must be formed, which again will be broken by as many lines of cliff, as there shall be periods of rest in the action of the subterranean forces. Now this is the structure of the plains of Patagonia; and such gradual changes harmonize well with the undisturbed strata, extending over so many hundred miles.

I must here observe, that I am far from supposing that the entire coast of this part of the continent has ever been lifted up, to the height of even a foot, at any one moment of time; but, drawing our analogies from the shores of the Pacific, that the whole may have been insensibly rising, with every now and then a paroxysmal or accelerated movement in certain spots. With respect to the alternation of the periods of such continued rise and those of quiescence, we may grant that they are probable, because such alternation agrees with what we see in the action, not only of a single volcano, but likewise of the disturbances affecting whole regions of the earth. At the present day, to the north of the parallel 44°, the subterranean forces are constantly manifesting their power over a space of more than 1,000 miles. But to the southward of that line, as far as Cape Horn, an earthquake is seldom or never experienced, and there is not a single point of eruption; yet in former ages, as we shall hereafter show, deluges of lava flowed from that very part. It is in conformity with our hypothesis that this southern region of repose, is at present suffering from the inroads of the ocean, as attested by the long line of cliff on the Patagonian coast. Such we believe to have been the causes of this singular configuration of the land. Nevertheless, we confess that it at first appears startling, that the most marked intervals between the heights of the successive plains should, instead of some great and sudden action of the subterranean forces, only indicate a longer period of repose.

In explaining the widely spread bed of gravel, we must first suppose a great mass of shingle to be collected by the action of innumerable torrents, and the swell of an open ocean, at the submarine basis of the Andes, prior to the elevation of the plains of Patagonia. If such a mass should then be lifted up, and left exposed during one of the periods of subterranean repose; a certain breadth, for instance a mile, would be washed down, and spread out over the bottom of the invading waters. (That the sea near the coast can carry out pebbles, we may feel sure from the circumstance of their gradual decrease in size, according to the distance from the coast-line.)

If this part of the sea should now be elevated, we should have a bed of gravel, but it would be of less thickness than in the first mass, both because it is spread over a larger area, and because it has been much reduced by attrition. This process being repeated, we might carry beds of gravel, always decreasing in thickness (as happens in Patagonia) to a considerable distance from the line of parent rock.* For instance, on the banks of the St Cruz at the

---

* It is needless to point out to the geologist, that this view, if correct, will account, without the necessity of any sudden rush of water, for the general covering of mixed shingle, so common in many parts of Europe, and likewise for the occurrence of widely extended strata of conglomerate; for the superficial beds might, during a period of subsidence, be covered by fresh deposits.

distance of 100 miles above the mouth of the river, the bed of gravel is 212 feet thick, whereas, near the coast, it seldom exceeds 25 or 30 feet; the thickness being thus reduced to nearly one-eighth.

I have already stated that the gravel is separated from the fossiliferous strata by some white beds of a friable substance, singularly resembling chalk, but which cannot be compared, as far as I am aware, with any formation in Europe. With respect to its origin, I may observe that the well-rounded pebbles all consist of various felspathic porphyrics; and that, from their prolonged attrition, during the successive re-modellings of the whole mass, much sediment must have been produced. I have already remarked that the white earthy matter more closely resembles decomposed felspar, than any other substance. If such is its origin it would always, from its lightness, be carried further to seaward than the pebbles. But as the land was elevated, the beds would be brought nearer the coast-line, and so become covered by the fresh masses of gravel which were travelling outwards. When these white beds were themselves elevated, they would hold a position intermediate between the gravel and the common foundation, or the fossiliferous strata. To explain my meaning more clearly, let us suppose the bottom of the present sea covered to a certain distance from the coast-line, with pebbles gradually decreasing in size, and beyond it by the white sediment. Let the land rise, so that the beach-line, by the fall of the water, may be carried outwards; then likewise the gravel, by the same agency as before, will be transported so much further from the coast, and will cover the white sediment, and these beds again will invade the more distant parts of the bottom of the sea. By this outward progress, the order of superposition must always be gravel, white sediment, and the fossiliferous strata.

Such is the history of the changes by which the present condition of Patagonia has, I believe, been determined. These changes all result from the assumption of a steady but very gradual elevation, extending over a wide area, and interrupted at long intervals by periods of repose. But we must now return to Port St Julian. On the south side of the harbour, a cliff of about 90 feet in height intersects a plain constituted of the formations above described; and its surface is strewed over with recent marine shells. The gravel, however, differently from that in every other locality, is covered by a very irregular and thin bed of a reddish loam, containing a few small calcareous concretions. The matter somewhat resembles that of the Pampas, and probably owes its origin either to a small stream having formerly entered the sea at that spot, or to a mud-bank similar to those now existing at the head of the harbour. In one spot this earthy matter filled up a hollow, or gully, worn quite through the gravel, and in this mass a group of large bones was embedded. The animal to which they belonged, must have lived, as in the case at Bahia Blanca, at a period long subsequent to the existence of the shells now inhabiting the coast. We may feel sure of this, because the formation of the lower terrace or plain, must necessarily have been posterior to those above it, and on the surface of the two higher ones, sea-shells of recent species are scattered. From the small physical change, which the last 100 feet elevation of the continent could have produced, the climate, as well as the general condition of Patagonia, probably was nearly the same, at the time when the animal was embedded, as it now is. This conclusion is moreover supported by the identity of the shells belonging to the two ages. Then immediately occurred the difficulty, how could any large quadruped have subsisted on these wretched deserts in lat. 49° 15′? I had no idea at the time, to what kind of animal these remains belonged. The puzzle, however, was soon solved when Mr Owen examined them; for he considers that they formed part of an animal allied to the guanaco or llama, but fully as large as the true camel. As all the existing members of the family of Camelidæ are inhabitants of the most sterile countries, so may we suppose was this extinct kind. The structure of the cervical vertebræ, the transverse processes not being perforated for the vertebral artery, indicates its affinity: some other parts, however, of its structure, probably are anomalous.

The most important result of this discovery, is the confirmation of the law that existing animals have a close relation in form with extinct species. As the guanaco is the characteristic quadruped of Patagonia, and the vicuna of the snow-clad summits of the Cordillera, so in bygone days, this gigantic species of the same family must have been conspicuous on the southern plains. We see this same relation of type between the existing and fossil Ctenomys, between the capybara (but less plainly, as shown by Mr Owen) and the gigantic Toxodon; and lastly, between the living and extinct Edentata. At the present day, in South America, there exist probably nineteen species of this order, distributed into several genera; while throughout the rest of the world there are but five. If, then, there is a relation between the living and the dead, we should expect that the Edentata would be numerous in the fossil state. I need only reply by enumerating the megatherium, and the three or four other great species, discovered at Bahia Blanca; the remains of some of which are also abundant over the whole immense territory of La Plata. I have already pointed out the singular relation between the armadilloes and their great prototypes, even in a point apparently of so little importance as their external covering.

The order of rodents at the present day, is most conspicuous in South America, on account of the vast number* and size of the species, and the multitude of individuals: according to the same law, we should expect to find their representatives in a fossil state. Mr Owen has shown how far the Toxodon is thus related; and it is moreover not improbable that another large animal has likewise a similar affinity.

The teeth of the rodent nearly equalling in size those of the Capybara, which were discovered near Bahia Blanca, must also be remembered.

The law of the succession of types, although subject to some remarkable exceptions, must possess the highest interest to every philosophical naturalist, and was first clearly observed in regard to Australia, where fossil remains of a large and extinct species of Kangaroo and other marsupial animals were discovered buried in a cave. In America the most marked change among the mammalia has been the loss of several species of Mastodon, of an elephant, and of the horse. These Pachydermata appear formerly to have had a range over the world, like that which deer and antelopes

---

* In my collection Mr Waterhouse distinguishes twenty-seven species of mice; to these must be added about thirteen more, known from the works of Azara, and other naturalists; so that we have forty species, which have actually been described as coming from between the Tropic and Cape Horn.

now hold. If Buffon had known of these gigantic armadilloes, llamas, great rodents, and lost pachydermata, he would have said with a greater semblance of truth, that the creative force in America had lost its vigour, rather than that it had never possessed such powers.

It is impossible to reflect without the deepest astonishment, on the changed state of this continent. Formerly it must have swarmed with great monsters, like the southern parts of Africa, but now we find only the tapir, guanaco, armadillo, and capybara; mere pigmies compared to the antecedent races. The greater number, if not all, of these extinct quadrupeds lived at a very recent period; and many of them were contemporaries of the existing molluscs. Since their loss, no very great physical changes can have taken place in the nature of the country. What then has exterminated so many living creatures? In the Pampas, the great sepulchre of such remains, there are no signs of violence, but on the contrary, of the most quiet and scarcely sensible changes. At Bahia Blanca I endeavoured to show the probability that the ancient Edentata, like the present species, lived in a dry and sterile country, such as now is found in that neighbourhood. With respect to the camel-like llama of Patagonia, the same grounds which, before knowing more than the size of the remains, perplexed me, by not allowing any great change of climate, now that we can guess the habits of the animal, are strangely confirmed. What shall we say of the death of the fossil horse? Did those plains fail in pasture, which afterwards were overrun by thousands and tens of thousands of the successors of the fresh stock introduced with the Spanish colonist? In some countries, we may believe, that a number of species subsequently introduced, by consuming the food of the antecedent races, may have caused their extermination; but we can scarcely credit that the armadillo has devoured the food of the immense Megatherium, the capybara of the Toxodon, or the guanaco of the camel-like kind. But granting that all such changes have been small, yet we are so profoundly ignorant concerning the physiological relations, on which the life, and even health (as shown by epidemics) of any existing species depends, that we argue with still less safety about either the life or death of any extinct kind.

One is tempted to believe in such simple relations, as variation of climate and food, or introduction of enemies, or the increased numbers of other species, as the cause of the succession of races. But it may be asked whether it is probable that any such cause should have been in action during the same epoch over the whole northern hemisphere, so as to destroy the *Elephas primigenus*, on the shores of Spain, on the plains of Siberia, and in Northern America; and in a like manner, the *Bos urus*, over a range of scarcely less extent? Did such changes put a period to the life of *Mastodon angustidens*, and of the fossil horse, both in Europe and on the Eastern slope of the Cordillera in Southern America? If they did, they must have been changes common to the whole world; such as gradual refrigeration, whether from modifications of physical geography, or from central cooling. But on this assumption, we have to struggle with the difficulty that these supposed changes, although scarcely sufficient to affect mollus-cous animals either in Europe or South America, yet destroyed many quadrupeds in regions now characterized by *frigid, temperate,* and *warm** climates! These cases of extinction forcibly recall the idea (I do not wish to draw any close analogy) of certain fruit-trees, which, it has been asserted, though grafted on young stems, planted in varied situations, and fertilized by the richest manures, yet at one period, have all withered away and perished. A fixed and determined length of life has in such cases been given to thousands and thousands of buds (or individual germs), although produced in long succession. Among the greater num-ber of animals, each individual appears nearly independent of its kind; yet all of one kind may be bound together by common laws, as well as a certain number of individual buds in the tree, or polypi in the Zoophyte.

I will add one other remark. We see that whole series of animals, which have been created with peculiar kinds of organiz-ation, are confined to certain areas; and we can hardly suppose

---

* The *Elephas primigenus* is thus circumstanced, having been found in Yorkshire (associated with recent shells: Lyell, vol. i, chap. vi), in Siberia, and in the warm regions of lat. 31°, in North America. The remains of the Mastodon occur in Paraguay (and I believe in Brazil, in lat. 12°), as well as in the temperate plains south of the Plata.

these structures are only adaptations to peculiarities of climate or country; for otherwise, animals belonging to a distinct type, and introduced by man, would not succeed so admirably, even to the extermination of the aborigines. On such grounds it does not seem a necessary conclusion, that the extinction of species, more than their creation, should exclusively depend on the nature (altered by physical changes) of their country. All that at present can be said with certainty, is that, as with the individual, so with the species, the hour of life has run its course, and is spent.

CHAPTER X

Santa Cruz – Expedition up river – Indians – Character of
Patagonia – Basaltic platform – Immense streams of lava –
Non-transport of blocks by river – Excavation of valley –
Condor, range and habits – Cordillera – Erratic boulders of
great size – Indian relics – Return to the ship

SANTA CRUZ – PATAGONIA

APRIL 13TH – The *Beagle* anchored within the mouth of the Santa
Cruz. This river is situated about 60 miles south of Port St Julian.
During the last voyage, Captain Stokes proceeded 30 miles up,
but then, from the want of provisions, was obliged to return.
Excepting what was discovered at that time, scarcely any thing
was known about this large river. Captain FitzRoy now determined to follow its course as far as time would allow. On the
18th, three whale-boats started, carrying three weeks' provisions;
and the party consisted of twenty-five souls – a force which would
have been sufficient to have defied a host of Indians. With a strong
flood-tide, and a fine day, we made a good run, soon drank some
of the fresh water, and were at night nearly above the tidal
influence.

The river here assumed a size and appearance, which, even at
the highest point we ultimately reached, was scarcely diminished.
It was generally from 300 to 400 yards broad, and in the middle
about 17 feet deep. The rapidity of the current, which in its whole
course runs at the rate of from 4 to 6 knots an hour, is perhaps its
most remarkable feature. The water is of a fine blue colour, but
with a slight milky tinge, and not so transparent as at first sight
would have been expected. It flows over a bed of pebbles, like
those which compose the beach and surrounding plains.

Although its course is winding, it runs through a valley which extends in a direct line to the westward. This valley varies from 5 to 10 miles in breadth; it is bounded by step-formed terraces, which rise in most parts one above the other to the height of 500 feet, and have on the opposite sides a remarkable correspondence.

APRIL 19TH – Against so strong a current, it was of course quite impossible to row or sail. Consequently the three boats were fastened together head and stern, two hands left in each, and the rest came on shore to track. As the general arrangements, made by Captain FitzRoy, were very good for facilitating the work of all, and as all had a share of it, I will describe the system. The party, including everyone, was divided into two spells, each of which hauled at the tracking line alternately for an hour and a half. The officers of each boat lived with, ate the same food, and slept in the same tent with their crew, so that each boat was quite independent of the others. After sunset, the first level spot where any bushes were growing, was chosen for our night's lodging. Each of the crew took it in turns to be cook. Immediately the boat was hauled up, the cook made his fire; two others pitched the tent; the coxswain handed the things out of the boat; the rest carried them up to the tents, and collected firewood. By this order, in half an hour, every thing was ready for the night. A watch of two men and an officer was always kept, whose duty it was to look after the boats, keep up the fire, and guard against Indians. Each in the party had his one hour every night.

During this day we tracked but a short distance, for there were many islets, covered by thorny bushes, and the channels between them were shallow.

APRIL 20TH – We passed the islands and set to work. Our regular day's march, although it was hard enough, carried us on an average only 10 miles in a straight line, and perhaps 15 or 20 altogether. Beyond the place where we slept last night the country is completely *terra incognita*, for it was there that Captain Stokes turned back. We saw in the distance a great smoke, and found the skeleton of a horse, so we knew that Indians were in the neighbourhood. On the next morning (21st) tracks of a party of horse, and marks left by the trailing of the *chuzos* were observed on the ground. It was generally thought they must have reconnoitred us during the night. Shortly afterwards we came to a spot, where from the fresh footsteps of men, children, and horses, it was evident the party had crossed the river.

* * *

APRIL 29TH – From some high land we hailed with joy the white summits of the Cordillera, as they were seen occasionally peeping through their dusky envelope of clouds. During the few succeeding days, we continued to get on slowly, for we found the river-course very tortuous, and strewed with immense fragments of various ancient slaty rocks, and of granite. The plain bordering the valley had here attained an elevation of about 1,100 feet, and its character was much altered. The well-rounded pebbles of porphyry were in this part mingled with many immense angular fragments of basalt and of the rocks above mentioned. The first of these erratic blocks which I noticed, was 67 miles distant from the nearest mountain; another which had been transported to rather a less distance, measured 5 yards square, and projected 5 feet above the gravel. Its edges were so angular, and its size so great, that I at first mistook it for a rock *in situ*, and took out my compass to observe the direction of its cleavage. The plains here were not quite so level as those nearer the coast, but yet, they betrayed little signs of any violent action. Under these circumstances, it would be difficult, as it appears to me, to explain this phenomenon on any theory, excepting through that of transport by ice while the country was under water. But this is a subject to which I shall again recur.

During the two last days we met with signs of horses, and with several small articles which had belonged to the Indians, – such as parts of a mantle and a bunch of ostrich feathers – but they appeared to have been lying long on the ground. Between the place where the Indians had so lately crossed the river and this neighbourhood, though so many miles apart, the country appears to be quite unfrequented. At first, considering the abundance of the guanacoes, I was surprised at this; but it is explained by the stony nature of the plains, which would soon disable an unshod horse from taking part in the chase. Nevertheless, in two places in this very central region, I found small heaps of stones, which I do not think could have been accidentally thrown together. They were placed on points, projecting over the edge of the highest lava cliff, and they resembled, but on a small scale, those near Port Desire.

MAY 4TH – Captain FitzRoy determined to take the boats no higher. The river had a winding course, and was very rapid; and the appearance of the country offered no temptation to proceed any further. Every where we met with the same productions, and the same dreary landscape. We were now 140 miles distant from the Atlantic, and about 60 from the nearest arm of the Pacific. The valley in this upper part expanded into a wide basin, bounded on the north and south by the basaltic platforms, and fronted by the long range of the snow-clad Cordillera. But we viewed these grand mountains with regret, for we were obliged to imagine their form and nature, instead of standing, as we had hoped, on their crest, and looking down on the plain below. Besides the useless loss of time which an attempt to ascend any higher would have cost us, we had already been for some days on half allowance of bread. This, although really enough for any reasonable men, was, after our hard day's march, rather scanty food. Let those alone who have never tried it, exclaim about the comfort of a light stomach and an easy digestion.

5TH – Before sunrise we commenced our descent. We shot down the stream with great rapidity, generally at the rate of 10 knots an hour. In this one day we effected what had cost us five-and-a-half hard days' labour in ascending. On the 8th, we reached the *Beagle* after our twenty-one days' expedition. Every one excepting myself had cause to be dissatisfied; but to me the ascent afforded a most interesting section of the great tertiary formation of Patagonia.

CHAPTER XI

TIERRA DEL FUEGO

DECEMBER 17TH, 1832 – Having now finished with Patagonia, I
will describe our first arrival in Tierra del Fuego. A little after
noon we doubled Cape St Diego, and entered the famous strait of
Le Maire. We kept close to the Fuegian shore, but the outline of
the rugged, inhospitable Staten land was visible amidst the
clouds. In the afternoon we anchored in the Bay of Good Success.
While entering we were saluted in a manner becoming the
inhabitants of this savage land. A group of Fuegians partly
concealed by the entangled forest, were perched on a wild point
overhanging the sea; and as we passed by, they sprang up, and
waving their tattered cloaks sent forth a loud and sonorous shout.
The savages followed the ship, and just before dark we saw their
fire, and again heard their wild cry. The harbour consists of a fine
piece of water half surrounded by low rounded mountains of
clay-slate, which are covered to the water's edge by one dense
gloomy forest. A single glance at the landscape was sufficient to
show me, how widely different it was from any thing I had ever
beheld. At night it blew a gale of wind, and heavy squalls from
the mountains swept past us. It would have been a bad time out
at sea, and we, as well as others, may call this Good Success
Bay.

In the morning, the Captain sent a party to communicate with the Fuegians. When we came within hail, one of the four natives who were present advanced to receive us, and began to shout most vehemently, wishing to direct us where to land. When we were on shore the party looked rather alarmed, but continued talking and making gestures with great rapidity. It was without exception the most curious and interesting spectacle I had ever beheld. I could not have believed how wide was the difference, between savage and civilized man. It is greater than between a wild and domesticated animal, in as much as in man there is a greater power of improvement. The chief spokesman was old, and appeared to be the head of the family; the three others were powerful young men, about 6 feet high. The women and children had been sent away. These Fuegians are a very different race from the stunted miserable wretches further to the westward. They are much superior in person, and seem closely allied to the famous Patagonians of the Strait of Magellan. Their only garment consists of a mantle made of guanaco skin, with the wool outside; this they wear just thrown over their shoulders, as often leaving their persons exposed as covered. Their skin is of a dirty coppery red colour.

The old man had a fillet of white feathers tied round his head, which partly confined his black, coarse, and entangled hair. His face was crossed by two broad transverse bars; one painted bright red reached from ear to ear, and included the upper lip; the other, white like chalk, extended parallel and above the first, so that even his eyelids were thus coloured. Some of the other men were ornamented by streaks of black powder, made of charcoal. The party altogether closely resembled the devils which come on the stage in such plays as Der Freischutz.

Their very attitudes were abject, and the expression of their countenances distrustful, surprised, and startled. After we had presented them with some scarlet cloth, which they immediately tied round their necks, they became good friends. This was shown by the old man patting our breasts, and making a chuckling kind of noise, as people do when feeding chickens. I walked with the old man, and this demonstration of friendship was repeated several times; it was concluded by three hard slaps, which were given me on the breast and back at the same time. He then bared his bosom for me to return the compliment, which being done, he seemed highly pleased. The language of these people, according to our notions, scarcely deserves to be called articulate. Captain Cook has compared it to a man clearing his throat, but certainly no European ever cleared his throat with so many hoarse, guttural, and clicking sounds.

They are excellent mimics: as often as we coughed or yawned, or made any odd motion, they immediately imitated us. Some of our party began to squint and look awry; but one of the young Fuegians (whose whole face was painted black, excepting a white band across his eyes) succeeded in making far more hideous grimaces. They could repeat with perfect correctness, each word in any sentence we addressed them, and they remembered such words for some time. Yet we Europeans all know how difficult it is to distinguish apart the sounds in a foreign language. Which of us, for instance, could follow an American Indian through a sentence of more than three words? All savages appear to possess, to an uncommon degree, this power of mimicry. I was told almost in the same words, of the same ludicrous habits among the Caffres: the Australians, likewise, have long been notorious for being able to imitate and describe the gait of any man, so that he may be recognized. How can this faculty be explained? is it a consequence of the more practised habits of perception and keener senses, common to all men in a savage state, as compared to those long civilized?

When a song was struck up by our party, I thought the Fuegians would have fallen down with astonishment. With equal surprise they viewed our dancing; but one of the young men, when asked, had no objection to a little waltzing. Little accustomed to Europeans as they appeared to be, yet they knew, and dreaded our fire-arms; nothing would tempt them to take a gun in their hands. They begged for knives, calling them by the Spanish word 'cuchilla'. They explained also what they wanted, by acting as if they had a piece of blubber in their mouth, and then pretending to cut instead of tear it.

It was interesting to watch the conduct of these people towards

Jemmy Button (one of the Fuegians* who had been taken, during the former voyage, to England): they immediately perceived the difference between him and the rest, and held much conversation between themselves on the subject. The old man addressed a long harangue to Jemmy, which it seems was to invite him to stay with them. But Jemmy understood very little of their language, and was, moreover, thoroughly ashamed of his countrymen. When York Minster (another of these men) came on shore, they noticed him in the same way, and told him he ought to shave; yet he had not twenty dwarf hairs on his face, whilst we all wore our untrimmed beards. They examined the colour of his skin, and compared it with ours. One of our arms being bared, they expressed the liveliest surprise and admiration at its whiteness. We thought that they mistook two or three of the officers, who were rather shorter and fairer (though adorned with large beards), for the ladies of our party. The tallest amongst the Fuegians was evidently much pleased at his height being noticed. When placed back to back with the tallest of the boat's crew, he tried his best to edge on higher ground, and to stand on tiptoe. He opened his mouth to show his teeth, and turned his face for a side view; and all this was done with such alacrity, that I dare say he thought himself the handsomest man in Tierra del Fuego. After the first feeling on our part of grave astonishment was over, nothing could be more ludicrous or interesting than the odd mixture of surprise and imitation which these savages every moment exhibited.

The next day I attempted to penetrate some way into the country. Tierra del Fuego may be described as a mountainous country, partly submerged in the sea, so that deep islets and bays occupy the place where valleys should exist. The mountain sides (except on the exposed western coast) are covered from the water's edge upwards by one great forest. The trees reach to an elevation of between 1,000 and 1,500 feet; and are succeeded by a band of peat, with minute alpine plants; and this again is succeeded by the line of perpetual snow, which, according to Captain King, in the Strait of Magellan descends to between 3,000 and

---

* Captain FitzRoy has given a history of these people. Four were taken to England; one died there, and the three others (two men and one woman) were now brought back and settled in their own country.

4,000 feet. To find an acre of level land in any part of the country is most rare. I recollect only one little flat near Port Famine, and another of rather larger extent near Goeree Road. In both these cases, and in all others, the surface was covered by a thick bed of swampy peat. Even within the forest the ground is concealed by a mass of slowly putrefying vegetable matter, which, from being soaked with water, yields to the foot.

Finding it nearly hopeless to push my way through the wood, I followed the course of a mountain torrent. At first, from the waterfalls and number of dead trees, I could hardly crawl along; but the bed of the stream soon became a little more open, from the floods having swept the sides. I continued slowly to advance for an hour along the broken and rocky banks; and was amply repaid by the grandeur of the scene. The gloomy depth of the ravine well accorded with the universal signs of violence. On every side were lying irregular masses of rock and up-torn trees; other trees, though still erect, were decayed to the heart and ready to fall. The entangled mass of the thriving and the fallen reminded me of the forests within the tropics; – yet there was a difference; for in these still solitudes, Death, instead of Life, seemed the predominant spirit. I followed the water-course till I came to a spot where a great slip had cleared a straight space down the mountain side. By this road I ascended to a considerable elevation, and obtained a good view of the surrounding woods. The trees all belong to one kind, the *Fagus betuloides*, for the number of the other species of beech, and of the Winter's bark, is quite inconsiderable. This tree keeps its leaves throughout the year; but its foliage is of a peculiar brownish-green colour, with a tinge of yellow. As the whole landscape is thus coloured, it has a sombre, dull appearance; nor is it often enlivened by the rays of the sun.

* * *

DECEMBER 21ST – The *Beagle* got under way: and on the succeeding day, favoured to an uncommon degree by a fine easterly breeze, we closed in with the Barnevelts, and, running past Cape Deceit with its stony peaks, about three o'clock doubled the weatherbeaten Cape Horn. The evening was calm and bright, and we enjoyed a fine view of the surrounding isles. Cape Horn, however, demanded his tribute, and before night sent us a gale of wind directly in our teeth. We stood out to sea, and on the second day again made the land, when we saw on our weather-bow this notorious promontory in its proper form – veiled in a mist, and its dim outline surrounded by a storm of wind and water. Great black clouds were rolling across the heavens, and squalls of rain, with hail, swept by us with extreme violence so that the captain determined to run into Wigwam Cove. This is a snug little harbour, not far from Cape Horn; and here, at Christmas-eve, we anchored in smooth water. The only thing which reminded us of the gale outside, was every now and then a puff from the mountains, which seemed to wish to blow us out of the water.

DECEMBER 25TH – Close by the cove, a pointed hill, called Kater's Peak, rises to the height of 1,700 feet. The surrounding islands all consist of conical masses of greenstone, associated sometimes with less regular hills of baked and altered clay-slate. This part of Tierra del Fuego may be considered as the extremity of the submerged chain of mountains already alluded to. The cove takes its name of 'Wigwam' from some of the Fuegian habitations; but every bay in the neighbourhood might be so called with equal propriety. The inhabitants living chiefly upon shell-fish, are obliged constantly to change their place of residence; but they return at intervals to the same spots, as is evident from the pile of old shells, which must often amount to some tons in weight. These heaps can be distinguished at a long distance by the bright green colour of certain plants, which invariably grow on them. Among these may be enumerated the wild celery and scurvy grass, two very serviceable plants, the use of which has not been discovered by the natives.

The Fuegian wigwam resembles, in size and dimensions, a haycock. It merely consists of a few broken branches stuck in the ground, and very imperfectly thatched on one side with a few tufts of grass and rushes. The whole cannot be so much as the work of an hour, and it is only used for a few days. At Goeree Roads I saw a place where one of these naked men had slept, which absolutely offered no more cover than the form of a hare. The man was evidently living by himself, and York Minster said he was 'very bad man', and that probably he had stolen some-thing. On the west coast, however, the wigwams are rather better, for they are covered with seal-skins. We were detained here several days by the bad weather. The climate is certainly wretched; the summer solstice was now passed, yet every day snow fell on the hills, and in the valleys there was rain, accom-panied by sleet. The thermometer generally stood about 45°, but in the night fell to 38° or 40°. From the damp and boisterous state of the atmosphere, not cheered by a gleam of sunshine, one fancied the climate even worse than it really was.

At a subsequent period the *Beagle* anchored for a couple of days under Wollaston Island, which is a short way to the northward. While going on shore we pulled alongside a canoe with six Fuegians. These were the most abject and miserable creatures I any where beheld.* On the east coast the natives, as we have seen, have guanaco cloaks, and on the west, they possess seal-skins. Amongst these central tribes the men generally possess an otter-skin, or some small scrap about as large as a pocket-handkerchief, which is barely sufficient to cover their backs as low down as their loins. It is laced across the breast by strings, and according as the wind blows, it is shifted from side to side. But these Fuegians in the canoe were quite naked, and even one full-grown woman was absolutely so. It was raining heavily, and the fresh water, together with the spray, trickled down her body. In another harbour not far distant, a woman, who was suckling a recently-born child, came one day alongside the vessel, and remained there whilst the sleet fell and thawed on her naked bosom, and on the skin of her naked child. These poor wretches were stunted in their growth,

---

* I believe, in this extreme part of South America, man exists in a lower state of improvement than in any other part of the world. The South Sea islander of either race is comparatively civilized. The Esquimaux, in his subterranean hut, enjoys some of the comforts of life, and in his canoe, when fully equipped, manifests much skill. Some of the tribes of Southern Africa, prowling about in search of roots, and living concealed on the wild and arid plains, are sufficiently wretched. But the Australian, in the simplicity of the arts of life, comes nearest the Fuegian. He can, however, boast of his boomerang, his spear and throwing-stick, his method of climbing trees, tracking animals, and scheme of hunting. Although thus superior in acquirements, it by no means follows that he should likewise be so in capabilities. Indeed, from what we saw of the Fuegians, who were taken to England, I should think the case was the reverse.

their hideous faces bedaubed with white paint, their skins filthy and greasy, their hair entangled, their voices discordant, their gestures violent and without dignity. Viewing such men, one can hardly make oneself believe they are fellow-creatures, and inhabitants of the same world. It is a common subject of conjecture what pleasure in life some of the less gifted animals can enjoy: how much more reasonably the same question may be asked with respect to these barbarians. At night, five or six human beings, naked and scarcely protected from the wind and rain of this tempestuous climate, sleep on the wet ground coiled up like animals. Whenever it is low water, they must rise to pick shellfish from the rocks; and the women, winter and summer, either dive to collect sea eggs, or sit patiently in their canoes, and, with a baited hair-line, jerk out small fish. If a seal is killed, or the floating carcass of a putrid whale discovered, it is a feast: such miserable food is assisted by a few tasteless berries and fungi. Nor are they exempt from famine, and, as a consequence, cannibalism accompanied by parricide.

The tribes have no government or head, yet each is surrounded by other hostile ones, speaking different dialects; and the cause of their warfare would appear to be the means of subsistence. Their country is a broken mass of wild rock, lofty hills, and useless forests: and these are viewed through mists and endless storms. The habitable land is reduced to the stones which form the beach; in search of food they are compelled to wander from spot to spot, and so steep is the coast, that they can only move about in their wretched canoes. They cannot know the feeling of having a home, and still less that of domestic affection; unless indeed the treatment of a master to a laborious slave can be considered as such. How little can the higher powers of the mind be brought into play! What is there for imagination to picture, for reason to compare, for judgment to decide upon? to knock a limpet from the rock does not even require cunning, that lowest power of the mind. Their skill in some respects may be compared to the instinct of animals; for it is not improved by experience: the canoe, their most ingenious work, poor as it is, has remained the same, for the last 250 years.

Whilst beholding these savages, one asks, whence have they

come? What could have tempted, or what change compelled a tribe of men to leave the fine regions of the north, to travel down the Cordillera or backbone of America, to invent and build canoes, and then to enter on one of the most inhospitable countries within the limits of the globe? Although such reflections must at first occupy one's mind, yet we may feel sure that many of them are quite erroneous. There is no reason to believe that the Fuegians decrease in number; therefore we must suppose that they enjoy a sufficient share of happiness (of whatever kind it may be) to render life worth having. Nature by making habit omnipotent, and its effects hereditary, has fitted the Fuegian to the climate and the productions of his country.

JANUARY 15TH, 1833 – The *Beagle* anchored in Goeree Roads. Captain FitzRoy having determined to settle the Fuegians, according to their wishes, in Ponsonby Sound, four boats were equipped to carry them there through the Beagle channel. This channel which was discovered by Captain FitzRoy during the last voyage, is a most remarkable feature in the geography of this, or indeed of any other country. Its length is about 120 miles with an average breadth, not subject to any very great variations, of about 2 miles. It is throughout the greater part so extremely straight, that the view, bounded on each side by a line of mountains, gradually becomes indistinct in the perspective. This arm of the sea may be compared to the valley of Lochness in Scotland, with its chain of lakes and entering friths. At some future epoch the resemblance perhaps will become complete. Already in one part we have proofs of a rising of the land in a line of cliff, or terrace, composed of coarse sandstone, mud, and shingle, which forms both shores. The Beagle channel crosses the southern part of Tierra del Fuego in an east and west line; in its middle, it is joined on the south side by an irregular channel at right angles to it, which has been called Ponsonby Sound. This is the residence of Jemmy Button's tribe and family.

JANUARY 19TH – Three whale-boats and the yawl, with a party of twenty-eight, started under the command of Captain FitzRoy. In the afternoon we entered the eastern mouth of the channel, and shortly afterwards found a snug little cove, concealed by some surrounding islets. Here we pitched our tents, and lighted our fires. Nothing could look more comfortable than this scene. The glassy water of the little harbour, with the trees sending their branches over the rocky beach, the boats at anchor, the tents supported by the crossed oars, and the smoke curling up the wooded valley, formed a picture of quiet retirement. The next day (20th) we smoothly glided onwards in our little fleet, and came to a more inhabited district. Few if any of these natives could ever have seen a white man; certainly nothing could exceed their astonishment at the apparition of the four boats. Fires were lighted on every point (hence the name of the land), both to attract our attention, and to spread far and wide the news. Some of the men ran for miles along the shore. As we passed under one cliff, four or five men suddenly appeared above us, forming the most wild and savage group that can be imagined. They were absolutely naked, with long streaming hair, and with rugged staffs in their hands: springing from the ground they waved their arms around their heads, and sent forth the most hideous yells.

At dinner-time we landed among a party of Fuegians. At first they were not inclined to be friendly; for until the captain pulled in ahead of the other boats, they kept their slings in their hands. We soon, however, delighted them by trifling presents, such as tying red tape round their heads. It was as easy to please, as it was difficult to satisfy these savages. Young and old, men and children, never ceased repeating the word 'yammerschooner', which means 'give me'. After pointing to almost every object, one after the other, even to the buttons on our coats, and saying their favourite word in as many intonations as possible, they would then use it in a neuter sense, and vacantly repeat 'yammerschooner'. After yammerschoonering for any article very eagerly, they would by a simple artifice point to their young women or little children, as much as to say, 'If you will not give it me, surely you will to such as these?'

At night we endeavoured in vain to find an uninhabited cove; and at last were obliged to bivouac not far from a party of natives. They were very inoffensive as long as they were few in numbers, but in the morning (21st) being joined by others they showed symptoms of hostility. An European labours under great disadvantages, when treating with savages like these, who have not the least idea of the power of fire-arms. In the very act of levelling his musket, he appears to the savage far inferior to a man armed with a bow and arrow, a spear, or even a sling. Nor is it easy to teach them our superiority except by striking a fatal blow. Like wild beasts they do not appear in all cases to compare numbers; for each individual if attacked, instead of retiring, will endeavour to dash your brains out with a stone, as certainly as a tiger under similar circumstances would tear you. Captain FitzRoy on one occasion, being very anxious from good reasons to frighten away a small party, twice fired his pistol close by the side of a native. The man both times looked astounded, and carefully but quickly rubbed his head; he then stared awhile, and gabbled to his companions; but he never seemed to think of running away. We can hardly put ourselves in the position of these savages, to understand their actions. In the case of the Fuegian, the possibility of such a sound as the report of a gun close to his ear, could never have entered his mind. He perhaps literally did not for a second know whether it was a sound or a blow, and therefore very naturally rubbed his head. In a similar manner, when a savage sees a mark struck by a bullet, it may be some time before he is able at all to understand how it is effected; for the fact of a body being invisible from its velocity, would perhaps be to him an idea totally inconceivable. Moreover, the extreme force of a bullet, that penetrates a hard substance without tearing it, may convince the savage that it has no force at all. Certainly I believe that many savages of the lowest grade, such as these of Tierra del Fuego, have seen objects struck, and even small animals killed by the musket, without being in the least aware how deadly an instrument it was.

22ND – After having passed an unmolested night, in what would appear to be neutral territory between Jemmy's tribe and the people we saw yesterday, we sailed pleasantly along. The scenery in this part had a peculiar and very magnificent character; although the effect was lessened from the lowness of the point of view in a boat, and from looking down the valley and hence losing all the beauty of a succession of ridges. The mountains attained an elevation of about 3,000 feet, and were terminated by sharp and jagged points. They rose in one unbroken sweep from the water's edge, and were covered to the height of 1,400 or 1,500 feet by the dusky-coloured forest. It was most curious to observe, how level and truly horizontal the line on the mountain-side was, as far as the eye could range, at which trees ceased to grow. It precisely resembled the high-water mark of drift weed on a sea-beach.

At night we slept close to the junction of Ponsonby Sound with the Beagle channel. A small family of Fuegians, who were living in the cove, were very quiet and inoffensive, and soon joined our party round the blazing fire. We were well clothed, and though sitting close to the fire, were far from too warm; yet these naked savages, though further off, were observed to our great surprise, to be streaming with perspiration at undergoing such a roasting. They seemed, however, very well pleased, and all joined in the chorus of the seamen's songs: but the manner in which they were invariably a little behindhand was quite ludicrous.

During the night the news had spread, and early in the morning (23rd) a fresh party arrived. Several of them had run so fast that their noses were bleeding, and their mouths frothed from the rapidity with which they talked, and with their naked bodies all bedaubed with black, white, and red, they looked like so many demoniacs who had been fighting. We then proceeded down Ponsonby Sound to the spot where poor Jemmy expected to find his mother and relations. We stayed there five days. Captain FitzRoy has given an account of all the interesting events which there happened.

During the succeeding year we paid another visit to the Fuegians, and the *Beagle* herself followed the same course which I have just described as having been taken in the boats. I was amused by finding what a difference the circumstance of being quite superior in force made, in the interest of beholding these savages. While in the boats I got to hate the very sound of their voices, so much trouble did they give us. The first and last word was 'yammerschooner'. When, entering some quiet little cove, we have looked round and thought to pass a quiet night, the odious word 'yammerschooner' has shrilly sounded from some gloomy nook, and then the little signal smoke has curled upwards to spread the news. On leaving some place we have said to each

other, 'Thank Heaven, we have at last fairly left these wretches!'
when one more faint halloo from an all-powerful voice, heard at a
prodigious distance, would reach our ears, and clearly could we
distinguish – 'yammerschooner'. But on the latter occasion, the
more Fuegians the merrier; and very merry work it was. Both
parties laughing, wondering, gaping at each other; we pitying
them, for giving us good fish and crabs for rags, &c.; they
grasping at the chance of finding people so foolish as to exchange
such splendid ornaments for a good supper. It was most amusing
to see the undisguised smile of satisfaction with which one young
woman, with her face painted black, tied with rushes several bits
of scarlet cloth round her head. Her husband, who enjoyed the
very universal privilege in this country of possessing two wives,
evidently became jealous of all the attention paid to his young
wife; and, after a consultation with his naked beauties, was
paddled away by them.

Some of the Fuegians plainly showed that they had a fair idea of
barter. I gave one man a large nail (a most valuable present)
without making any signs for a return; but he immediately picked
out two fish, and handed them up on the point of his spear. If any
present was designed for one canoe, and it fell near another, it was
invariably given to the right owner. We were always much
surprised at the little notice, or rather none whatever, which was
evinced respecting many things, even such as boats, the use of
which must have been evident. Simple circumstances, – such as
the whiteness of our skins, the beauty of scarlet cloth or blue
beads, the absence of women, our care in washing ourselves –
excited their admiration far more than any grand or complicated
object, such as the ship. Bougainville has remarked concerning these very people that they treat the 'chef d'œuvres de
l'industrie humaine, comme ils traitent les loix de la nature et ses
phénomènes'.

The perfect equality among the individuals composing these
tribes, must for a long time retard their civilization. As we see
those animals, whose instinct compels them to live in society and
obey a chief, are most capable of improvement, so is it with the
races of mankind. Whether we look at it as a cause or a consequence, the more civilized always have the most artificial

governments. For instance, the inhabitants of Otaheite, who, when first discovered, were governed by hereditary kings, had arrived at a far higher grade than another branch of the same people, the New Zealanders – who although benefited by being compelled to turn their attention to agriculture, were republicans in the most absolute sense. In Tierra del Fuego, until some chief shall arise with power sufficient to secure any acquired advantages, such as the domesticated animals or other valuable presents, it seems scarcely possible that the political state of the country can be improved. At present, even a piece of cloth is torn into shreds and distributed; and no one individual becomes richer than another. On the other hand, it is difficult to understand how a chief can arise till there is property of some sort by which he might manifest and still increase his authority.

*   *   *

CHAPTER XII

Falkland Islands – Excursion round island – Aspect – Cattle, horses, rabbit, wolf-like fox – Fire made of bones – Art in making fire – Manner of hunting wild cattle – Geology, fossil shells – Valleys filled with great fragments, scenes of violence – Penguin – Geese – Eggs of doris – Zoophytes, coralline phosphorescent – Compound animals

FALKLAND ISLANDS

MARCH 16TH, 1834 – The *Beagle* anchored in Berkeley Sound, in East Falkland Island.* This archipelago is situated in nearly the same latitude as the mouth of the Strait of Magellan. It covers a space of about 120 by 60 geographical miles, and is a little more than half the size of Ireland. After the possession of these miserable islands had been contested by France, Spain, and England, they were left uninhabited. The government of Buenos Ayres then sold them to a private individual, but likewise used them, as old Spain had done before, for a penal settlement. England claimed her right and seized them. The Englishman who was left in charge of the flag was consequently murdered. A British officer was next sent, unsupported by any power: and when we arrived, we found him in charge of a population, of which rather more than half were runaway rebels and murderers.

The theatre is worthy of the scenes acted on it. An undulating land, with a desolate and wretched aspect, is every where covered by a peaty soil and wiry grass, of one monotonous brown colour. Here and there a peak or ridge of gray quartz rock, breaks through the smooth surface. Every one has heard of the climate of these regions; it may be compared to that which is experienced at the height of between 1,000 and 2,000 feet, on the mountains of

* In the same month, also, of the previous year, the *Beagle* visited these islands.

North Wales; having however less sunshine and less frost, but more wind and rain.

* * *

During our whole ride we only saw one troop of wild horses. These animals, as well as the cattle, were introduced by the French in 1764, since which time they have greatly increased. It is a curious fact, that the horses have never left the eastern end of the island, although there is no natural boundary to prevent them from roaming, and that part of the island is not more tempting than the rest. The Gauchos, though asserting this to be the case, are unable to account for the circumstance. The horses appear to thrive well, yet they are small sized, and have lost so much strength, that they are unfit to be used in taking wild cattle with the lazo. In consequence, it is necessary to go to the great expense of importing fresh horses from the Plata. At some future period the southern hemisphere probably will have its breed of Falkland ponies, as the northern has that of Shetland.

The rabbit is another animal which has been introduced, and has succeeded very well; so that they abound over large parts of the island. Yet, like the horses, they are confined within certain limits; for they have not crossed the central chain of hills; nor would they have extended even so far as the base, if, as the Gauchos informed me, small colonies had not been carried there. I should not have supposed that these animals, natives of northern Africa, could have existed in a climate so extremely humid as this, and which enjoys so little sunshine that even wheat ripens only occasionally. It is asserted that in Sweden, which any one would have thought a more favourable climate, the rabbit cannot live out of doors. The first few pair moreover had here to contend against pre-existing enemies, in the fox, and some large hawks. The French naturalists have considered the black variety a distinct species, and called it *Lepus Magellanicus.** They imagined that

---

* Lesson's Zoology of the Voyage of the Coquille, vol. i, p. 168. All the early voyagers, and especially Bougainville, distinctly state that the wolf-like fox was the only native animal on the island. The distinction of this rabbit as a species, is taken from peculiarities in the fur, from the shape of the head, and from the shortness of the ears. I may here observe that the difference between the Irish and English hare, rests upon nearly similar characters, only more strongly marked.

Magellan, when talking of an animal under the name of 'conejos', in the Strait of Magellan, referred to this species; but he was alluding to a small cavy, which to this day is thus called. The Gauchos laughed at the idea of the black kind being different from the gray, and they said that at all events it had not extended its range any further than the other; that the two were never found separate; and that they readily bred together, and produced piebald offspring. Of the latter I now possess a specimen, and it is marked about the head, differently from the French specific description. This circumstance shows how cautious naturalists should be in making species; for even Cuvier, on looking at the skull of one of these rabbits, thought it was probably distinct.

The only quadruped native to the island, is a large wolf-like fox,* which is common to both East and West Falkland. I have no doubt it is a peculiar species, and confined to this archipelago; because many sealers, Gauchos, and Indians, who have visited these islands, all maintain that no such animal is found in any part of South America. Molina, from a similarity in habits, thought this was the same with his 'culpeu';† but I have seen both, and they are quite distinct. These wolves are well known, from Byron's account of their tameness and curiosity; which the sailors, who ran into the water to avoid them, mistook for fierceness. To this day their manners remain the same. They have been observed to enter a tent, and actually pull some meat from beneath the head of a sleeping seaman. The Gauchos, also, have frequently killed them in the evening, by holding out a piece of meat in one hand, and in the other a knife ready to stick them. As far as I am aware, there is no other instance in any part of the world, of so small a mass of broken land, distant from a continent, possessing so large a quadruped peculiar to itself. Their numbers have rapidly decreased; they are already banished from that half of the island which lies to the eastward of the neck of land between St Salvador Bay and Berkeley Sound. Within a very few years after

---

* I have reason to believe there is likewise a field-mouse. The common European rat and mouse have roamed from the habitations, and have settled themselves at various points. The common hog has also run wild.
† The 'culpeu' is the *Vulpes Magellanicus* brought home by Captain King from the Strait of Magellan. It is common in Chile.

these islands shall have become regularly settled, in all probability this fox will be classed with the dodo, as an animal which has perished from the face of the earth. Mr Lowe, an intelligent person who has long been acquainted with these islands, assured me, that all the foxes from the western island were smaller and of a redder colour than those from the eastern. In the four specimens which were brought to England in the *Beagle** there was some variation, but the difference with respect to the islands could not be perceived. At the same time the fact is far from improbable.

*   *   *

In many parts of the island, the bottoms of the valleys are covered in an extraordinary manner, by myriads of great angular fragments of the quartz rock. These have been mentioned with surprise by every voyager since the time of Pernety. The whole may be called 'a stream of stones'. The blocks vary in size, from that of a man's chest to ten or twenty times as large, and occasionally they altogether exceed such measures. Their edges show no signs of being waterworn, but are only a little blunted. They do not occur thrown together in irregular piles, but are spread out into level sheets, or great streams. It is not possible to ascertain their thickness, but the water of small streamlets could be heard trickling through the stones many feet below the surface. The actual depth is probably much greater, because the crevices between the lower fragments must long ago have been filled up with sand, and the bed of the rivulet thus raised. The width of these beds varies from a few hundred feet to a mile; but the peaty soil daily encroaches on the borders, and even forms islets wherever a few fragments happen to lie close together. In a valley south of Berkeley Sound, which some of our party called the 'great valley of fragments', it was necessary to cross an uninterrupted band half a mile wide, by jumping from one pointed stone to another. So large were the fragments, that being overtaken by a shower of rain, I readily found good shelter beneath one of them.

---

* Captain FitzRoy has presented two of these foxes to the British Museum, where Mr Gray had the kindness to compare them in my presence.

Their little inclination is the most remarkable circumstance in these 'streams of stones'. On the hill-sides I have seen them sloping at an angle of 10 degrees with the horizon; but in some of the level, broad-bottomed valleys, the inclination is only just sufficient to be clearly perceived. On so rugged a surface there was no means of measuring the angle; but to give a common illustration, I may say that the slope alone would not have checked the speed of an English mail-coach. In some places, a continuous stream of these fragments followed up the course of a valley, and even extended to the very crest of the hill. On these crests huge masses, exceeding in dimensions any small building, seemed to stand arrested in their headlong course: there, also, the curved strata of the archways lay piled over each other, like the ruins of some vast and ancient cathedral. In endeavouring to describe these scenes of violence, one is tempted to pass from one simile to another. We may imagine, that streams of white lava had flowed from many parts of the mountains into the lower country, and that, when consolidated, they had been rent by some enormous convulsion into myriads of fragments. The expression, 'streams of stones', which immediately occurred to every one, conveyed the same idea. These scenes are, on the spot, rendered more striking, by the contrast of the low, rounded forms of the neighbouring hills.

I was much interested by finding on the highest peak of one range (about 700 feet above the sea) a great arched fragment, lying on its convex or upper surface. Must we believe that it was fairly pitched up in the air, and thus turned? Or, with more probability, that there existed formerly a part of the same range more elevated than the point on which this monument of a great convulsion of nature now lies. As the fragments in the valleys are neither rounded nor the crevices filled up with sand, we must infer that the period of violence was subsequent to the land having been raised above the waters of the sea. In a transverse section within these valleys the bottom is nearly level, or rises but very little towards either side. Hence the fragments appear to have travelled from the head of the valley; but in reality it seems most probable, either that they have been hurled down from the nearest slopes, or that masses of rock were broken up in the position they formerly occupied; and that since, by a vibratory movement of over-whelming force,* the fragments have been levelled into one continuous sheet. If during the earthquake† which in 1835 over-threw Concepcion, in Chile, it was thought wonderful that small bodies should have been pitched a few inches from the ground, what must we say to a movement which has caused fragments, many tons in weight (like so much sand on a vibrating board), to move onwards and find their level? I have seen, in the Cordillera of the Andes, the evident marks where stupendous mountains have been broken into pieces like so much thin crust, and the strata thrown on their vertical edges; but never did any scene, like the 'streams of stones', so forcibly convey to my mind the idea of a convulsion of which in historical records we might in vain seek for any counterpart.

* * *

In Tierra del Fuego, as well as at the Falkland Islands, I made many observations on the lower marine animals,‡ but they are of little general interest. I will only mention one class of facts, relating to certain zoophytes in the more highly organized div-ision of that class. Several genera (*flustra, eschara, cellaria, crisia*, and others) agree in having singular moveable organs, like those

---

* 'Nous n'avons pas été moins saisis d'étonnement à la vûe de l'innombrable quantité de pierres de toutes grandeurs, bouleversées les unes sur les autres, et cependant rangées, comme si elles avoient été amoncelées négligemment pour remplir des ravins. On ne se lassoit pas d'admirer les effets prodigieux de la nature.' *Pernety*, p. 526.

† An inhabitant of Mendoza, and hence well capable of judging, assured me that, during the several years he had resided on these islands, he had never felt the slightest shock of an earthquake.

‡ While at the Falklands, during the autumn of the southern hemisphere, most of the lower marine animals were breeding. I was surprised to find on counting the eggs of a large white Doris (this sea-slug was 3½ inches long) how extraordi-narily numerous they were. From two to five eggs (each three-thousandths of an inch in diameter) were contained in a spherical little case. These were arranged two deep in transverse rows forming a ribbon. The ribbon adhered by its edge to the rock in an oval spire. One, which I found, measured nearly 20 inches in length and half in breadth. By counting how many balls were contained in a tenth of an inch in the row, and how many rows in an equal length of the ribbon, on the most moderate computation there were 600,000 eggs. Yet this Doris was certainly not very common: although I was often searching under the stones I saw only seven individuals.

of *Flustra avicularia* (found in the European seas), attached to their cells. The organ, in the greater number of cases, very closely resembles the head of a vulture; but the lower mandible can be opened much wider, so as to form even a straight line with the upper. The head itself possesses considerable powers of movement, by means of a short neck. In one zoophyte the head itself was fixed, but the lower jaw free: in another it was replaced by a triangular hood, with a beautifully fitted trap-door, which evidently answered to the lower mandible. A species of stony eschara had a structure somewhat similar. In the greater number of species, each shell was provided with one head, but in others each had two.

The young cells at the end of the branches necessarily contained quite immature polypi, yet the vulture-heads attached to them, though small, were in every respect perfect. When the polypus was removed by a needle from any of the cells, these organs did not appear in the least affected. When one of the latter was cut off from a cell, the lower mandible retained its power of opening and closing. Perhaps the most singular part of their structure is, that when there were more rows of cells than two, both in a Flustra and an Eschara, the central cells were furnished with these appendages, of only one-fourth the size of the lateral ones. Their movements varied according to the species: – in some I never saw the least motion; while others, with the lower mandible generally wide open, oscillated backwards and forwards at the rate of about five seconds each turn; others moved rapidly and by starts. When touched with a needle the beak generally seized the point so firmly, that the whole branch might be shaken.

These bodies have no relation whatever with the production of the gemmules. I could not trace any connexion between them and the polypus. From their formation being completed before that of the latter; from the independence of their movements; from the difference of their size in different parts of the branch; I have little doubt that in their functions they are related rather to the axis than to any of the polypi. In a similar manner, the fleshy appendage at the extremity of the sea-pen forms part of the zoophyte as a whole, as much as the roots of a tree do of the whole and not of the individual buds. Without doubt this is a very curious variation in the structure of a zoophyte: for the growing part in most other cases does not manifest the least irritability or power of movement.

I will mention one other kind of structure quite as anomalous. A small and elegant Crisia is furnished, at the corner of each cell, with a long and slightly-curved bristle, which is fixed at the lower end by a joint. It terminates in the finest point, and has its outer or convex side serrated with delicate teeth or notches. Having placed a *small* piece of a branch under the microscope, I was exceedingly surprised to see it suddenly start from the field of vision by the movement of these bristles, which acted as oars. Irritation generally produced this motion, but not always. When the coralline was laid flat on that side from which the toothed bristles projected, they were necessarily all pressed together and entangled. This scarcely ever failed to excite a considerable movement among them, and evidently with the object of freeing themselves. In a small piece, which was taken out of water and placed on blotting-paper, the movement of these organs was clearly visible for a few seconds by the naked eye.

In the case of the vulture-heads, as well as in that of the bristles, all that were on one side of a branch, moved sometimes co-instantaneously, sometimes in regular order one after the other; at other times the organs on both sides the branch moved together; but generally all were independent of each other, and entirely so of the polypi. In the Crisia, if the bristles were excited to move by irritation in any one branch, generally the whole zoophyte was affected. In the instance where the branch started from the simultaneous movement of these appendages, we see as perfect a transmission of will as in a single animal. The case, indeed, is not different from that of the sea-pen, which when touched drew itself into the sand. I will state one other instance of uniform action, though of a very different nature, in a zoophyte* closely allied to Clytia, and therefore very simply organized. Having kept a large tuft of it in a basin of salt water, when it was dark I found that as often as I rubbed any part of a branch, the whole became strongly phosphorescent with a green light: I do not think

---

* This coralline emitted a very strong and disagreeable odour, when freshly taken from the sea.

I ever saw any object more beautifully so. But the remarkable circumstance was, that the flashes of light always proceeded up the branches, from the base towards the extremities.

The examination of these compound animals was always very interesting to me. What can be more remarkable than to see a plant-like body producing an egg, furnished with setæ, and having independent movements, which soon becomes fixed, branches into numberless arms, and these, though crowded with polypi, yet in some cases possessing independent organs of movement, and obeying uniform impulses of will? The polypi are frequently animals of no simple organization; and in most respects certainly are to be considered as true individuals. It is therefore more curious to observe, in the young and terminal cells, their gradual formation, from the growth of the simple horny substance of which so many zoophytes are composed. The known organization of a tree should remove all surprise at the union of many individuals together, and their relation to a common body. Indeed we might expect, according to the apparent law, that any structure which prevails in one class will be produced in a lesser degree in some others – that since so many plants are compound, so would some animals be thus constructed. It requires, however, a greater effort of reason to view a bud as an individual, than a polypus furnished with a mouth and intestines; and therefore the union does not appear so strange.

Our conception of a compound animal,* where in some respects the individuality of each is not completed, may be aided, by reflecting on the production of two distinct creatures by bisecting one with a knife, or where nature herself performs the task. We may consider the polypi in a zoophyte, or the buds in a tree, as cases where the division of the individual has not been completely effected. In this kind of generation, the individuals seem produced only with relation to the present time; their numbers are multiplied, but their life is not extended beyond a fixed period. By the

---

* With regard to associated life, animals of other classes besides the mollusca and radiata present obscure instances of it. The bee could not live by itself. And in the neuter, we see an individual produced which is not fitted for the reproduction of its kind – that highest point at which the organization of all animals, especially the lower ones, tends – therefore such neuters are born as much for the good of the community, as the leaf-bud is for the tree.

other, and more artificial kind, through intermediate steps or ovules, the relation is kept up through successive ages. By the latter method many peculiarities, which are transmitted by the former, are obliterated, and the character of the species is limited; while on the other hand, certain peculiarities (doubtless adaptations) become hereditary and form races. We may fancy that in these two circumstances we see a step towards the final cause of the shortness of life.

CHAPTER XIII

STRAIT OF MAGELLAN

* * *

JUNE 1ST – We anchored in the fine bay of Port Famine. It was now the beginning of winter, and I never saw a more cheerless prospect; the dusky woods piebald with snow, could be only indistinctly seen through a drizzling hazy atmosphere. We were, however, lucky in getting two fine days. On one of these, Mount Sarmiento, a distant mountain 6,800 feet high, presented a very noble spectacle. I was frequently surprised, in the scenery of Tierra del Fuego, at the little apparent elevation of mountains really lofty. I suspect it is owing to a cause, which would not at first be imagined, namely, that the whole mass, from the summit to the water's edge, is generally in full view. I remember having seen a mountain, first from the Beagle channel, where the whole sweep from the summit to the base was full in view, and then from Ponsonby Sound across several successive ridges; and it was curious to observe in the latter case, as each fresh step afforded means of judging of the distance, how the mountain appeared to rise in height.

The Fuegians twice came and plagued us. As there were many instruments, clothes, and men on shore, it was thought necessary to frighten them away. The first time, a few great guns were fired, when they were far distant. It was most ludicrous to watch through a glass the Indians, as often as the shot struck the water, take up stones, and as a bold defiance, throw them towards the ship, though about a mile and a half distant! A boat was then sent with orders to fire a few musket-shot wide of them. The Fuegians hid themselves behind the trees; and for every discharge of the musket they fired their arrows: all, however, fell short of the boat, and the officer as he pointed at them laughed. This made the Fuegians frantic with passion, and they shook their mantles in vain rage. At last seeing the balls cut and strike the trees, they ran away; and we were left in peace and quietness.

On a former occasion, when the *Beagle* was here in the month of February, I started one morning at four o'clock to ascend Mount Tarn, which is 2,600 feet high, and is the most elevated point in this immediate neighbourhood. We went in a boat to the foot of the mountain (but not to the best part), and then began our ascent. The forest commences at the line of high-water mark, and during the two first hours I gave over all hopes of reaching the summit. So thick was the wood, that it was necessary to have constant recourse to the compass; for every landmark, though in a mountainous country, was completely shut out. In the deep ravines, the death-like scene of desolation exceeded all description; outside it was blowing a gale, but in these hollows, not even a breath of wind stirred the leaves of the tallest trees. So gloomy, cold, and wet was every part, that not even the fungi, mosses, or ferns, could flourish. In the valleys it was scarcely possible to crawl along, they were so completely barricaded by the great mouldering trunks, which had fallen down in every direction. When passing over these natural bridges, one's course was often arrested by sinking knee deep into the rotten wood; at other times, when attempting to lean against a firm tree, one was startled by finding a mass of decayed matter ready to fall at the slightest touch. We at last found ourselves among the stunted trees, and then soon reached the bare ridge, which conducted us to the summit. Here was a view characteristic of Tierra del Fuego; – irregular chains of hills, mottled with patches of snow, deep yellowish-green valleys, and arms of the sea intersecting the land in many directions. The strong wind was piercingly cold, and the atmosphere rather hazy, so that we did not stay long on the top of the mountain. Our descent was not quite so laborious as our ascent; for the weight of the body forced a passage, and all the slips and falls were in the right direction.

* * *

The perfect preservation of the Siberian animals, perhaps presented, till within a few years, one of the most difficult problems which geology ever attempted to solve. On the one hand it was granted, that the carcasses had not been drifted from any great distance by any tumultuous deluge, and on the other it was assumed as certain, that when the animals lived, the climate must have been so totally different, that the presence of ice in the vicinity was as incredible, as would be the freezing of the Ganges. Mr Lyell in his *Principles of Geology** has thrown the greatest light on this subject, by indicating the northerly course of the existing rivers with the probability that they formerly carried carcasses in the same direction; by showing (from Humboldt) how far the inhabitants of the hottest countries sometimes wander; by insisting on the caution necessary in judging of habits between animals of the same genus, when the species are not identical; and especially by bringing forward in the clearest manner the probable change from an insular to an extreme climate, as the consequence of the elevation of the land, of which proofs have lately been brought to light.†

In a former part of this volume, I have endeavoured to prove, that as far as regards the *quantity* of food, there is no difficulty in supposing that these large quadrupeds inhabited sterile regions, producing but a scanty vegetation. With respect to temperature,

* In the fourth and subsequent editions.
† Wrangel's Voyage in the Icy Sea, in the years 1821, 1822, and 1823. Edited by Professor Parrot, of Dorpat, Berlin, 1826.

the woolly covering both of the elephant and the rhinoceros seems at once to render it at least probable (although it has been argued that some animals living in the hottest regions are thickly clothed) that they were fitted for a cold climate. I suppose no reason can be assigned why, during a former epoch, when the pachydermata abounded over the greater part of the world, some species should not have been fitted for the northern regions, precisely as now happens with deer and several other animals. * If, then, we believe that the climate of Siberia, anteriorly to the physical changes above alluded to, had some resemblance with that of the southern hemisphere at the present day – a circumstance which harmonizes well with other facts,† as I think has been shown by the imaginary case, when we transported existing phenomena from one to the other hemisphere – the following conclusions may be deduced as probable: first, that the degree of cold formerly was not excessive; secondly, that snow did not for a long time together cover the ground (such not being the case at the extreme parts 55°–56° of S. America); thirdly, that the veg-

* Dr Fleming first brought this notion forward in two papers published in the Edinburgh Philosoph. Journ. (April, 1829, and Jan. 1830). He adduces the case of allied species of the bear, fox, hare, and ox, living under widely different climates.

† Since writing the above, I have been much interested by reading an account by Professor Esmark, which proves that formerly, glaciers in Norway descended to a lower altitude than at present; and therefore, that they came down to the .level of the sea in a lower latitude. This, according to generally received ideas, would indicate a colder climate, and so it was considered to do by Professor Esmark; for he argues from it in favour of Whiston's hypothesis, that the 'earth in its aphelion was covered with ice and snow'. Professor Esmark describes a glacier-dike, in lat. 58° 57′, as 'lying close to the level of the sea, in a district, where you find only a few heaps of perpetual snow in the hollows of the mountains'. He says, 'Not only the dike itself, but the whole horizontal surface, exhibits proofs that there has been a glacier here, for the plain exactly resembles those which I found adjoining the glaciers presently existing between Londfiord and Lomb.' (See Ed. New Phil. Journal, p. 117, October 1826.) These facts afford a very strong and admirable confirmation of the view, that the climate of Europe has been gradually changing, from a character resembling that of the southern hemisphere, to its present condition. For on this hypothesis, we might have anticipated, that proofs would have been discovered, that glaciers formerly descended to a lower altitude than they now do; and yet, that the organic remains of that epoch, instead of a former period of refrigeration, would have indicated a climate of a more tropical character – a conclusion, which may be deduced from plain geological evidence.

etation partook of a more tropical character than it now does in the same latitudes; and lastly, that at but a short distance to the northward of the country thus circumstanced (even not so far as where Pallas found the entire rhinoceros), the soil might be perpetually congealed: so that if the carcass of any animal should once be buried a few feet beneath the surface, it would be preserved for centuries.

Both Humboldt* and Lyell have remarked, that at the present day, the bodies of any animals, wandering beyond the line of perpetual congelation which extends as far south as 62°, if once embedded by any accident a few feet beneath the surface, would be preserved for an indefinite length of time: the same would happen with carcasses drifted by the rivers; and by such means the extinct mammalia may have been entombed. There is only one small step wanting, as it appears to me, and the whole problem would be solved with a degree of simplicity very striking, compared with the several theories first invented. From the account given by Mr Lyell of the Siberian plains, with their innumerable fossil bones, the relics of many successive generations, there can be little doubt that the beds were accumulated either in a shallow sea, or in an estuary. From the description given in Beechey's voyage of Eschscholtz Bay, the same remark is applicable to the north-west coast of America: the formation there appears identical with the common littoral deposits† recently elevated, which I have seen on the shores of the southern part of the same continent. It seems also well established, that the Siberian remains are only exposed where the rivers intersect the plain. With this fact, and the proofs of recent elevation, the whole case appears to be precisely similar to that of the Pampas: namely, that the carcasses were formerly floated into the sea, and the remains covered up in the deposits which were then accumulating. These beds have since been elevated; and as the rivers excavate their channels the entombed skeletons are exposed.

Here then, is the difficulty: how were the carcasses preserved at

* See Humboldt, Fragmens Asiatiques, vol. ii, pp. 385–395.
† See some remarks by Dr Buckland on the similarity of this formation with the deposits so commonly found over a great part of Europe. Appendix to Beechey's Voyage, p. 609.

the bottom of the sea? I do not think it has been sufficiently noticed, that the preservation of the animal with its flesh was an occasional event, and not directly consequent on its position far northward. Cuvier* refers to the voyage of Billing as showing that the *bones* of the elephant, buffalo, and rhinoceros, are nowhere so abundant as on the islands between the mouths of the Lena and Indigirska. It is even said that excepting some hills of rock, the whole is composed of sand, ice, and bones. These islands lie to the northward of the place where Adams found the mammoth with its flesh preserved, and even 10° north of the Wiljui, where the rhinoceros was discovered in a like condition. In the case of the *bones* we may suppose that the carcasses were drifted into a deeper sea, and there remaining at the bottom, the flesh decomposed.† But in the second and more extraordinary case, where putrefaction seems to have been arrested, the body probably was soon covered up by deposits which were then accumulating. It may be asked, whether the mud a few feet deep, at the bottom of a shallow sea which is annually frozen, has a temperature higher than 32°? It must be remembered how intense a degree of cold is required to freeze salt water; and that the mud at some depth below the surface, would have a low mean temperature, precisely in the same manner as the subsoil on the land is frozen in countries which enjoy a short but hot summer. If this be possible,‡ the entombment of these extinct quadrupeds is

---

* Ossemens Fossiles, vol. i, p. 151.

† Under these circumstances of slow decomposition, the surrounding deposits would probably be impregnated with much animal matter; and thus the peculiar odour perceived in the neighbourhood of the strata containing fossil bones at Eschscholtz Bay, may be accounted for. See Appendix to Beechey's Voyage.

‡ With respect to the possibility of even ice accumulating at the bottom of the sea, I shall only refer to the following passage taken from the English translation of the *Expedition to the East Coast of Greenland, by Captain W. Graah, Danish Royal Navy.* 'Nor is this the only danger to be apprehended: the ice off this blink, even to a considerable distance from it, being said to shoot up from the bottom of the sea in such a manner, and in such masses, as in many years to make it utterly impassable. How to account for the phenomenon to which I have just adverted I know not, unless by supposing that the bottom of the sea itself is hereabouts like the dry land covered with a thick crust of ice. But whether this crust is formed upon the spot, or is the remains of icebergs and the heavy drift-ice frozen to the

rendered very simple; and with regard to the conditions of their former existence, the principal difficulties have, I think, already been removed.

* * *

There is one marine production, which from its importance is worthy of a particular history. It is the kelp or *Fucus giganteus* of Solander. This plant grows on every rock from low-water mark to a great depth, both on the outer coast and within the channels. I believe, during the voyages of the *Adventure* and *Beagle*, not one rock near the surface was discovered, which was not buoyed by this floating weed. The good service it thus affords to vessels navigating near this stormy land is evident; and it certainly has saved many a one from being wrecked. I know few things more surprising than to see this plant growing and flourishing amidst those great breakers of the western ocean, which no mass of rock, let it be ever so hard, can long resist. The stem is round, slimy, and smooth, and seldom has a diameter of so much as an inch. A few taken together are sufficiently strong to support the weight of the large loose stones to which in the inland channels they grow attached; and some of these stones are so heavy, that when drawn to the surface they can scarcely be lifted into a boat by one person.

Captain Cook, in his second voyage, says, that at Kerguelen Land 'some of this weed is of a most enormous length, though the stem is not much thicker than a man's thumb. I have mentioned,

---

bottom during severe winters, or a portion of the land-ice, which loaded with stones and fragments of the crumbling hill has protruded itself into the sea, is a problem impossible, perhaps to solve.' Again he says: 'We passed it without any accident, and without having observed any thing of that upheaving of the ice off it, to which allusion has been made, though the fact of its occurrence cannot be doubted, the very name of the place. Puisortok, being thence derived.' It seems fully established on excellent testimony (see Journ. of Geograph. Soc., vol. v, p. 12, and vol. vi, p. 416; also a collection of notices in Edinburgh Journal of Nat. and Geograph. Soc., vol. ii, p. 55), that fresh-water rivers in Russia and Siberia, and even in England, often freeze at the bottom, and that the flakes of ice when they rise to the surface, often 'bring with them large stones'. All that seems to be required in producing ground-ice, is, that there should be sufficient movement in the fluid, so that the whole is cooled down to the freezing point, and then the water *crystallizes*, wherever there is a point of attachment.

that on some of the shoals upon which it grows, we did not strike ground with a line of 24 fathoms. The depth of water, therefore, must have been greater. And as this weed does not grow in a perpendicular direction, but makes a very acute angle with the bottom, and much of it afterwards spreads many fathoms on the surface of the sea, I am well warranted to say that some of it grows to the length of sixty fathoms and upwards.' Certainly at the Falkland Islands, and about Tierra del Fuego, extensive beds frequently spring up from 10- and 15-fathom water. I do not suppose the stem of any other plant attains so great a length as 360 feet, as stated by Captain Cook. Its geographical range is very considerable; it is found from the extreme southern islets near Cape Horn, as far north, on the eastern coast (according to information given me by Mr Stokes), as lat. 43° – and on the western it was tolerably abundant, but far from luxuriant, at Chiloe, in lat. 42°. It may possibly extend a little further northward, but is soon succeeded by a different species. We thus have a range of 15° in latitude; and as Cook, who must have been well acquainted with the species, found it at Kerguelen Land, no less than 140° in longitude.

The number of living creatures of all orders, whose existence intimately depends on the kelp, is wonderful. A great volume might be written, describing the inhabitants of one of these beds of sea-weed. Almost every leaf, excepting those that float on the surface, is so thickly incrusted with coral-lines, as to be of a white colour. We find exquisitely-delicate structures, some inhabited by simple hydra-like polypi, others by more organized kinds, and beautiful compound Ascidiæ.* On the flat surfaces of the leaves various patelliform shells, Trochi, uncovered molluscs, and some bivalves are attached. Innumerable crustacea frequent every part of the plant. On shaking the great entangled roots, a pile of small fish, shells, cuttle-fish, crabs of all orders, sea-eggs, star-fish, beautiful Holuthuriæ (some taking the external form of the nudibranch molluscs), Planariæ, and crawling nereidous animals of a multitude of forms, all fall out together. Often as I recurred to

---

* I have reason to believe that many of these animals are exclusively confined to this station.

a branch of the kelp, I never failed to discover animals of new and curious structures. In Chiloe, where, as I have said, the kelp did not thrive very well, the numerous shells, corallines, and crustacea were absent; but there yet remained a few of the flustraceæ, and some compound Ascidiæ; the latter, however, were of different species from those in Tierra del Fuego. We here see the fucus possessing a wider range than the animals which use it as an abode.

I can only compare these great aquatic forests of the southern hemisphere with the terrestrial ones in the intertropical regions. Yet if the latter should be destroyed in any country, I do not believe nearly so many species of animals would perish, as, under similar circumstances, would happen with the kelp. Amidst the leaves of this plant numerous species of fish live, which nowhere else would find food or shelter; with their destruction the many cormorants, divers, and other fishing birds, the otters, seals, and porpoises, would soon perish also; and lastly, the Fuegian savage, the miserable lord of this miserable land, would redouble his cannibal feast, decrease in numbers, and perhaps cease to exist.

JUNE 8TH – We weighed anchor early in the morning, and left Port Famine. Captain FitzRoy determined to leave the Strait of Magellan by the Magdalen channel, which had not long been discovered. Our course lay due south, down that gloomy passage which I have before alluded to, as appearing to lead to another and worse world. The wind was fair, but the atmosphere was very thick; so that we missed much curious scenery. The dark ragged clouds were rapidly driven over the mountains, from their summits nearly to their bases. The glimpses which we caught through the dusky mass were highly interesting: jagged points, cones of snow, blue glaciers, strong outlines marked on a lurid sky, were seen at different distances and heights. In the midst of such scenery we anchored at Cape Turn, close to Mount Sarmiento, which was then hidden in the clouds. At the base of the lofty and almost perpendicular sides of our little cove, there was one deserted wigwam, and it alone reminded us that man sometimes wandered in these desolate regions. But it would be difficult to imagine a scene where he seemed to have less claims, or less authority. The inanimate works of nature – rock, ice, snow, wind, and water – all warring with each other, yet combined against man – here reigned in absolute sovereignty.

JUNE 9TH – In the morning we were delighted by seeing the veil of mist gradually rise from Sarmiento, and display it to our view. This mountain, which is one of the highest in Tierra del Fuego, has an elevation of 6,800 feet. Its base, for about an eighth of its total height, is clothed by dusky woods, and above this a field of snow extends to the summit. These vast piles of snow, which never melt, and seem destined to last as long as the world holds together, present a noble and even sublime spectacle. The outline of the mountain was admirably clear and defined. Owing to the abundance of light reflected from the white and glittering surface, no shadows are cast on any part; and those lines which intersect the sky can alone be distinguished: hence the mass stood out in the boldest relief. Several glaciers descended in a winding course, from the snow to the sea-coast: they may be likened to great frozen Niagaras; and perhaps these cataracts of blue ice are to the full as beautiful as the moving ones of water. By night we reached the western part of the channel; but the water was so deep that no anchorage could be found. We were in consequence obliged to stand off and on, in this narrow arm of the sea, during a pitch-dark night of fourteen hours long.

JUNE 10TH – In the morning we made the best of our way into the open Pacific. The Western coast generally consists of low, rounded, quite barren, hills of granite and greenstone. Sir John Narborough called one part South Desolation, because it is 'so desolate a land to behold'; and well indeed might he say so. Outside the main islands there are numberless scattered rocks, on which the long swell of the open ocean incessantly rages. We passed out between the East and West Furies, and a little further northward there are so many breakers that the sea is called the Milky Way. One sight of such a coast is enough to make a landsman dream for a week about shipwreck, peril, and death; and with this sight, we bade farewell for ever to Tierra del Fuego.

CHAPTER XIV

===

CENTRAL CHILE

JULY 23RD – The *Beagle* anchored late at night in the bay of
Valparaiso, the chief seaport of Chile. When morning came,
every thing appeared delightful. After Tierra del Fuego, the
climate felt quite delicious – the atmosphere so dry, and the
heavens so clear and blue, with the sun shining brightly, that all
nature seemed sparkling with life. The view from the anchorage is
very pretty. The town is built at the very foot of a range of hills,
about 1,600 feet high, and rather steep. From its position, it
consists of one long, straggling street, which runs parallel to the
beach, and wherever a ravine comes down, the houses are piled up
on each side of it. The rounded hills, being only partially protected by a very scanty vegetation, are worn into numberless little
gullies, which expose a singularly bright red soil. From this cause,
and from the low whitewashed houses with tile roofs, the view
reminded me of St Cruz in Teneriffe. In a north-easterly direction
there are some fine glimpses of the Andes: but these mountains
appear much grander when viewed from the neighbouring hills;
the great distance at which they are situated can then more readily
be perceived. The volcano of Aconcagua is particularly magnificent. This huge and irregularly conical mass has an elevation
greater than that of Chimborazo; for, from measurements made

by the officers in the *Beagle*, its height is no less than 23,000 feet. The Cordillera, however, viewed from this point, owe the greater part of their beauty to the atmosphere through which they are seen. When the sun was setting in the Pacific, it was admirable to watch how clearly their rugged outlines could be distinguished, yet how varied and how delicate were the shades of their colour.

I had the good fortune to find living here Mr Richard Corfield, an old schoolfellow and friend, to whose hospitality and kindness I was greatly indebted, in having afforded me a most pleasant residence during the *Beagle*'s stay in Chile. The immediate neighbourhood of Valparaiso is not very productive to the naturalist. The surrounding hills consist of a granitic formation, which sometimes assumes the character of gneiss, and sometimes of granite. Their summits are flat-topped, and their flanks rounded. I have before stated, that forests cover that side of the Cordillera which fronts the prevailing winds. Here, during the summer, which forms the longer portion of the year, the winds blow steadily from the southward, and a little off shore, so that rain never falls: during the three winter months it is however sufficiently abundant. The vegetation in consequence is very scanty. Except in some deep valleys, trees nowhere occur, and only a little grass and a few low bushes are scattered over the less steep parts of the hills. When we reflect that, at the distance of 350 miles to the southward, this side of the Andes is completely hidden by one impenetrable forest, the contrast is very remarkable.

I took several long walks while collecting objects of natural history. The country is pleasant for exercise. There are many very beautiful flowers, and as in most other dry climates, the plants and shrubs possess strong and peculiar odours; even one's clothes by brushing through them became scented. I did not yet cease from wonder, at finding each succeeding day as fine as the foregoing. What a difference does climate make in the enjoyment of life! How opposite are the sensations when viewing black mountains half enveloped in clouds, and seeing another range through the light blue haze of a fine day! The one for a time may be very sublime; the other is all gaiety and happy life.

AUGUST 14TH – I set out on a riding excursion, for the purpose of geologizing the basal parts of the Andes, which alone at this time of the year were not shut up by the winter snow. Our first day's ride was northward along the sea-coast. After dark we reached the Hacienda of Quintero, the estate which formerly belonged to Lord Cochrane. My object in coming here was to see the great beds of shells, which are elevated some yards above the level of the sea. They nearly all consist of one species of Erycina; and these shells at the present day live together in great numbers, on the sandy flats. So wonderfully numerous are those forming the beds, that for years they have been quarried, and burnt for the lime, with which the large town of Valparaiso is supplied. As any change of level, even in this neighbourhood, has often been disputed, I may add, that I saw dead barnacles adhering to points of solid rock which were now so much elevated, that even during gales of wind they would scarcely be wetted by the spray.

15TH – We returned towards the valley of Quillota. The country was exceedingly pleasant; just such as poets would call pastoral: green open lawns, separated by small valleys with rivulets, and the cottages, we will suppose of the shepherds, scattered on the hill-sides. We were obliged to cross the ridge of the Chilecauquen. At its base there were many fine evergreen forest-trees, but these only flourished in the ravines, where there was running water. Any person who had seen only the country near Valparaiso, would never have imagined that there had been such picturesque spots in Chile. As soon as we reached the brow of the Sierra, the valley of Quillota was immediately under our feet. The prospect was one of remarkable artificial luxuriance. The valley is very broad and quite flat, and is thus easily irrigated in all parts. The little square gardens are crowded with orange and olive trees, and every sort of vegetable. On each side huge bare mountains rise, and this from the contrast renders the patchwork valley the more pleasing. Whoever called 'Valparaiso' the 'Valley of Paradise' must have been thinking of Quillota. We crossed over to the Hacienda de San Isidoro, situated at the very foot of the Bell mountain.

Chile, as may be seen in the maps, is a narrow strip of land between the Cordillera and the Pacific; and this strip is itself traversed by several mountain-lines, which in this part run parallel to the great range. Between these outer lines, and the main Cordillera, a succession of level basins, generally opening into each other by narrow passages, extend far to the southward. In these the principal towns are situated, as San Felipe, Santiago, S. Fernando. These basins or plains, together with the transverse flat valleys (like that of Quillota) which connect them with the coast, I have little doubt, are the bottoms of ancient inlets and deep bays, such as at the present day intersect every part of Tierra del Fuego, and the west coast of Patagonia. Chile must formerly have resembled the latter country, in the configuration of its land and water. This resemblance was occasionally seen with great force, when a level fog-bank covered, as with a mantle, all the lower parts of the country: the white vapour curling into the ravines, beautifully represented little coves and bays; and here and there a solitary hillock peeping up, showed that it had formerly stood there as an islet. The contrast of these flat valleys and basins with the irregular mountains, gave the scenery a character which to me was novel and very interesting.

From the natural slope to seaward of these plains, they are very easily irrigated, and in consequence singularly fertile. Without this process the land would produce scarcely any thing; for during the whole summer the sky is cloudless. The mountains and hills are dotted over with bushes and low trees, and excepting these, the vegetation is very scanty. Each landowner in the valley possesses a certain portion of hill-country, where his half-wild cattle, in considerable numbers, manage to find sufficient pasture. Once every year there is a grand 'rodeo', when all the cattle are driven down, counted, and marked, and a certain number separated to be fattened in the irrigated fields. Wheat is extensively cultivated, and also a good deal of Indiàn corn: a kind of bean is, however, the staple article of food for the common labourers. The orchards produce an overflowing abundance of peaches, figs, and grapes. With all these advantages, the inhabitants of the country ought to be much more prosperous than they are.

AUGUST 16TH – The mayor-domo of the Hacienda was good enough to give me a guide and fresh horses; and in the morning we set out to ascend the Campana, or Bell mountain, which is 6,400 feet high. The paths were very bad, but both the geology and scenery amply repaid the trouble. We reached, by the evening, a spring called the Agua de Guanaco, which is situated at a great height. This must be an old name, for it is very many years since a guanaco has drunk its waters. During the ascent I noticed that nothing grew on the northern slope but bushes, whilst on the southern there was a sort of bamboo, about 15 feet high. In a few places there were palms, and I was surprised to see one at an elevation of at least 4,500 feet. These palms are, for their family, ugly trees. Their stem is very large, and of a curious form, being thicker in the middle than at the base or top. They are excessively numerous in some parts of Chile, and valuable on account of a sort of treacle made from the sap. On one estate near Petorca, they tried to count them, but failed, after having numbered several hundred thousand. Every year in August (early spring time) very many are cut down, and when the trunk is lying on the ground, the crown of leaves is lopped off. The sap then immediately begins to flow from the upper end, and continues so doing for some months: it is, however, necessary that a thin slice should be shaved off from that end every morning, so as to expose a fresh surface. A good tree will give 90 gallons, and all this must have been contained in the vessels of the apparently dry trunk. It is said that the sap flows much more quickly on those days when the sun is powerful; and likewise, that it is absolutely necessary to take care, in cutting down the tree, that it should fall with its head upwards on the side of the hill; for if it falls down the slope, scarcely any sap will flow; although in that case, one would have thought that the action would have been aided, instead of checked, by the force of gravity. The sap is concentrated by boiling, and is then called treacle, which it very much resembles in taste.

We unsaddled our horses near the spring, and prepared to pass the night. The evening was fine, and the atmosphere so clear, that the masts of the vessels at anchor in the bay of Valparaiso, although no less than 26 geographical miles distant, could be distinguished clearly, as little black streaks. A ship doubling the point under sail appeared as a bright white speck. Anson expresses much surprise, in his voyage, at the distance his vessels were discovered from the coast; but he did not sufficiently allow for the height of the land, and the great transparency of the air.

The setting of the sun was glorious; the valleys being black, whilst the snowy peaks of the Andes yet retained a ruby tint. When it was dark, we made a fire beneath a little arbour of bamboos, fried our *charqui* (or dried strips of beef), took our maté, and were quite comfortable. There is an inexpressible charm in thus living in the open air. The evening was calm and still; the shrill noise of the mountain bizcacha, and the faint cry of the goatsucker, were only occasionally to be heard. Besides these, few birds, or even insects, frequent these dry, parched mountains.

AUGUST 17TH – In the morning we climbed up the rough mass of greenstone which crowns the summit. This rock, as frequently happens, was much shattered and broken into huge angular fragments. I observed, however, one remarkable circumstance, namely, that many of the surfaces presented every degree of freshness – some appearing as if broken the day before, whilst on others lichens had either just become, or had long grown, attached. I so fully believed that this was owing to the frequent earthquakes, that I felt inclined to hurry from beneath every pile of the loose masses. As this is an observation in which one would be very apt to be deceived, I doubted its accuracy, until ascending Mount Wellington, near Hobart Town. The summit of that mountain is similarly composed, and similarly shattered; but all the blocks appeared as if they had been hurled into their present position thousands of years ago.

We spent the day on the summit, and I never enjoyed one more thoroughly. Chile, bounded by the Andes and the Pacific, was seen as in a map. The pleasure from the scenery, in itself beautiful, was heightened by the many reflections which arose from the mere view of the grand range, with its lesser parallel ones, and of the broad valley of Quillota directly intersecting the latter. Who can avoid admiring the wonderful force which has upheaved these mountains, and even more so the countless ages which it must have required, to have broken through, removed, and levelled whole masses of them? It is well in this case to call to mind the vast shingle and sedimentary beds of Patagonia, which, if heaped on the Cordillera, would increase by so many thousand feet its height. When in that country, I wondered how any mountain-chain could have supplied such masses, and not have been utterly obliterated. We must not now reverse the wonder, and doubt whether all-powerful time can grind down mountains – even the gigantic Cordillera – into gravel and mud.

The appearance of the Andes was different from that which I had expected. The lower line of the snow was of course horizontal, and to this line the even summits of the range seemed quite parallel. Only at long intervals a mass of points, or a single cone, showed where a volcano had existed, or does now exist. Hence the range resembled a great solid wall, surmounted here and there by a tower, and thus made a most complete barrier to the country.

* * *

AUGUST 18TH – We descended the mountain, and passed some beautiful little spots, with rivulets and fine trees. Having slept at the same hacienda as before, we rode during the two succeeding days up the valley, and passed through Quillota, which is more like a collection of nursery-gardens than a town. The orchards were beautiful, presenting one mass of peach-blossoms. I saw also in one or two places the date-palm. It is a most stately tree; and I should think a group of them in their native Asiatic and African deserts must be superb. We passed likewise San Felipe, a pretty straggling town like Quillota. The valley in this part expands into one of those great bays or plains, reaching to the foot of the Cordillera, which have been mentioned as forming so curious a part of the scenery of Chile.

In the evening we reached the mines of Jajuel, situated in a ravine at the flank of the great chain. I stayed here five days. My host, the superintendent of the mine, was a shrewd but rather ignorant Cornish miner. He had married a Spanish woman, and did not mean to return home; but his admiration for the mines of Cornwall remained unbounded. Amongst many other questions, he asked me, 'Now that George Rex is dead, how many more of the family of Rexes are yet alive?' This Rex certainly must be a relation of the great author Finis, who wrote all books!

These mines are of copper, and the ore is all shipped to

Swansea, to be smelted. Hence the mines have an aspect singularly quiet, as compared to those in England: here no smoke, furnaces, or great steam-engines, disturb the solitude of the surrounding mountains.

The Chilian government, or rather the old Spanish law, encourages by every method the searching for mines. The discoverer may work a mine on any ground, by paying five shillings; and before paying this he may try, even in the garden of another man, for twenty days.

It is now well known that the Chilian method of mining is the cheapest. My host says the two principal improvements introduced by foreigners have been, first, reducing by previous roasting the copper pyrites – which, being the common ore in Cornwall, the English miners were astounded on their arrival to find thrown away as useless: secondly, stamping and washing the scoriæ from the furnaces – by which process particles of metal are recovered in abundance. I have actually seen mules carrying to the coast, for transportation to England, a cargo of such cinders. But the first case is much the most curious. The Chilian miners were so convinced that copper pyrites contained not a particle of copper, that they laughed at the Englishmen for their ignorance, who laughed in turn, and bought their richest veins for a few dollars. It is very odd that, in a country where mining had been extensively carried on for many years, so simple a process as gently roasting the ore, to expel the sulphur previous to smelting it, had never been discovered. A few improvements have likewise been introduced in some of the simple machinery; but even to the present day, water is removed from some mines by men carrying it up the shaft in leathern bags!

The labouring men work very hard. They have little time allowed for their meals, and during summer and winter they begin when it is light, and leave off at dark. They are paid one pound sterling a month, and their food is given them: this for breakfast consists of sixteen figs and two small loaves of bread; for dinner boiled beans; for supper broken roasted wheatgrain. They scarcely ever taste meat; as, with the twelve pounds per annum, they have to clothe themselves, and support their families. The miners who work in the mine itself, have twenty-five shillings per month, and are allowed a little *charqui*. But these men come down from their bleak habitations only once in every fortnight or three weeks.

During my stay here I thoroughly enjoyed scrambling about these huge mountains. The geology, as might have been expected, was very interesting. The shattered and baked rocks, traversed by innumerable dykes of greenstone, showed what commotions had formerly taken place there. The scenery was much the same as that near the Bell of Quillota – dry barren mountains, dotted at intervals by bushes with a scanty foliage. The cactuses, or rather opuntias, were here very numerous. I measured one of aspherical figure, which, including the spines, was 6 feet and 4 inches in circumference. The height of the common cylindrical, branching kind is from 12 to 15 feet, and the girth (with spines) of the branches between 3 and 4.

A heavy fall of snow on the mountains prevented me, during the last two days, from making some interesting excursions. I attempted to reach a lake, which the inhabitants, from some unaccountable reason, believe to be an arm of the sea. During a very dry season, it was proposed to attempt cutting a channel from it, for the sake of the water; but the padre, after a consultation, declared it was too dangerous, as all Chile would be inundated, if, as generally supposed, the lake was connected with the Pacific. We ascended to a great height, but becoming involved in the snow-drifts, failed in reaching this wonderful lake, and had some difficulty in returning. I thought we should have lost our horses; for there was no means of guessing how deep the drifts were, and the animals, when led, could only move by jumping. The black sky showed that a fresh snow-storm was gathering, and we therefore were not a little glad when we escaped. By the time we reached the base, the storm commenced, and it was lucky for us that this did not happen three hours earlier in the day.

* * *

SEPTEMBER 14TH – This morning we rode to the mines, which are situated at the distance of some leagues, near the summit of a lofty hill. On the way we had a glimpse of the lake Tagua-tagua, celebrated for its floating islands, which have been described by M. Gay.* They are composed of the stalks of various dead plants intertwined together, and on the surface of which other living ones take root. Their form is generally circular, and their thickness from 4 to 6 feet, of which the greater part is immersed in the water. As the wind blows they pass from one side of the lake to the other, and often carry cattle and horses as passengers.

When we arrived at the mine, I was struck by the pale appearance of many of the men, and inquired from Mr Nixon respecting their condition. The mine is 450 feet deep, and each man brings up about 200 pounds† weight of stone. With this load they have to climb up the alternate notches cut in the trunks of trees, placed in a zigzag line up the shaft. Even beardless young men, eighteen and twenty years old, with little muscular development of their bodies (they are quite naked excepting drawers) ascend with this great load from nearly the same depth. A strong man, who is not accustomed to this labour, perspires most profusely, with merely carrying up his own body. With this very severe labour, they live entirely on boiled beans and bread. They would prefer having the latter alone; but their masters, finding they cannot work so hard upon this, treat them like horses, and make them eat the beans. Their pay is here rather more than at the mines of Jajuel, being from twenty-four to twenty-eight shillings per month. They leave the mine only once in three weeks; when they stay with their families for two days.‡ One of the rules in this mine sounds very

---

* Annales des Sciences Naturelles, March, 1833. M. Gay, a zealous and able naturalist, is now occupied in studying every branch of natural history throughout the kingdom of Chile.

† In another mine, as will hereafter be mentioned, I picked out a load by hazard, and weighed it: it was 197 pounds.

‡ Bad as all the above treatment appears, it is gladly accepted of by the miners; for the condition of the labouring agriculturists is much worse. The wages of the latter are lower, and they live almost exclusively on beans. This poverty must be chiefly owing to the feudal-like system on which the land is tilled. The landowner gives a small plot of ground to the labourer, for building and cultivating, and in return has his services (or that of a proxy) for every day of his life, without any wages. Until a father has a grown-up son who can by his labour pay the rent, there is no one, except on chance days, to take care of the patch of ground. Hence extreme poverty is very common among the labouring classes in this country.

harsh, but answers pretty well for the master. The only method of stealing gold, is to secrete pieces of the ore, and take them out as occasion may offer. Whenever the major–domo finds a lump thus hidden, its full value is stopped out of the wages of all the men; who thus, without they all combine, are obliged to keep watch over each other.

When the ore is brought to the mill, it is ground into an impalpable powder; the process of washing removes all the lighter particles, and amalgamation finally secures the gold dust. The washing, when described, sounds a very simple process; but it is beautiful to see how the exact adaption of the current of water to the specific gravity of the gold, so easily separates the powdered matrix from the metal. The mud which passes from the mills is collected into pools, where it subsides, and every now and then is cleared out, and thrown into a common heap. A great deal of chemical action then commences, salts of various kinds effloresce on the surface, and the mass becomes hard. In the heap which I examined, an angulo–concretionary structure was also superinduced, and what was very remarkable, these pseudo-fragments possessed an even and well–defined slaty structure; but the laminæ were not inclined at any uniform angle. The mud, after having been left for a year or two, and then rewashed, yields gold; and this process may be repeated even six or seven times; but the gold each time becomes less in quantity, and the intervals required (as the inhabitants say to generate the metal) are longer. There can be no doubt that the chemical action, already mentioned, each time liberates fresh gold from some combination. The discovery of a method to effect this before the first grinding, would without doubt raise the value of gold ores many fold.

It is curious to find how the minute particles of gold, after being scattered about, and from not corroding, at last accumulate in some quantity. A short time since a few miners, being out of work, obtained permission to scrape the ground round the house and mill: they washed the earth thus got together, and so procured thirty dollars' worth of gold. This is an exact counterpart of what takes place in nature. Mountains suffer degradation and wear away, and with them the metallic veins which they contain. The hardest rock is worn into impalpable mud, the ordinary metals oxidate, and both are removed; but gold, platina, and a few others, are nearly indestructible, and from their weight, sinking to the bottom, are left behind. After whole mountains have passed through this grinding mill, and have been washed by the hand of nature, the residue becomes metalliferous, and man finds it worth his while to complete the task of separation.

There are some old Indian ruins in this neighbourhood, and I was shown one of the perforated stones which Molina* mentions, as being found in many places in considerable numbers. They are of a circular flattened form, from 5 to 6 inches in diameter, and with a hole passing quite through the centre. It has generally been supposed, that they were used as heads to clubs, although their form does not appear at all well adapted for that purpose. Burchell† states that some of the tribes in Southern Africa dig up roots, by the aid of a pointed stick, the force and weight of which is increased by a round stone with a hole, into which the stick is firmly wedged. It appears probable that the Indians of Chile formerly used some such rude agricultural instrument.

One day, a German collector in natural history, of the name of Renous, called, and nearly at the same time an old Spanish lawyer. I was amused by afterwards hearing the conversation which took place between them. Renous speaks Spanish so well, that the old lawyer mistook him for a fellow-countryman. Renous, alluding to myself, asked him what he thought of the King of England sending out a collector to their country, to pick up lizards and beetles, and to break stones? The old gentleman thought seriously for some time, and then said, 'It is not well, – *hay un gato encerrado aqui* (there is a cat shut up here). No man is so rich as to send out people to pick up such rubbish. I do not like it: if one of us were to go and do such things in England, do not you think the King of England would very soon send us out of his country?' And this old gentleman, from his profession, belongs to the better informed and more intelligent classes! Renous himself, two or three years before, left in a house at S. Fernando some caterpillars, under charge of a girl to feed, that they might turn into butterflies.

* Molina, Compendio de la Historia, &c. del Reyno de Chile, vol. i, p. 81.
† Burchell's Travels, vol. ii, p. 45.

This was rumoured through the town, and at last the Padres and Governor consulted together, and agreed it must be some heresy. Accordingly, when Renous returned, he was arrested.

SEPTEMBER 19TH – We left Yaquil, and followed the flat valley, formed like that of Quillota, in which the Rio Tinderidica flows. Even at these few miles south of Santiago the climate is much damper; in consequence there were fine tracks of pasturage, which were not irrigated. (20th) We followed this valley till it expanded into a great plain, which reaches from the sea to the mountains west of Rancagua. We shortly lost all trees and even bushes; so that the inhabitants are nearly as badly off for firewood as those in the Pampas. Never having heard of these plains, I was much surprised at meeting with such scenery in Chile. The plains belong to more than one series of different elevations, and they are traversed by broad flat-bottomed valleys; both of which circumstances, as in Patagonia, bespeak the gentle retreat of the ocean. In the steep cliffs bordering these valleys, there are some large caves, which no doubt were formed by the waters of the ancient bays and channels. One of these which I visited is celebrated under the name of Cueva del Obispo; having formerly been consecrated. During the day I felt very unwell, and from that time till the end of October did not recover.

SEPTEMBER 22ND – We continued to pass over green plains without a tree. The next day we arrived at a house near Navedad, on the sea-coast, where a rich Haciendero gave us lodgings. I stayed here the two ensuing days, and although very unwell, managed to collect from the tertiary formation some marine shells, many of which turn out to be quite new forms. (24th) Our course was now directed towards Valparaiso, which with great difficulty I reached on the 27th, and was there confined to my bed till the end of October. During this time I was an inmate in Mr Corfield's house, whose kindness to me I do not know how to express.

*   *   *

CHAPTER XV

CHILOE AND CHONOS ISLANDS

NOVEMBER 10TH – The *Beagle* sailed from Valparaiso to the
southward, for the purpose of surveying the southern part of
Chile, the island of Chiloe, and the broken land called the Chonos
Archipelago, as far south as the Peninsula of Tres Montes. On the
21st we anchored in the bay of S. Carlos, the capital of Chiloe.

This island is about 90 miles long, with a breadth of rather less
than 30. The land is hilly, but not mountainous, and is every
where covered by one great forest, excepting a few scattered
green patches, which have been cleared round the thatched
cottages. From a distance the view somewhat resembles Tierra
del Fuego; but the woods, when seen nearer, are incomparably
more beautiful. Many kinds of fine evergreen trees, and plants
with a tropical character, here take the place of the gloomy beech
of the southern shores. In winter the climate is detestable, and in
summer it is only a little better. I should think there are few parts
of the world, within the temperate regions, where so much rain
falls. The winds are very boisterous, and the sky almost always
clouded: to have a week of fine weather is something wonderful.
It is even difficult to get a single glimpse of the Cordillera: during
our first visit only one opportunity occurred, and that was before

sunrise, when the volcano of Osorno stood out in bold relief; and it was curious to watch, as the sun rose, the outline gradually fading away in the glare of the eastern sky.

The inhabitants, from their complexion and low stature, appear to have three-fourths of Indian blood in their veins. They are an humble, quiet, industrious set of men. Although the fertile soil, resulting from the decomposition of volcanic rocks, supports a rank vegetation, yet the climate is not favourable to any production which requires much sunshine to ripen it. There is very little pasture for the larger quadrupeds; and in consequence, the staple articles of food are pigs, potatoes, and fish. The people all dress in strong woollen garments, which each family makes for itself, and dyes with indigo of a dark blue colour. The arts, however, are in the rudest state; as may be seen in their strange fashion of ploughing, their method of spinning, grinding corn, and in the construction of their boats.

The forests are so impenetrable, that the land is nowhere cultivated except near the coast, and on the adjoining islets. Even where roads exist, they are scarcely passable from the soft and swampy state of the soil. The inhabitants, like those of Tierra del Fuego, chiefly move about on the beach, or in boats: in some cases the latter afford the only means of getting from one house to another. Although with plenty to eat, the people are very poor: there is no demand for labour, and consequently the lower orders cannot scrape together money sufficient to purchase even the smallest luxuries. There is also a great deficiency of a circulating medium. I have seen a man bringing on his back a bag of charcoal, with which to buy some trifle, and another a plank to exchange for a bottle of wine. Hence every tradesman must also be a merchant, and again sell the goods which he takes in exchange.

NOVEMBER 24TH – The yawl and whale-boat were sent under the command of Mr Sulivan, to survey the eastern or inland coast of Chiloe; and with orders to meet the *Beagle* at the southern extremity of the island; to which point she would proceed by the outside, so as thus to circumnavigate the island. I accompanied this expedition, but instead of going in the boats the first day, I hired horses to take me to Chacao, at the northern extremity of the island. The road followed the coast; every now and then crossing promontories covered by fine forests. In these shaded paths it is absolutely necessary that the whole road should be made of logs of wood, which are squared and placed by the side of each other. From the rays of the sun never penetrating the evergreen foliage, the ground is so damp and soft, that except by such means, neither man nor horse would be able to pass along. I arrived at the village of Chacao, shortly after the tents belonging to the boats had been pitched for the night.

The land in this neighbourhood had been extensively cleared, and there were many quiet and most picturesque nooks in the forest. Chacao was formerly the principal port; but many vessels having been lost, owing to the dangerous currents and rocks in the straits, the Spanish government burnt the church, and thus arbitrarily compelled the greater number of inhabitants to migrate to S. Carlos. In a short time the barefooted son of the governor came down to reconnoitre us. Seeing the English flag hoisted at the yawl's mast-head, he asked, with the utmost indifference, whether it was always to fly at Chacao. In several places, the inhabitants were much astonished at the appearance of men-of-war's boats, and hoped and believed it was the forerunner of a Spanish fleet, coming to recover the island from the patriot government of Chile. All the men in power had, however, been informed of our intended visit, and were exceedingly civil. While we were eating our supper, the governor paid us a visit. He had been a lieutenant-colonel in the Spanish service, but now was miserably poor. He gave us two sheep, and accepted in return two cotton handkerchiefs, some brass trinkets, and a little tobacco.

25TH – Torrents of rain: we managed, however, to run down the coast as far as Huapi-lenou. The whole of this eastern side of Chiloe has one aspect: it is a plain, broken by valleys, and divided into little islands, and the whole thickly covered with one impervious blackish-green forest. On the margins there are some cleared spaces, surrounding the high-roofed cottages.

26TH – The day rose splendidly clear. The volcano of Osorno was spouting out volumes of smoke. This most beautiful mountain, formed like a perfect cone, and white with snow, stands out in front of the Cordillera. Another great volcano, with a saddle-shaped summit, also emitted from its immense crater little jets of steam. Subsequently we saw the lofty-peaked Corcovado – well deserving the name of 'el famoso Corcovado'. Thus we beheld, from one point of view, three great active volcanoes, each of which had an elevation of about 7,000 feet. In addition to this, far to the south, there were other very lofty cones covered with snow, which although not known to be active, must have been in their origin volcanic. The line of the Andes is not, in this neighbourhood, nearly so elevated as in Chile; neither does it appear to form so perfect a barrier between the regions of the earth. This great range, although running in a direct north and south line, owing to an optical deception, always appeared more or less semicircular; for the extreme peaks being seen standing above the same horizon together with the nearer ones, their much greater distance was not so easily recognized.

*   *   *

DECEMBER 6TH – We reached Caylen, called 'el fin del Cristiandad'. In the morning we stopped for a few minutes at a house on the northern end of Laylec, which was the extreme point of South American Christendom, and a miserable hovel it was. The latitude is 43° 10′, which is two degrees further south than the Rio Negro on the Atlantic coast. These extreme Christians were very poor, and, under the plea of their situation, begged some tobacco. As a proof of the poverty of these Indians, I may mention that, shortly before this, we had met a man who had travelled three days and a half on foot, and had as many to return, for the sake of recovering the value of a small axe, and a few fish. How very difficult it must be to buy the smallest article, when such trouble is taken to recover so small a debt!

In the evening we reached the island of S. Pedro, where we found the *Beagle* at anchor. In doubling the point, two of the officers landed to take a round of angles with the theodolite. A fox, of a kind said to be peculiar to the island, and very rare in it, and which is an undescribed species, was sitting on the rocks. He was so intently absorbed in watching their manœuvres, that I was able, by quietly walking up behind, to knock him on the head with my geological hammer. This fox, more curious or more scientific, but less wise, than the generality of his brethren, is now mounted in the museum of the Zoological Society.

We staid three days in this harbour; on one of which Captain FitzRoy, with a party, attempted to ascend to the summit of San Pedro. The woods here had rather a different aspect from those on the northern parts of the island. The rock also being micaceous slate, there was no beach, but the steep sides dipped directly beneath the water. The general aspect in consequence was more like that of Tierra del Fuego than of Chiloe. In vain we tried to gain the summit: the forest was so impenetrable that no one, who has not beheld it, can imagine so entangled a mass of dying and dead trunks. I am sure that often, for more than ten minutes together, our feet never touched the ground, and we were frequently 10 or 15 feet above; it so that the seamen as a joke called out the soundings. At other times we crept one after another on our hands and knees, under the rotten trunks. In the lower part of the mountain, noble trees of the winter's bark, and a laurel like the sassafras with fragrant leaves, and others, the names of which I do not know, were matted together by a trailing bamboo or cane. Here we were more like fishes struggling in a net than any other animal. On the higher parts, brushwood takes the place of larger trees, with here and there a red cedar or an alerce pine. I was also pleased to see, at an elevation of a little less than 1,000 feet, our old friend the southern beech. They were, however, poor stunted trees; and I should think this must be nearly their northern limit. We ultimately gave up the attempt in despair.

* * *

JANUARY 1ST, 1835 – The new year is ushered in, with the ceremonies proper to it in these regions. She lays out no false hopes; a heavy NW gale, with steady rain, bespeaks the rising year. Thank God, we are not destined here to see the end of it, but hope then to be in the Pacific, where a blue sky tells one there is a heaven – a something beyond the clouds above our heads.

The NW winds prevailing for the next four days, we only managed to cross a great bay, and then anchored in another secure harbour. I accompanied the captain in a boat to the head of a deep creek. On the way the number of seals which we saw was quite astonishing; every bit of flat rock, and parts of the beach, were covered with them. They appeared to be of a loving disposition, and lay huddled together, fast asleep, like so many pigs; but even pigs would be ashamed of their dirt, and of the foul smell which came from them. Each herd was watched by the patient but inauspicious eyes of the turkey-buzzard. This disgusting bird, with its bald scarlet head, formed to wallow in putridity, is very common on the west coast, and their attendance on the seals shows that they are dependant on their mortality. We found the water (probably only that of the surface) nearly fresh; this was caused by the number of torrents, which, in the form of cascades, came tumbling over the bold granite mountains into the sea. The fresh water attracts the fish, and these bring many terns, gulls, and two kinds of cormorant. We saw also a pair of the beautiful black-necked swans, and several small sea-otters, the fur of which is held in such high estimation. In returning we were again amused by the impetuous manner in which the heap of seals, old and young, tumbled into the water as the boat passed. They did not remain long under, but rising, followed us with outstretched necks, expressing great wonder and curiosity.

7TH – Having run up the coast, we anchored near the northern end of the Chonos Archipelago, in Lowe's Harbour, where we remained a week. The islands were here, as in Chiloe, composed of a stratified littoral deposit of soft sandstone with shingle; and the vegetation in consequence was beautifully luxuriant. The woods came down to the sea-beach, just in the same manner as an evergreen shrubbery over a gravel walk. We also enjoyed from the anchorage a splendid view of four great snowy cones of the Cordillera, beginning with the saddle-topped volcano, next 'el famoso Corcovado', and then two others to the southward. The range itself had in this latitude so little elevation, that few parts of it appeared above the line of the neighbouring islets. We found here a party of five men from Caylen, 'el fin del Cristiandad', who had most adventurously crossed the open space of sea which separates Chonos from Chiloe, in their miserable boat-canoe, for the purpose of fishing. These islands will, in all probability, in a short time become peopled, like those adjoining the coast of Chiloe.

Humboldt,* in his Essay on the Kingdom of New Spain, has given a most interesting discussion on the history of the common potato. He believes that the plant described by Molina,† under the name of *maglia*, is the original stock of this useful vegetable, and that it grows in Chile in its native soil. He supposes that thence it was transported by the Indian population to Peru, Quito, New Granada, and the whole Cordillera, from 40° south to 5° north. He oberves that it is a remarkable circumstance, and in accordance with all records respecting the course of the stream of American population, that previously to the Spanish conquest, it was unknown in Mexico. Among the Chonos Islands, a wild potato grows in abundance, which in general habit is even more closely similar to the cultivated kind than is the *maglia* of Molina.

These potatoes grow near the sea-beach, in thick beds, on a sandy, shelly soil, wherever the trees are not too close together. In the middle of January they were in flower, but the tubers were small, and few in number; especially in those plants which grew in the shade, and had the most luxuriant foliage. Nevertheless, I found one which was of an oval form, with one diameter 2 inches in length. The raw bulbs had precisely the smell of the common potato of England, but when cooked they shrunk, and became watery and insipid. They had not a bitter taste, as, according to Molina, is the case with the Chilian kind; and they could be eaten with safety. Some plants measured from the ground to the tip of the upper leaf, not less than 4 feet.

So very close is the general resemblance with the cultivated species, that it is necessary to show that they have not been imported. The simple fact of their growth on the islands, and even small rocks, throughout the Chonos Archipelago, which has never been inhabited, and very seldom visited, is an argument of some weight. But the circumstance of the wildest Indian tribes being well acquainted with the plant, is stronger. Mr Lowe, a very intelligent and active sealer, informs me, that on showing

* Humboldt's New Spain, book iv, chap. ix.
† Molina's Chile, Spanish edition, vol. i, p. 136.

some potatoes to the naked savages in the Gulf of Trinidad (lat. 50°), they immediately reognized them, and calling them 'Aquina', wanted to take them away. The savages also pointed to a place where they grew: which fact was subsequently verified. The Indians of Chiloe, belonging to another tribe, also give them a name in their own language. The simple fact of their being known and named by distinct races, over a space of 400 or 500 miles on a most unfrequented and scarcely known coast, almost proves their native existence. Professor Henslow, who has examined the dried specimens which I brought home, says that they are the same with those described by Mr Sabine* from Valparaiso, but that they form a variety which by some botanists has been considered as specifically distinct. It is remarkable that the same plant should be found on the sterile mountains of central Chile, where a drop of rain does not fall for more than six months, and within the damp forests of the southern islands. From what we know of the habits of the potato, this latter situation would appear more congenial than the former, as its birthplace.

In the central parts of the Chonos Archipelago, in lat. 45° 30′, the forest has assumed very much the same character which is found along the whole west coast for 600 miles to Cape Horn. The arborescent grass of Chiloe has here ceased to exist; while the beech of Tierra del Fuego both grows to a good size, and forms a considerable proportion of the wood; not, however, in the same exclusive manner as it does further to the southward. Cryptogamic plants here find a most congenial climate. In the neighbourhood of the Strait of Magellan, I have before remarked that the country appears too cold and wet to allow of their arriving at perfection; but in these islands, within the forest, the number of species, and great abundance of mosses, lichens, and small ferns, is quite extraordinary.† In Tierra del Fuego trees grow only on

---

* Horticultural Transact., vol. v, p. 249. Mr Caldcleugh sent home two tubers, which being well manured, even the first season produced numerous potatoes and an abundance of leaves.

† By sweeping with my insect-net, I procured from these situations a considerable number of minute insects of the family of Staphylinidæ, and others allied to Pselaphus, and minute Hymenoptera. But the most characteristic family in number of both individuals and species, throughout the more open parts of Chiloe and Chonos, is that of the Telephoridæ.

the hill-sides; every level piece of land being invariably covered by a thick bed of peat; but in Chiloe the same kind of situation supports the most luxuriant forest. Here, within the Chonos Archipelago, the nature of the climate more closely approaches that of the southern, than that of the northern, of these two countries. Nearly every patch of level ground is covered by two species of plants (*Astelia pumila* of Brown,* and *Donatia magellanica*), which by their joint decay compose a thick bed of elastic peat.

In Tierra del Fuego, above the region of woodland, the former of these eminently sociable plants is the chief agent in the production of peat. Fresh leaves are always succeeding one to the other, round the central tap root; the lower ones soon decay; and in tracing a root downward in the peat, the leaves (yet holding their position) can be observed passing through every stage of decomposition, till the whole becomes blended in one confused mass. The Astelia is assisted by few other plants; here and there a small creeping one (*Myrtus nummularia*), with a woody stem like our cranberry, but with a sweet berry; another (*Empetrum rubrum*) like our heath, and a third (*Juncus grandiflorus*) a rush; are nearly the only ones that grow on the swampy surface. These plants, though possessing a very close general resemblance to the English kinds, are botanically different. In the more level parts of the country, the surface of the peat is broken up into little pools of water, which stand at different heights, and appear as if artificially excavated. Small streams of water, flowing under ground, complete the disorganization of the vegetable matter, and consolidate the whole.

* * *

The zoology of these broken islets of the Chonos Archipelago, is, as might have been expected, very poor. Of quadrupeds, two aquatic kinds are common. The *Myopotamus Coipus* (like a beaver, but with a round tail) is well known from its fine fur, which is an object of trade, throughout the tributaries of La Plata. It here,

---

* *Anthericum trifarium* of Solander.

however, exclusively frequents salt water; which same circumstance has been mentioned, as sometimes occurring with the great rodent, the Capybara. A small sea-otter is very numerous. This animal does not feed exclusively on fish, but, like the seals, draws a large supply from a small red crab, which swims in shoals near the surface of the water. Mr Bynoe saw one in Tierra del Fuego eating a cuttle-fish; and at Lowe's Harbour another was killed, in the act of carrying to its hole a large volute; and this was the only specimen of that shell which was procured. At one place I caught in a trap a singular little mouse; it appeared common on several of the islets, but the Chilotans at Lowe's Harbour said that it was not found in all. What a succession of chances,* or what changes of level, must have been brought into play, thus to spread these small animals throughout this broken archipelago!

* * *

---

* Many rapacious animals bring their prey alive to feed their young. Are there any instances on record of such a habit among owls or hawks? If so, in the course of centuries, every now and then, one might escape from the young birds. Some such agency is wanted, to account for the distribution of the smaller gnawing animals on islands near to each other.

CHAPTER XVI

CHILOE AND CONCEPCION

* * *

FEBRUARY 20TH – The day has been memorable in the annals of
Valdivia, for the most severe earthquake experienced by the
oldest inhabitant. I happened to be on shore, and was lying down
in the wood to rest myself. It came on suddenly, and lasted two
minutes; but the time appeared much longer. The rocking of the
ground was most sensible. The undulations appeared to my
companion and myself to come from due east; whilst others
thought they proceeded from south-west; which shows how
difficult it is in all cases to perceive the direction of these
vibrations. There was no difficulty in standing upright, but the
motion made me almost giddy. It was something like the movement of a vessel in a little cross ripple, or still more like that felt by
a person skating over thin ice, which bends under the weight of
his body.

A bad earthquake at once destroys the oldest associations: the
world, the very emblem of all that is solid, has moved beneath our

feet like a crust over a fluid; one second of time has conveyed to the mind a strange idea of insecurity, which hours of reflection would never have created. In the forest, as a breeze moved the trees, I only felt the earth tremble, but saw no consequences from it. Captain FitzRoy and the officers were at the town during the shock, and there the scene was more awful; for although the houses, from being built of wood, did not fall, yet they were so violently shaken that the boards creaked and rattled. The people rushed out of doors in the greatest alarm. I feel little doubt that it is these accompaniments which cause that horror of earthquakes, experienced by all those who have thus seen as well as felt their effects. Within the forest it was a deeply interesting, but by no means an awe-exciting phenomenon. The tides were very curiously affected. The great shock took place at the time of low water; and an old woman who was on the beach told me, that the water flowed very quickly, but not in big waves, to high-water mark, and then as quickly returned to its proper level; this was also evident by the line of wet sand. This same kind of quick but quiet movement in the tide happened a few years since at Chiloe, during a slight earthquake, and created much causeless alarm. In the course of the evening there were other weaker shocks, all of which seemed to produce in the harbour the most complicated currents, and some of great strength.

22ND – We sailed from Valdivia, and on the 4th of March, entered the harbour of Concepcion. While the ship was beating up to the anchorage, which is distant several miles, I was landed on the island of Quiriquina. The mayor–domo of the estate quickly rode down to tell us the terrible news of the great earthquake of the 20th; 'that not a house in Concepcion, or Talcuhano, (the port) was standing; that seventy villages were destroyed; and that a great wave had almost washed away the ruins of Talcuhano.' Of this latter fact I soon saw abundant proof; the whole coast being strewed over with timber and furniture, as if a thousand great ships had been wrecked. Besides chairs, tables, bookshelves, &c, in great numbers, there were several roofs of cottages, which had been drifted in an almost entire state. The storehouses at Talcuhano had burst open, and great bags of cotton, yerba, and other valuable merchandise, were scattered about on the shore.

During my walk round the island, I observed that numerous fragments of rock, which, from the marine productions adhering to them, must recently have been lying in deep water, had been cast up high on the beach. One of these was a slab 6 feet by 3, and about 2 feet thick.

The island itself as plainly showed the overwhelming power of the earthquake, as the beach did that of the consequent great wave. The ground was fissured in many parts, in north and south lines; which direction perhaps was caused by the yielding of the parallel and steep sides of the narrow island. Some of the fissures near the cliffs were a yard wide: many enormous masses had already fallen on the beach; and the inhabitants thought, that when the rains commenced, even much greater slips would happen. The effect of the vibration on the hard primary slate, which composes the foundation of the island, was still more curious: the superficial parts of some narrow ridges were as completely shivered, as if they had been blasted by gunpowder. This effect, which was rendered very evident by the fresh fractures and displaced soil, must, during earthquakes, be confined to near the surface, for othewise there would not exist a block of solid rock throughout Chile. This limited action is not improbable, as it is certain, that the surface of any body, when vibrating, is in a different condition from the central parts. It is, perhaps, owing to this same reason, that earthquakes do not cause quite such terrific havoc within deep mines, as would at first have been expected. I believe this convulsion has been more effectual in lessening the size of the island of Quiriquina, than the ordinary wear and tear of the weather and the sea during the course of an entire century.

The next day I landed at Talcuhano, and afterwards rode to Concepcion. Captain FitzRoy has given so detailed and accurate an account of the earthquake, that it is almost useless for me to say any thing on the subject; but I will extract a few passages from my journal. Both towns presented the most awful yet interesting spectacle I ever beheld. To a person who had formerly known the places, it possibly might have been still more impressive; for the ruins were so mingled together, and the whole scene possessed so little the air of a habitable place, that it was scarcely possible to imagine its former appearance or condition. The earthquake commenced at half-past eleven in the forenoon. If it had happened in the middle of the night the greater number of the inhabitants (which in this one province amount to many thousands),* instead of less than a hundred, must have perished. In Concepcion, each house, or row of houses, stood by itself, a heap or line of ruins; but in Talcuhano, owing to the great wave, little more than one layer of bricks, tiles, and timber, with here and there part of a wall left standing, could be distinguished. From this circumstance, Concepcion, although not so completely desolated, was a more terrible, and if I may so call it, picturesque sight. The first shock was very sudden. The invariable practice among the residents in these provinces, of running out of doors at the first trembling of the ground, alone saved them. The mayor-domo at Quiriquina told me, that the first notice he received of the earthquake, was finding both the horse he rode, and himself, rolling together on the ground. Rising up, he was again thrown down. He also told me that some cows, which were standing on the steep sides of the island, were rolled into the sea. The great wave, however, was far more destructive in this respect: on one low island near the head of the bay, seventy animals were washed off and drowned. It is generally thought that this has been the worst earthquake ever recorded in Chile; but as the very bad ones occur only after long intervals, this cannot easily be known; nor indeed would a much more severe shock have made any great difference, for the ruin is now complete.

After viewing Concepcion, I cannot understand how the greater number of inhabitants escaped unhurt. The houses in many parts fell outwards; thus forming in the middle of the streets little hillocks of brickwork and rubbish. Mr Rous, the English consul, told us that he was at breakfast when the first movement warned him to run out. He had scarcely reached the middle of the courtyard, when one side of his house came thundering down. He retained presence of mind to remember, that if he once got on the top of that part which had already fallen, he should be safe. Not

---

* Miers estimates them at 40,000; but the towns in some of the other provinces were likewise overthrown.

being able, from the motion of the ground, to stand, he crawled up on his hands and knees; and no sooner had he ascended this little eminence, than the other side of the house fell in, the great beams sweeping close in front of his head. With his eyes blinded, and his mouth choked with the cloud of dust which darkened the sky, at last he gained the street. As shock succeeded shock, at the interval of a few minutes, no one dared approach the shattered ruins; and no one knew whether his dearest friends and relations might not be perishing from the want of help. The thatched roofs fell over the fires, and flames burst forth in all parts. Hundreds knew themselves to be ruined, and few had the means of providing food for the day. Can a more miserable and fearful scene be imagined?

Earthquakes alone are sufficient to destroy the prosperity of any country. If, for instance, beneath England, the now inert subterranean forces should exert those powers which most assuredly in former geological ages they have exerted, how completely would the entire condition of the country be changed! What would become of the lofty houses, thickly packed cities, great manufactures, the beautiful public and private edifices? If the new period of disturbance were first to commence by some great earthquake in the dead of the night, how terrific would be the carnage! England would at once be bankrupt; all papers, records, and accounts would from that moment be lost. Government being unable to collect the taxes, and failing to maintain its authority, the hand of violence and rapine would go uncontrolled. In every large town famine would be proclaimed, pestilence and death following in its train.

Captain FitzRoy has given an account of the great wave, which, travelling from seaward, burst over Talcuhano. In the middle of the bay it was seen as one unbroken swell of the water; but on each side, meeting with resistance, it curled over, and tore up cottages and trees as it swept onwards with overwhelming force. At the head of the bay it is easy to imagine the fearful line of white breakers which three times rushed over, and almost obliterated, the ruins of the former town. Pools of salt water yet remained in the streets; and children, making boats with old tables and chairs, appeared as happy as their parents were miserable. It was, however, exceedingly interesting to observe how active and cheerful all appeared, after their heavy misfortune. It was remarked with much truth, that from the destruction being universal, no one individual was humbled more than another, or could suspect his friends of coldness; and this latter effect is perhaps the most grievous one of the loss of wealth. Mr Rous, and a large party whom he kindly took under his protection, lived for the first week in a garden beneath some apple-trees. At first they were as merry as if it had been a picnic; but soon afterwards heavy rain caused much discomfort, for they were absolutely without shelter.

In Captain FitzRoy's paper it is said that two explosions, one like a column of smoke, and another like the blowing of a great whale, were seen in the bay of Concepcion. The water also appeared every where to be boiling; and it 'became black, and exhaled a most disagreeable sulphureous smell'. I am informed by Mr Alison, that during the earthquake of 1822 these last-mentioned circumstances occurred in the bay of Valparaiso. The two great explosions in the first case must no doubt be connected with deep-seated changes; but the bubbling water, its black colour and fetid smell, the usual concomitants of a severe earthquake, may, I think, be attributed to the disturbance of mud containing organic matter in decay. In the bay of Callao, during a calm day, I noticed, that as the ship dragged her cable over the bottom, its course was marked by a line of bubbles.

The lower orders in Talcuhano thought that the earthquake was caused by some old Indian women, who two years ago having been offended, stopped the volcano of Antuco. This silly belief is curious, because it shows that experience has taught them to observe the constant relation between the suppressed activity of volcanoes, and the trembling of the ground. It was necessary to apply the witchcraft to the point where their knowledge stopped; and this was the closing of the volcanic vent. This saying is the more odd in this particular instance, because the result of Captain FitzRoy's investigation was to discountenance the belief that Antuco (whatever might have been the case with the volcanoes further northward) was any way affected.

The town of Concepcion was built in the usual Spanish fashion, with all the streets running at right angles to each other. One set ranged S W by W and N E by E, and the other N W by N and S E by S. The walls in the former direction certainly stood better than those in the other. Captain FitzRoy* has likewise remarked, that the greater number of the masses of brickwork were thrown down towards the N E. Both these circumstances perfectly agree with the general idea of the undulation having come from the S W; in which quarter subterranean noises were also sometimes heard. It is evident on this supposition, that the N W and S E walls, being nearly coincident with the line of undulation (or with the crests of the successive waves), would be much more likely to fall than those which had their extremities presented towards the point whence the vibration proceeded; for, in the first case, the whole wall would be thrown at the same moment out of its perpendicular. This may be illustrated by placing books edgewise on a carpet, and then, after the manner suggested by Michell, imitating the undulations of an earthquake: it will be found, that they fall with more or less readiness, according to their direction. The fissures in the ground, though not uniform, generally had a S E and N W direction;† and therefore they corresponded to the lines of principal flexure. Bearing in mind all these circumstances, which so clearly point to the S W as the chief focus of disturbance, it is a very interesting fact that the island of S. Maria,† situated in that quarter, was during the general uplifting of the land (to which I shall presently refer) raised to nearly three times the altitude of any other part of the coast.

The different resistance offered by the walls, according to their direction, was well exemplified in the case of the cathedral. The side which fronted the N E presented a grand pile of ruins, in the midst of which door-cases and masses of timber stood up, as if floating in a stream. Some of the angular blocks of brickwork were of great dimensions; and they had been rolled to a distance on the level plaza, like fragments of rock round the base of some high mountain. The side walls, though exceedingly fractured, yet

* 'Sketch of Surveying Voyages of *Adventure* and *Beagle*' by Captain FitzRoy, Royal Geograph. Journal, vol. vi, p. 320.
† Ditto, p. 327, *et passim*.

remained standing; but the vast buttresses (at right angles to them, and therefore parallel to the walls that fell) were in many cases cut clean off, as if by a chisel, and hurled to the ground.

Some square ornaments on the coping of these same walls were moved by the earthquake into a diagonal position. The buttresses of the church of La Merced, at Valparaiso, and some heavy pieces of furniture in the rooms were similarly affected by the shock of 1822.[*] Mr Lyell[†] has also given a drawing of an obelisk in Calabria, of which the separate stones were partially turned round. In these instances, the displacement at first appears to be owing to a vorticose movement beneath each point thus affected; but such can hardly be the case. May it not be caused by a tendency in each stone to arrange itself in some particular position, with respect to the lines of vibration – in a manner somewhat similar to pins on a sheet of paper, or on a board, when it is shaken? Generally speaking, arched doorways or windows stood much better than any other kind of building. Nevertheless, a poor lame old man, who had been in the habit, during trifling shocks, of crawling to a certain doorway, was this time crushed to pieces.

I have not attempted to give any detailed description of the appearance of Concepcion, for I feel it is quite impossible to convey the mingled feelings with which one beholds such a spectacle. Several of the officers visited it before me, but their strongest language failed to communicate a just idea of the desolation. It is a bitter and humiliating thing to see works, which have cost men so much time and labour, overthrown in one minute; yet compassion for the inhabitants is almost instantly forgotten, from the interest excited in finding that state of things produced in a moment of time, which one is accustomed to attribute to a succession of ages. In my opinion, we have scarcely beheld since leaving England, any other sight so deeply interesting.

In almost every severe earthquake which has been described, the neighbouring waters of the sea are said to have been greatly agitated. The disturbance seems generally, as in the case of Concepcion, to have been of two kinds: first, at the instant of the

---

[*] Miers's Chile. vol. i, p. 392.
[†] Lyell's Principles of Geology, chap. xv, book ii.

shock, the water swells high up on the beach, with a gentle motion, and then as quietly retreats; secondly, some little time afterwards, the whole body of the sea retires from the coast, and then returns in great waves of overwhelming force. The first and less regular movement seems to be an immediate consequence of the earthquake differently affecting a fluid and a solid, so that their respective levels are slightly deranged. But the second case is a far more important phenomenon, and at first appears of less easy explanation. In reading accounts of earthquakes, and especially of those on the west coast of America, as collated from various authors by Sir W. Parish,* it is certain that the first great movement of the waters has been that of retiring. Several hypotheses† have been invented to explain this fact. Some have supposed it owing to a vertical oscillation in the land, the water retaining its level: but this can hardly happen, even on a moderately shoal coast; for the water near the land must partake of the motion of the bottom. Moreover, as Mr Lyell has urged, a change of level in the land will not account for movements in the sea, of a similar nature, affecting islands distant from the line of uplifted coast. This occurred at Madeira during the famous Lisbon earthquake. Juan Fernandez also offers a parallel instance; for the sea was disturbed there much in the same manner as on the coast of Chile.

The whole phenomenon, it appears to me, is due to a common undulation in the water, proceeding from a line or point of disturbance, some little way distant. If the waves sent off from the paddles of a steam-vessel be watched breaking on the sloping shore of a still river, the water will be seen first to retire 2 or 3 feet, and then to return in little breakers, precisely analogous to those consequent on an earthquake. From the oblique direction in which the waves are sent off from the paddles, the vessel has proceeded a long way ahead, before the undulation reaches the shore; and hence it is at once manifest, that this movement bears no relation to the actual displacement of the fluid from the bulk of the vessel. Indeed, it seems a general circumstance, that in all cases where the equilibrium of an undulation is thus destroyed, the

---

* Sir W. Parish had the kindness to lend me the original manuscript, which was read before the Geological Society, March 5th, 1835.

† Lyell's Geology, book ii, ch. xvi.

water is drawn from the resisting surface to form the advancing breaker.* Considering then a wave produced by an earthquake as an ordinary undulation proceeding from some point or line in the offing, we can see the cause, first of its occurrence some time after the shock; secondly, of its affecting the shores of the mainland and of outlying islets in a uniform manner – namely, the water retiring first, and then returning in a mountainous breaker; and lastly, of its size being modified (as appears to be the case) by the form of the neighbouring coast. For instance Talcuhano and Callao are situated at the head of great shoaling bays, and they have always suffered from this phenomenon; whereas, the town of Valparaiso, which is seated close on the border of a profound ocean, though shaken by the severest earthquakes, has never been overwhelmed by one of these terrific deluges. On this view, we have only to imagine, in the case of Concepcion, a point of disturbance in the bottom of the sea in a south-west direction, whence the wave was seen to travel, and where the land was elevated to a greater height than any other part – and the whole phenomenon will be explained.

It is probable that near every coast, the chief line of disturbance would be situated at that distance in the offing, where the fluid which was most agitated, from overlying the shallow bottom near the land, joined on to that part which covered the depths (but slightly moved) of the ocean. In all distant parts of the coast the small oscillations of the sea, both at the moment of the great shock, and during the lesser following ones, would be confounded with the undulation propagated from the focus of disturbance, and hence the series of movements would be undistinguishable.

The most remarkable effect (or perhaps speaking more correctly, cause) of this earthquake was the permanent elevation of the land. Captain FitzRoy having twice visited the island of Santa Maria, for the purpose of examining every circumstance with extreme accuracy, has brought a mass of evidence in proof of such elevation, far more conclusive than that on which geologists on

---

* I am indebted to Mr Whewell for explaining to me the probable movements on the shore, of an undulation of which the equilibrium has been destroyed.

most other occasions place implicit faith. The phenomenon possesses an uncommon degree of interest, from this particular part of the coast of Chile having previously been the theatre of several earthquakes of the worst class. It is almost certain, from the altered soundings, together with the circumstance of the bottom of the bay near Penco, consisting of hard stone, that there has been an uplifting to the amount of 4 fathoms, since the famous convulsion of 1751. With this additional instance fresh before us, we may assume as probable, according to the principles laid down by Mr Lyell,* other small successive elevations, and may fearlessly maintain that the problem of the raised shells,† recorded by Ulloa, is explained.

Some of the consequences which may be deduced from the phenomena connected with this earthquake are most important in a geological point of view; but in the present work I cannot do more than simply allude to the results. Although it is known that earthquakes have been felt over enormous spaces, and strange subterranean noises likewise heard over nearly equal areas, yet few cases are on record of volcanoes, very far distant from each other, bursting out at the same moment of time. In this instance, however, at the same hour when the whole country around Concepcion was permanently elevated, a train of volcanoes situated in the Andes, in front of Chiloe, instantaneously spouted out a dark column of smoke, and during the subsequent year continued in uncommon activity. It is, moreover, a very interesting circumstance, that, in the immediate neighbourhood, these eruptions entirely relieved the trembling ground, although at a little distance, and in sight of the volcanoes, the island of Chiloe was strongly affected. To the northward, a volcano burst out at the bottom of the sea adjoining the island of Juan Fernandez, and several of the great chimneys in the Cordillera of central Chile commenced a fresh period of activity. We thus see a permanent elevation of the land, renewed activity through habitual vents, and a submarine outburst, forming parts of one great phenom-

---

* Lyell's Geology, book ii, chap. xvi.
† I saw these shells in very great quantities on the flanks of the island of Quiriquina.

enon. The extent of country throughout which the subterranean forces were thus unequivocally displayed, measures 700 by 400 geographical miles. From several considerations, which I have not space here to enter on, and especially from the number of intermediate points whence liquefied matter was ejected, we can scarcely avoid the conclusion, however fearful it may be, that a vast lake of melted matter, of an area nearly doubling in extent that of the Black Sea, is spread out beneath a mere crust of solid land.

The elevation of the land to the amount of some feet during these earthquakes, appears to be a paroxysmal movement, in a series of lesser and even insensible steps, by which the whole west coast of South America has been raised above the level of the sea. In the same manner, the most violent explosion from any volcano is merely one in a series of lesser eruptions: and we have seen that both these phenomena, which are in so many ways related, are parts of one common action, only modified by local circumstances. With respect to the cause of the paroxysmal convulsion in particular portions of the great area which is simultaneously affected, it can be shown to be extremely probable, that it is owing to the giving way of the superincumbent strata, (and this giving way probably is a consequence of the tension from the general elevation) and their interjection by fluid rock – one step in the formation of a mountain chain. On this view we are led to conclude, that the unstratified mass forming the axis of any mountain, has been pumped in when in a fluid state, by as many separate strokes as there were earthquakes. For instance, in the case of Concepcion, during the few months subsequent to the great shock, upwards of 300 tremours of the ground were felt, each of which indicated a fresh fracture, and injection of the fluid stone. It is a case precisely analogous to what happens in all bad eruptions, which are invariably followed by a succession of smaller ones: the difference is, that in the volcano the lava is ejected, while in the formation of a mountain chain it is injected. This view of the extremely gradual elevation of a line of mountains, will alone explain the difficulty (which, as far as I am aware, has never been attempted to be solved) of the axis consisting of rock which has become solid under the pressure of the super-incumbent strata, while yet these same strata, in their present inclined and vertical positions, cannot possibly cover more than a small portion of that axis.

CHAPTER XVII

PASSAGE OF CORDILLERA

MARCH 7TH, 1835 – We staid only three days at Concepcion, and then sailed for Valparaiso. The wind being northerly, we only reached the mouth of the harbour of Concepcion before it was dark. Being very near the land, and a fog coming on, the anchor was dropped. Presently a large American whaler appeared close alongside of us; and we heard the Yankee swearing at his men, to make them keep quiet, whilst he listened where the breakers were. Captain FitzRoy hailed him in a loud clear voice, to anchor where he then was. The poor man must have thought the voice came from the shore: such a Babel of cries issued at once from the ship – every one hallooing out, 'Let go the anchor! veer cable! shorten sail!' It was the most laughable thing I ever heard. If the ship's crew had been all captains, and no men, there could not have been a greater uproar of orders. We afterwards found that the mate stuttered. I suppose all hands were assisting him in giving his orders.

On the 11th we anchored at Valparaiso; and two days afterwards

I set out on an excursion to cross the Cordillera. I proceeded to Santiago, where Mr Caldcleugh most kindly assisted me in every possible way, in making the little preparations which were necessary. In this immediate part of Chile there are two passes across the Andes to Mendoza, and the plains on the opposite side. The one most commonly used, namely, that of Aconcagua, or Uspallata, is situated some way to the northward of the capital: the other, called the Portillo, is to the southward, and less distant. The latter is, however, rather more lofty, and from the double chain, more dangerous during a snow-storm. For these reasons it is but little used, especially late in the season.

MARCH 18TH – We set out for the Portillo pass. Leaving Santiago we crossed the wide burnt-up plain on which that city stands, and in the afternoon arrived at the Maypo, one of the principal rivers in Chile. The valley, at the point where it enters the first Cordillera, is bounded on each side by lofty barren mountains; and although not broad, it is very fertile. Numerous cottages were surrounded by vines, and by orchards of apple, nectarine, and peach trees; the boughs of the latter breaking with the weight of the beautiful ripe fruit.

In the evening we passed the custom-house, where our luggage was examined. The frontier of Chile is better guarded by the Cordillera, than by the waters of the sea. There are very few valleys which lead to the central ranges, and, except by these, the mountains are far too steep and lofty for any beast of burden to pass over them. The custom-house officers were very civil; which was perhaps partly owing to the passport which the President of the republic had given me; but I must also express my admiration at the natural politeness of almost every Chileno. In this instance the contrast with the same class of men in most other countries was strongly marked. I may mention an anecdote with which I was much pleased at the time. We met near Mendoza a little and very fat negress, riding astride on a mule. She had a *goître* so enormous, that it was scarcely possible to avoid gazing at her for a moment; but my two companions almost instantly, by way of apology, made the common salute of the country, by taking off their hats. Where would one of the lower classes in Europe have shown such feeling politeness to a poor and miserable object of a degraded race?

At night we slept at a cottage. Our manner of travelling was delightfully independent. In the inhabited parts we bought a little firewood, hired pasture for the animals, and bivouacked in the corner of the same field with them. Carrying an iron pot, we cooked and ate our supper under the cloudless sky, and knew no trouble. My companions were Mariano Gonzales, who had formerly accompanied me, and an 'arriero', with his ten mules and a 'madrina'.

The madrina (or godmother) is a most important personage. She is an old steady mare, with a little bell round her neck; and wheresoever she goes, the mules, like good children, follow her. If several large troops are turned into one field to graze, in the morning the muleteer has only to lead the madrinas a little apart, and tinkle their bells; and, although there may be 200 or 300 mules together, each immediately knows its own bell, and separates itself from the rest. The affection of these animals for their madrinas saves infinite trouble. It is nearly impossible to lose an old mule; for if detained for several hours by force, she will, by the power of smell, like a dog, track out her companions, or rather the madrina; for, according to the muleteer, she is the chief object of affection. The feeling, however, is not of an individual nature; for I believe I am right in saying, that any animal with a bell will serve as madrina. In a troop each animal carries, on a level road, a cargo weighing 416 pounds (more than 29 stone); but in a mountainous country 100 pounds less.* Yet with what delicate slim limbs, without any proportional bulk of muscle, these animals support so great a burden! The mule always appears to me a most surprising animal. That a hybrid should possess more reason, memory, obstinacy, social affection, and powers of muscular endurance, than either of its parents, seems to indicate that art has here outmastered nature. Of our ten animals, six were intended for riding and four for carrying cargoes, each taking turn

* Throughout Chile, except between Santiago and Valparaiso, every thing is conveyed on mules. This is an expensive method of transport, but unavoidable without good roads and improved waggons. In a troop of mules, there is generally a muleteer to each six animals.

about. We carried a good deal of food, in case we should be snowed up, as the season was rather late for passing the Portillo.

* * *

MARCH 20TH – As we ascended the valley, the vegetation, with the exception of a few pretty alpine flowers, became exceedingly scanty; and of birds, animals, or insects, scarcely one could be seen. The lofty mountains, their summits marked with a few patches of snow, stood well separated from each other; the valleys being filled up with an immense thickness of stratified alluvium. I may here briefly remark, without detailing the reasons on which the opinion is grounded, that in all probability this matter was accumulated at the bottoms of deep arms of the sea, which running from the inland basins, penetrated to the axis of the Cordillera – in a similar manner to what now happens in the southern part of this same great range. This fact, in itself most curious, as preserving a record of a very ancient state of things, possesses a high theoretical interest, when considered in relation to the kind of elevation by which the present great altitude of these mountains has been attained.

The features in the scenery of the Andes which struck me most, as contrasted with the few other mountain chains with which I am acquainted, were, the flat fringes sometimes expanding into narrow plains on each side the valleys, the bright colours, chiefly red and purple, of the utterly bare and precipitous hills, the grand and continuous wall-like dikes, the strongly-marked strata which, when nearly vertical, form the most picturesque and wild pinnacles, but where less inclined, great massive mountains – the latter occupying the outskirts of the range, and the former the more lofty and central parts – lastly, the smooth conical piles of fine and brightly coloured detritus, which slope at a high angle from the flanks of the mountains to their bases, some of the piles having a height of more than 2,000 feet.

* * *

I will here give a very brief sketch of the geological structure of these mountains: first, of the Peuquenes, or western line; for the constitution of the two ranges is totally different. The lowest stratified rock is a dull red or purple claystone porphyry, of many varieties, alternating with conglomerates, and breccia composed of a similar substance: this formation attains a thickness of more than a mile. Above it there is a grand mass of gypsum, which alternates, passes into, and is replaced by, red sandstone, conglomerates, and black calcareous clay-slate. I hardly dare venture to guess the thickness of this second division; but I have already said some of the beds of gypsum alone attain a thickness of at least 2,000 feet. Even at the very crest of the Peuquenes, at the height of 13,210 feet, and above it, the black clay-slate contained numerous marine remains, amongst which a gryphæa is the most abundant, likewise shells, resembling turritellæ, terebratulæ, and an ammonite. It is an old story, but not the less wonderful, to hear of shells, which formerly were crawling about at the bottom of the sea, being now elevated nearly 14,000 feet above its level. The formation probably is of the age of the central parts of the secondary series of Europe.

These great piles of strata have been penetrated, upheaved, and overturned, in the most extraordinary manner, by masses of injected rock, equalling mountains in size. On the bare sides of the hills, complicated dikes, and wedges of variously-coloured porphyries and other stones, are seen traversing the strata in every possible form and direction; proving also by their intersections, successive periods of violence. The rock which composes the axis of these great lines of dislocation, at a distance very closely resembles granite, but on examination, it is found rarely to contain any quartz; and instead of ordinary felspar, albite.

The metamorphic action has been very great, as might have been expected from the close proximity of such grand masses of rock, which were injected when in a liquefied state from heat. When it is known, first, that the stratified porphyries have flowed as streams of submarine lava under an enormous pressure, and that the mechanical beds separating them owe their origin to explosions from the same submarine craters; secondly, that the whole mass in the lower part has generally been so completely fused into one solid rock by metamorphic action, that the lines of division can only be traced with much difficulty; and thirdly, that masses of porphyry, undistinguishable by their mineralogical characters from the two first kinds, have been subsequently injected; the extreme complication of the whole will readily be believed.

We now come to the second range, which is of even greater altitude than the first. Its nucleus in the section seen in crossing the Portillo pass, consists of magnificent pinnacles of coarsely crystallized red granite. On the eastern flank, a few patches of mica slate still adhere to the unstratified mass; and at the foot a stream of basaltic lava has burst forth at some remote period – perhaps when the sea covered the wide surface of the Pampas. On the western side of the axis, between the two ranges, laminated fine sandstone has been penetrated by immense granitic dikes proceeding from the central mass, and has thus been converted into granular quartz rock. The sandstone is covered by other sedimentary deposits, and these again by a coarse conglomerate, the vast thickness of which I will not attempt even to estimate. All these coarse mechanical beds dip from the red granite directly towards the Peuquenes range, as if they passed beneath it; though such is not the case. On examining the pebbles composing this conglomerate (which, to my surprise, betrayed no signs of metamorphic action), I was astonished to find perfectly rounded masses of the black calcareous clay-slate with organic remains – the same rock which I had just crossed *in situ* on the Peuquenes. These phenomena compel us to arrive at the following conclusion: that the Peuquenes existed as dry land for a long period anterior to the formation of the second range, and that, during this period, immense quantities of shingle were accumulated at its submarine flank. The action of a disturbing force then commenced: these more modern deposits were injected by dikes, altered by heat, and tilted towards the line whence, in the form of sediment and pebbles, they had originally proceeded, thus making the offspring at first appear older than its parent. This second, grand, and subsequent line of elevation is parallel to the first and more ancient one.

* * *

About noon we began the tedious ascent of the Peuquenes, and then for the first time experienced some little difficulty in our respiration. The mules would halt every 50 yards, and then the poor willing animals after a few seconds started of their own accord again. The short breathing from the rarefied atmosphere is called by the Chilenos 'puna'; and they have most ridiculous notions concerning its origin. Some say, 'all the waters here have puna'; others that 'where there is snow there is puna'; — and this no doubt is true. It is considered a kind of disease, and I was shown the crosses over the graves of some who had died 'punado'. Excepting perhaps in the case of a person suffering from some organic disease of the heart or chest, I should think this must be an erroneous conclusion. A person near death, would probably at this elevation experience a more unusual difficulty in breathing than others; and hence the effect might be assumed as the cause. The only sensation I felt was a slight tightness over the head and chest; a feeling which may be experienced by leaving a warm room and running violently on a frosty day. There was much fancy even in this; for upon finding fossil shells on the highest ridge, I entirely forgot the puna in my delight. Certainly the exertion of walking was extreme, and the respiration became deep and laborious. It is incomprehensible to me, how Humboldt and others were able to ascend to the elevation of 19,000 feet. No doubt a residence of some months in the lofty region of Quito would prepare the constitution for such an exertion; yet I am told that in Potosi (about 13,000 feet), strangers do not become quite accustomed to the atmosphere for an entire year. The inhabitants all recommend onions for the puna; as this vegetable has some-times been given in Europe for pectoral complaints, it may possibly be of real service: for my part, I found nothing so good as the fossil shells!

* * *

When nearly on the crest of the Portillo, we were enveloped in a cloud which was falling, under the form of minute frozen spicula. This was very unfortunate, as it continued the whole day, and quite intercepted our view. The pass takes its name of Portillo from a narrow cleft or doorway on the highest ridge, through which the road passes. From this point, on a clear day, those vast plains which extend from the base of the mountains towards the Atlantic can be seen. We descended to the upper limit of vegetation, and found good quarters for the night under the shelter of some large fragments of rock. We here met some passengers, who made anxious inquiries about the state of the road. Shortly after it was dark, the clouds suddenly cleared away; and the effect was quite magical. The great mountains, bright with the full moon, seemed impending over us on all sides, as if we had been buried at the bottom of some deep crevice. One morning also, very early, I witnessed the same striking effect. As soon as the clouds were dispersed, it froze severely; but as there was no wind, we slept very comfortably.

The increased brilliancy of the moon and stars at this elevation, owing to the perfect transparency of the atmosphere, was very remarkable. Travellers having observed the difficulty of judging heights and distances amidst lofty mountains, have generally attributed it to the absence of objects of comparison. It appears to me that it is fully as much owing to this transparency, confounding different distances, and partly, likewise, to the novelty of an unusual degree of fatigue arising from a little exertion, – habit being thus opposed to the evidence of the senses. I am sure that this extreme clearness of the air gives a peculiar character to the landscape; all objects appearing to be brought nearly into one plane, as in a drawing or panorama. The transparency is, I presume, owing to the equable and nearly perfect state of atmospheric dryness. The latter quality was shown by the manner in which woodwork shrunk (as I soon found by the trouble my geological hammer gave me); by articles of food, such as bread and sugar, becoming extremely hard; and by the preservation of the skin and parts of the flesh of the beasts, which perish on the road. To the same cause we must attribute the singular facility with which electricty is excited. My flannel-waistcoat, when rubbed in the dark, appeared as if it had been washed with phosphorus; every hair on a dog's back crackled; even the linen sheets, and leathern straps of the saddle, when handled, emitted sparks.

MARCH 23 RD – The descent on the eastern side of the Cordillera is much shorter or steeper than on the Pacific side; in other words, the mountains rise more abruptly from the plains, than from the alpine country of Chile. A level and brilliantly white sea of clouds was extended beneath our feet, and thus shut out the view of the equally level Pampas. We soon entered the band of clouds, and did not again emerge from it that day. About noon, finding pasture for the animals and bushes for firewood, in a part of the valley called Los Arenales, we stopped for the night. This was near the uppermost limit of bushes, and the elevation, I suppose, was between 7,000 and 8,000 feet.

I was very much struck with the marked difference between the vegetation of these eastern valleys and that of the opposite side: yet the climate, as well as the kind of soil, is nearly identical, and the difference of longitude very trifling. The same remark holds good with the quadrupeds, and in a lesser degree with the birds and insects. We must except certain species which habitually or occasionally frequent elevated mountains; and in the case of the birds, certain kinds, which have a range as far south as the Strait of Magellan. This fact is in perfect accordance with the geological history of the Andes; for these mountains have existed as a great barrier, since a period so remote that whole races of animals must subsequently have perished from the face of the earth. Therefore, unless we suppose the same species to have been created in two different countries, we ought not to expect any closer similarity between the organic beings on opposite sides of the Andes, than on shores separated by a broad strait of the sea. In both cases we must leave out of the question those kinds which have been able to cross the barrier, whether of salt water or solid rock. *

A great number of the plants and animals were absolutely the same, or most closely allied with those of Patagonia. We here have the agouti, bizcacha, three species of armadillo, the ostrich, certain kinds of partridges, and other birds, none of which are

---

* This is merely an illustration of the admirable laws first laid down by Mr Lyell of the geographical distribution of animals as influenced by geological changes. The whole reasoning, of course, is founded on the assumption of the immutability of species. Otherwise the changes might be considered as superinduced by different circumstances in the two regions during a length of time.

ever seen in Chile, but are the characteristic animals of the desert plains of Patagonia. We have likewise many of the same (to the eyes of a person who is not a botanist) thorny stunted bushes, withered grass, and dwarf plants. Even the black slowly crawling beetles are closely similar, and some, I believe, on rigorous examination, absolutely identical. It had always been a subject of regret to me, that we were unavoidably compelled to give up the ascent of the St Cruz river before reaching the mountains. I always had a latent hope of meeting with some great change in the features of the country; but I now feel sure, that it would only have been following the plains of Patagonia up an ascent.

MARCH 24TH – Early in the morning I climbed up a mountain on one side of the valley, and enjoyed a far-extended view over the Pampas. This was a spectacle to which I had always looked forward with interest, but I was disappointed. At the first glance there was a strong resemblance to a distant view of the ocean, but in the northern parts many irregularities in the surface were soon distinguishable. The most striking feature in the scene consisted of the rivers, which, facing the rising sun, glittered like silver threads, till lost in the immensity of the distance.

In the middle of the day, we descended the valley, and reached a hovel, where an officer and three soldiers were posted to examine passports. One of these men was a thorough-bred Pampas Indian. He was kept much for the same purpose as a bloodhound, to track out any person who might pass by secretly, either on foot or horseback. Some years ago, a passenger had endeavoured to escape detection, by making a long circuit over a neighbouring mountain; but this Indian, having by chance crossed his track, followed it for the whole day, over dry and very stony parts, till at last he came on his prey hidden in a gully. We here heard that the silvery clouds, which we had admired from the bright region above, had poured down torrents of rain. The valley from this point gradually opened, and the hills became mere water-worn hillocks compared to the giants behind. It then expanded into a gently-sloping plain of shingle, covered with low trees and bushes. This talus, although it looked of little breadth, must be nearly ten miles wide, before it blends into the apparently dead level Pampas. We had already passed the only house in this neighbourhood, the Estancia of Chaquaio; and at sunset we pulled up in the first snug corner, and there bivouacked.

* * *

We crossed the Luxan, which is a river of considerable size, though its course towards the sea-coast is very imperfectly known. It is even doubtful whether, in passing over the plains, it is evaporated, or whether it forms a tributary of the Sauce or Colorado. We slept in the village, which is a small place surrounded by gardens, and forms the most southern part, that is cultivated, of the province of Mendoza; it is 5 leagues south of the capital. At night I experienced an attack (for it deserves no less a name) of the *Benchuca* (a species of Reduvius) the great black bug of the Pampas. It is most disgusting to feel soft wingless insects, about an inch long, crawling over one's body. Before sucking they are quite thin, but afterwards become round and bloated with blood, and in this state they are easily crushed. They are also found in the northern parts of Chile and in Peru. One which I caught at Iquique was very empty. When placed on the table, and though surrounded by people, if a finger was presented, the bold insect would immediately draw its sucker, make a charge, and if allowed, draw blood. No pain was caused by the wound. It was curious to watch its body during the act of sucking, as it changed in less than ten minutes, from being as flat as a wafer to a globular form. This one feast, for which the *benchuca* was indebted to one of the officers, kept it fat during four whole months; but, after the first fortnight, the insect was quite ready to have another suck.

* * *

MARCH 29TH — We set out on our return to Chile by the Uspallata pass to the northward of Mendoza. We had to cross a long and most sterile traversia of 15 leagues. The soil in parts was absolutely bare, in others covered by numberless dwarf cacti, armed with formidable spines, and called by the inhabitants 'little lions'. There were also a few low bushes. Although the plain is elevated about 3,000 feet above the sea, the sun was very powerful; this, and the clouds of impalpable dust, rendered the travelling extremely irksome. Our course during the day lay nearly parallel to the mountains, but gradually approaching them. Before sunset we entered one of the wide valleys, or rather bays, which open on the plain: this soon narrowed into a ravine, and a little higher up the house of the Villa Vicencio was situated. As we had ridden all day without a drop of water, both ourselves and our animals were very thirsty, and we looked out anxiously for the stream which flows down this valley. It was curious to observe how gradually the water made its appearance: on the plain the course was quite dry; by degrees it became a little damper; then puddles of water were formed; these soon became connected, and at Villa Vicencio there was a nice little rivulet.

30TH – The solitary hovel which bears the imposing name of Villa Vicencio, has been mentioned by every traveller who has crossed the Andes. I stayed here, and at some neighbouring mines, during the two succeeding days. The geology of the surrounding country is very curious. The Uspallata range is separated from the true Cordillera by a long narrow plain or basin, like those so often mentioned in Chile, but with an altitude of about 6,000 feet. The range consists of various kinds of submarine lava, alternating with volcanic sandstones and other remarkable sedimentary deposits; the whole having a very close resemblance to some of the newer horizontal beds on the shores of the Pacific. From this resemblance I expected to find silicified wood, which is generally characteristic of those formations. I was gratified in a very extraordinary manner. In the central part of the range, at an elevation probably of 7,000 feet, on a bare slope, I observed some snow-white projecting columns. These were petrified trees, eleven being silicified, and from thirty to forty converted into coarsely-crystallized white calcareous spar. They were abruptly broken off; the upright stumps projecting a few feet above the ground. The trunks measured from 3 to 5 feet each in circumference. They stood a little way apart from each other, but the whole formed one distinct group. Mr Robert Brown has been kind enough to examine the wood: he says it is coniferous, and that it partakes of the character of the Araucarian tribe (to which the common South Chilian pine belongs), but with some curious points of affinity with the yew. The volcanic sandstone in which they were embedded, and from the lower part of which they must have sprung, had accumulated in successive thin layers around their trunks; and the stone yet retained the impression of the bark.

It required little geological practice to interpret the marvellous story, which this scene at once unfolded; though I confess I was at first so much astonished that I could scarcely believe the plainest evidence of it. I saw the spot where a cluster of fine trees had once waved their branches on the shores of the Atlantic, when that ocean (now driven back 700 miles) approached the base of the Andes. I saw that they had sprung from a volcanic soil which had been raised above the level of the sea, and that this dry land, with its upright trees, had subsequently been let down to the depths of the ocean. There it was covered by sedimentary matter, and this again by enormous streams of submarine lava – one such mass alone attaining the thickness of 1,000 feet; and these deluges of melted stone and aqueous deposits had been five times spread out alternately. The ocean which received such masses must have been deep; but again the subterranean forces exerted their power, and I now beheld the bed of that sea forming a chain of mountains more than 7,000 feet in altitude. Nor had those antagonist forces been dormant, which are always at work to wear down the surface of the land to one level: the great piles of strata had been intersected by many wide valleys; and the trees now changed into silex were exposed projecting from the volcanic soil now changed into rock, whence formerly in a green and budding state they had raised their lofty heads. Now, all is utterly irreclaimable and desert; even the lichen cannot adhere to the stony casts of former trees. Vast, and scarcely comprehensible as such changes must ever appear, yet they have all occurred within a period recent when compared with the history of the Cordillera; and that Cordillera itself is modern as compared with some other of the fossiliferous strata of South America.

APRIL 1ST – We crossed the Uspallata range; and at night slept at the custom-house – the only inhabited spot on the plain. Shortly before leaving the mountains, there was a very extra-ordinary view: red, purple, green and quite white sedimentary rocks, alternating with black lavas, were broken up and thrown into all kinds of disorder, by masses of porphyry, of every shade, from dark brown to the brightest lilac. It was the first view I ever saw, which really resembled those pretty sections which geologists make of the inside of the earth.

* * *

I have certain proofs that this part of the continent of South America has been elevated, near the coast, at least from 400 to 500 feet, since the epoch of existing shells; and further inland the rise possibly may have been greater. As the peculiarly arid character of the climate is evidently a consequence of the height of the great range of mountains, we may feel almost sure, that prior to the latter elevations, the atmosphere was not so completely drained of its moisture as at the present day. At a remote geological era, it is probable that the Andes consisted of a chain of islands, which were covered by luxuriant forests; and many of the trees, in a silicified state, may now be seen embedded in the upper conglomerates. Of these I measured one which was cylindrical, with a circumference of 15 feet. As it is nearly certain that the mountains have risen slowly, so would the climate likewise become deteriorated slowly. We need not feel greatly surprised at walls of stone and hardened mud here lasting for many ages, when we remember how many centuries the Druidical mounds have withstood even the climate of England. The only question is, whether the amount of change, since the introduction of man into South America, has been sufficient to cause a sensible effect on the atmospheric moisture, and therefore on the fertility of the valleys in the upper Cordillera. From the extreme slowness with which there is reason to believe the continent is rising, the longevity of man as a species, required to allow of sufficient change, is the most valid objection to the above speculations: for on the eastern shores of this continent, we have seen that several animals, belonging to the same class of mammalia with man, have passed away, while the change of level between land and water, in that part at least, has been so small, that it can scarcely have caused any sensible difference in the climate. I may add, however, that at

Lima, the elevation, within the human epoch, certainly has amounted to between 70 and 80 feet.

* * *

8TH – We left the valley of the river of Aconcagua, by which we had descended, and reached in the evening a cottage near the Villa de St Rosa. The fertility of the plain was extremely delightful. The autumn being well advanced, the leaves of many of the fruit-trees were falling; and of the labourers – some were busy in drying figs and peaches on the roofs of their cottages; while others were gathering the grapes from the vineyards. It was a pretty scene; but that pensive stillness was absent, which makes the autumn in England indeed the evening of the year.

On the 10th we reached Santiago, where I experienced a very kind and hospitable reception from Mr Caldcleugh. My excursion only cost me twenty-four days, and never did I more deeply enjoy an equal space of time. A few days afterwards I returned to Mr Corfield's house at Valparaiso.

CHAPTER XVIII

Bell mountain – Miners – Great loads carried by the Apires –
Coquimbo – Earthquake – Geology – Terraces – Excursion up
valley – Road to Guasco – Desert country – Valley of Copiapó –
Rain and earthquakes, meteorolites – Hydrophobia – Copiapó
– Excursion to Cordillera – Dry valley – Cold gales of wind –
Noises from a hill – Iquique, complete desert – Salt alluvium –
Nitrate of soda – Lima – Unhealthy country – Ruins of Callao,
overthrown by earthquake – Elevated shells on island of San
Lorenzo – Plain with embedded fragments of pottery

NORTHERN CHILE AND PERU

APRIL 27TH – I set out on a journey to Coquimbo, and thence
through Guasco to Copiapó, where Captain FitzRoy kindly
offered to pick me up in the *Beagle*. The distance in a straight line
along the shore northward is only 420 miles; but my mode of
travelling caused me to find it a very long journey. I bought four
horses and two mules, the latter carrying the cargo on alternate
days. The six animals together only cost the value of £25 sterling,
and at Copiapó I sold again for £23. We travelled in the same
independent manner as before, cooking our own meals, and
sleeping in the open air. As we rode towards the Vino del Mar, I
took a farewell view of Valparaiso, and admired its picturesque
appearance. For geological purposes I made a detour from the
high road to the foot of the Bell mountain. We passed through a
highly auriferous district to the neighbourhood of Limache,
where we slept. The country is covered with much alluvium, and
by the side of each little rivulet it has been washed for gold. This
employment supports the inhabitants of numerous scattered
hovels; but, like all those who gain by chance, they are unthrifty
in their habits.

28TH—In the afternoon we arrived at a cottage at the foot of the Bell mountain. The inhabitants were freeholders, which is not very usual in Chile. They supported themselves on the produce of a garden and a little field, but were very poor. Capital is so deficient in this part, that the people are obliged to sell their green corn while it is standing in the field, in order to buy necessaries for the ensuing year. Wheat in consequence was dearer in the very district of its production, than at Valparaiso, where the contractors live. The next day we joined the main road to Coquimbo. At night there was a very light shower of rain: this was the first drop that had fallen since the heavy rain of September 11th and 12th, which detained me a prisoner at the baths of Cauquenes. The interval was seven and a half months; but the rain this year in Chile was rather later than usual. The Andes were now covered by a thick mass of snow; and they presented, in the distance, a very glorious sight.

* * *

We proceeded to Los Hornos, another mining district, where the principal hill was drilled with holes, like a great ants' nest. The Chilian miners are in their habits a peculiar race of men. Living for weeks together in the most desolate spots, when they descend to the villages on feast-days, there is no excess or extravagance into which they do not run. They sometimes gain a considerable sum, and then, like sailors with prize-money, they try how soon they can contrive to squander it. They drink excessively, buy quantities of clothes, and in a few days return penniless to their miserable abodes, there to work harder than beasts of burden. This thoughtlessness, as with sailors, is evidently the result of a similar manner of life. Their daily food is found them, and they acquire no habitual care as to the means of subsistence: moreover, at the same moment that temptation is offered, the means of enjoying it is placed in their power. On the other hand, in Cornwall, and some other parts of England, where the system of selling part of the vein is followed, the miners, from being obliged to act for themselves, and to judge with clearness, are a singularly intelligent and well-conducted set of men.

The dress of the Chilian miner is peculiar and rather

picturesque. He wears a very long shirt, of some dark-coloured baize, with a leathern apron; the whole being fastened round his waist by a brightly coloured sash. His trousers are very broad, and his small cap of scarlet cloth is made to fit the head closely. We met a party of these miners in full costume, carrying the body of one of their companions to be buried. They marched at a very quick trot, four men supporting the corpse. One set having run as hard as they could for about 200 yards, were relieved by four others, who had previously dashed on ahead on horseback. Thus they proceeded, encouraging each other by wild cries: altogether the scene formed a most strange funeral.

We continued travelling northward, in a zigzag line; sometimes stopping a day to geologize. The country was so thinly inhabited, and the track so obscure, that we often had difficulty in finding our way. On the 12th I stayed at some mines. The ore in this case was not considered particularly good, but from being abundant it was supposed the mine would sell for about $30,000 or $40,000 (that is £6,000 or £8,000 sterling); yet it was bought by one of the English Associations for an ounce of gold (3*l*. 8*s*.). The ore is yellow pyrites, which as I have already remarked, before the arrival of the English, was not supposed to contain a particle of copper. On a scale of profits, nearly as great as in the above instance, piles of scoriæ abounding with minute globules of metallic copper were purchased; yet with these advantages, the mining associations, as is well known, contrived to lose immense sums of money. The folly of the greater number of commissioners and shareholders, amounted to infatuation: £1,000 per annum given in some cases to entertain the authorities; libraries of well-bound geological books; bringing out miners for particular metals (as tin) which were soon found not to exist in the country; contracts to supply the miners with milk, in parts where there were no cows; machinery, where such could not possibly be used; and a hundred similar arrangements, bore witness to our absurdity, and to this day afford amusement to the natives. Yet there can be no doubt, that the same capital well employed in these mines would have yielded an immense return: a confidential man of business, a practical miner and assayer, would have been all that was required.

Captain Head has described the wonderful load which the 'Apires', truly beasts of burden, carry up from deep mines. I confess I thought the account exaggerated; so that I was glad to take the opportunity of weighing one of the loads, which I picked out by hazard. It required considerable exertion on my part, when standing directly over it, to lift it from the ground. The load was considered under weight when found to be 197 pounds. The apire had carried this up 80 perpendicular yards – part of the way by a steep passage, but the greater part up notched poles, placed in a zigzag line in the shaft. According to the general regulation, the apire is not allowed to halt for breath, except the mine is 600 feet deep. The average load is considered as rather more than 200 pounds, and I have been assured that one of 300 pounds (twenty-two stone and a half) by way of a trial has been brought up from the deepest mine! At this time the apires were bringing up the usual load twelve times in the day; that is, 2,400 pounds from 80 yards deep; and they were employed in the intervals in breaking and picking ore.

These men, excepting from accidents, are healthy, and appear cheerful. Their bodies are not very muscular. They rarely eat meat once a week, and never oftener, and then only the hard dry *charqui*. Although with a knowledge that the labour is voluntary, it was nevertheless quite revolting to see the state in which they reached the mouth of the mine; their bodies bent forward, leaning with their arms on the steps, their legs bowed, the muscles quivering, the perspiration streaming from their faces over their breasts, their nostrils distended, the corners of their mouth forcibly drawn back, and the expulsion of their breath most laborious. Each time, from habit, they utter an articulate cry of 'ay-ay', which ends in a sound rising from deep in the chest, but shrill like the note of a fife. After staggering to the pile of ores, they emptied the 'carpacho'; in two or three seconds recovering their breath, they wiped the sweat from their brows, and apparently quite fresh descended the mine again at a quick pace. This appears to me a wonderful instance of the amount of labour which habit (for it can be nothing else), will enable a man to endure.

In the evening, talking with the mayor-domo of these mines,

about the number of foreigners now scattered over the whole country, he told me that, though quite a young man, he remembers when a boy at school at Coquimbo, a holiday being given, to see the captain of an English ship, who was brought to the city to speak to the governor. He believes that nothing would have induced any boy in the school, himself included, to have gone close to the Englishman; so deeply had they been impressed with an idea of the heresy, contamination, and evil to be derived from contact with such a person. To this day they relate the atrocious actions of the bucaniers; and especially of one man, who took away the figure of the Virgin Mary, and returned the year after for that of St Joseph, saying it was a pity the lady should not have a husband. I heard also of an old lady who, at a dinner in Coquimbo, remarked how wonderfully strange it was that she should have lived to dine in the same room with an Englishman; for she remembered as a girl, that twice, at the mere cry of 'Los Ingleses', every soul, carrying what valuables they could, had taken to the mountains.

MAY 14TH – We reached Coquimbo, where we staid a few days. The town is remarkable for nothing but its extreme quietness. It is said to contain from 6,000 to 8,000 inhabitants. On the morning of the 17th it rained lightly (the first time this year) for about five hours. With this shower, the farmers, who plant corn near the sea-coast where the atmosphere is more humid, would break up the ground; with a second, put the seed in: and if a third should fall, they would reap in the spring a good harvest. It was interesting to watch the effect of this trifling amount of moisture. Twelve hours afterwards the ground appeared as dry as ever; yet after an interval of ten days, all the hills were faintly tinged with green patches; the grass being sparingly scattered in hair-like fibres a full inch in length. Before this shower every part of the surface was bare as on a high road.

In the evening, Captain FitzRoy and myself were dining with Mr Edwards, an English resident well known for his hospitality by all who have visited Coquimbo, when a sharp earthquake happened. I heard the forecoming rumble, but from the screams of the ladies, the running of servants, and the rush of several of the gentlemen to the doorway, I could not distinguish the motion.

Some of the women afterwards were crying with terror, and one person said he should not be able to sleep all night, or if he did, it would only be to dream of falling houses. The father of this gentleman had lately lost all his property at Talcuhano, and he himself only just escaped a falling roof at Valparaiso, in 1822. He mentioned a curious coincidence which then happened: he was playing at cards, when a German, one of the party, got up, and said he would never sit in a room in these countries with the door shut, since, owing to his having done so, he had nearly lost his life at Copiapó. Accordingly he opened the door; and no sooner had he done this, than he cried out, 'Here it comes again!' and the famous shock commenced. The whole party escaped. The danger in an earthquake is not from the time lost in opening a door, but from the chance of its becoming jammed by the movement of the walls.

It is impossible to be much surprised at the fear which natives and old residents, though some of them known to be men of great command of mind, so generally experience during earthquakes. I think, however, this excess of panic may be partly attributed to a want of habit in governing their fears; the usual restraint of shame being here absent. Indeed, the natives do not like to see a person indifferent. I heard of two Englishmen who, sleeping in the open air, during a smart shock, knowing there was no danger, did not rise. The natives cried out indignantly, 'Look at those heretics, they will not even get out of their beds!'

I spent two or three days in examining the step-formed terraces of shingle first described by Captain Basil Hall, in his work, so full of spirited descriptions, on the west coast of America. Mr Lyell concluded from the account, that they must have been formed by the sea during the gradual rising of the land. Such is the case: on some of the steps which sweep round from within the valley, so as to front the coast, shells of existing species both lie on the surface, and are embedded in a soft calcareous stone. This bed of the most modern tertiary epoch passes downward into another, containing some living species associated with others now lost. Amongst the latter may be mentioned shells of an enormous perna and an oyster, and the teeth of a gigantic shark, closely allied to, or identical with the *Carcharias Megalodon* of ancient Europe; the bones of which, or of some cetaceous animal, are also present, in a silicified state, in great numbers. At Guasco, the phenomenon of the parallel terraces is very strikingly seen: no less than seven perfectly level, but unequally broad plains, ascending by steps, occur on one or both sides the valley. So remarkable is the contrast of the successive horizontal lines, corresponding on each side, with the irregular outline of the surrounding mountains, that it attracts the attention of even those who feel no interest regarding the causes, which have modelled the surface of the land. The origin of the terraces of Coquimbo is precisely the same, according to my view, with that of the plains of Patagonia; the only difference is that the plains are rather broader than the terraces, and that they front the Atlantic ocean instead of a valley – which valley, however, was formerly occupied by an arm of the sea, but now by a fresh-water river. In every case it must be remembered, that the successive cliffs do not mark so many distinct elevations, but on the contrary, periods of comparative repose during the gradual and perhaps scarcely sensible rise of the land. In the valley of Guasco we have the record of seven such nights of rest, in the action of the subterranean powers.

MAY 21ST – I set out in company with Don Jose Edwards to the silver-mine of Arqueros, and thence up the valley of Elque or Coquimbo. Passing through a mountainous country, we reached by nightfall the mines belonging to Mr Edwards. I enjoyed my night's rest here from a cause which will not be fully understood in England, namely, the absence of fleas! The rooms in Coquimbo swarm with them; but they will not live at the elevation of 3,000 or 4,000 feet, even if brought there, as is constantly occurring at these mines. It can scarcely be the trifling diminution of temperature, but some other cause which is here destructive to these troublesome insects.

* * *

28TH – We continued gradually ascending as we followed the valley, which now had assumed the character of a ravine. During the day we saw several guanacoes, and the track of the closely allied species, called the Vicuna. This latter animal is pre-eminently alpine in its habits; it seldom descends much below the limit of perpetual snow, and therefore haunts even a more lofty and sterile situation than the guanaco. The only other animal which we saw in any number was a small fox. I suppose this animal preys on the mice and other small rodents, which, as long as there is the least vegetation subsist in considerable numbers in very desert places. In Patagonia, even on the borders of the salinas, where a drop of fresh water can never be found, these little animals swarm. Next to lizards, mice appear to be able to support existence on the smallest and driest portions of the earth – even on islets in the midst of great oceans. I believe it will be found, that several islands, which possess no other warmblooded quadruped, have small rodents peculiar to themselves.

The scenery on all sides showed desolation, brightened, and made palpable, by a clear, unclouded sky. Custom excludes the feeling of sublimity, and when this is wanting, such scenery is rather the reverse of interesting. We bivouacked at the 'primera linea', or the first line of the partition of the waters. The streams, however, on the east side do not flow to the Atlantic, but into an elevated district, in the middle of which there is a large salina, or salt lake; thus forming a little Caspian sea at the elevation, perhaps, of 10,000 feet. Where we slept, there were some considerable patches of snow, but they do not remain throughout the year. The winds in these lofty regions obey very regular laws: everyday a fresh breeze blows up the valley, and at night, an hour or two after sunset, the air from the cold regions above descends, as through a funnel. This night it blew a gale of wind, and the temperature must have been considerably below the freezing-point, for water in a short time became a block of ice. No clothes seemed to oppose any obstacle to the air; I suffered very much from the cold, so that I could not sleep, and in the morning rose with my body quite dull and benumbed.

In the Cordillera further southward, people lose their lives from snow-storms; here, this sometimes happens from another cause. My guide, when a boy of fourteen years old, was passing the Cordillera with some others, in the month of May; and while in the central parts, a furious gale of wind arose so that the men could hardly stick on their mules, and stones were flying along the ground. The day was cloudless and not a speck of snow fell, but the temperature was low. It is probable that the thermometer would not have stood very many degrees below the freezing-point, but the effect on their bodies, ill protected by clothing, would be in proportion to the rapidity of the current of cold air. The gale lasted for more than a day; the men began to lose all their strength, and the mules would not move onwards. My guide's brother tried to return, but he perished, and his body was found two years afterwards, lying by the side of his mule near the road, with the bridle still in his hand. Two other men in the party lost their fingers and toes, and out of 200 mules and thirty cows, only fourteen of the former escaped alive. Many years ago the whole of a large party are supposed to have perished from a similar cause, but their bodies to this day have never been discovered. The union of a cloudless sky, low temperature, and a furious gale of wind, must be I should think, in all parts of the world, an unusual occurrence.

JUNE 29TH – We gladly travelled down the valley to our former night's lodging, and thence to near the Agua amarga. On July 1st, we reached the valley of Copiapó. The smell of the fresh clover was quite delightful, after the scentless air of the dry sterile Despoblado.

* * *

Lima stands on a plain in a valley, formed during the gradual retreat of the sea. It is distant 7 miles from Callao, and is elevated 500 feet above it; but from the slope being very gradual, the road appears absolutely level; so that when at Lima it is difficult to believe one has ascended some 100 feet. Humboldt has remarked on this singularly deceptive case. Steep, barren hills rise like islands from the plain, which is divided, by straight mud-walls, into large green fields. In these scarcely a tree grows excepting a few willows; and the presence of an occasional clump of bananas and of oranges, alone reminded one that the landscape of a country in latitude 12° might have boasted of a far more splendid vegetation. The city of Lima is now in a wretched state of decay: the streets are nearly unpaved, and heaps of filth are piled up in all directions; where the black gallinazos, tame as poultry, pick up bits of carrion. The houses have generally an upper story, built, on account of the earthquakes, of plastered woodwork; but some of the old ones, which are now used by several families, are immensely large, and would rival in suites of apartments the most magnificent in any place. Lima, the City of the Kings, must formerly have been a splendid town. The extraordinary number of churches, even at the present day, gives it a peculiar and striking character, especially when viewed from a short distance.

On day I went out with some merchants to hunt in the immediate vicinity of the city. Our sport was very poor; but I had an opportunity of seeing the ruins of one of the ancient Indian villages, with its hill-like mound in the centre. The remains of houses, enclosures, irrigating streams, and burial mounds, scattered over this plain, cannot fail to give one a high idea of the condition and number of the ancient population. When their earthenware, woollen clothes, utensils of elegant forms cut out of the hardest rocks, tools of copper, ornaments of precious stones, palaces and hydraulic works, are considered, it is impossible not to respect the considerable advance made by them in the arts of civilization. The burial mounds, called Huacas, are really stupendous; although in some places it is only a natural hill which appears to have been incased and modelled.

There is also another and very different class of ruins, which possesses some interest, namely, those of old Callao, overwhelmed by the great earthquake of 1746, and its accompanying wave. The destruction must have been more complete even than at Concepción. Quantities of shingle almost conceal the foundations of the walls, and vast masses of brickwork appear to have been whirled about by the retiring waves like pebbles. It has been stated that the land subsided during this memorable shock: I could not discover any proof of this; yet it seems far from improbable, for the form of the coast must certainly have undergone some change since the foundation of the old town; as no people in their senses would willingly have chosen for their building place the narrow spit of shingle on which the ruins now stand. On the island of San Lorenzo, there are very satisfactory proofs of elevation within the recent period: this of course would not contravene the belief of a small subsidence, if any signs of such movement could be discovered. The side of the mountain fronting the bay on that island, is worn into three obscure terraces, which are covered by masses of shells many hundred tons in weight, of species now existing on the beach. Several of the univalves had serpulæ and small balani attached on their insides; proving that they must have remained some time, after the animal had died, at the bottom of the sea. In such cases we may feel sure that they had not been carried up, as has sometimes been believed, either by birds or men for food.

When examining the beds of shells, which have been raised above the level of the sea, on other parts of the coast, I often felt curious to trace their final disappearance from decay. On the island of San Lorenzo, this could be done in the most satisfactory manner: at a small height the shells were quite perfect; on a terrace, 85 feet above the sea, they were partially decomposed and coated by a soft scaly substance; at double this altitude a thin layer of calcareous powder beneath the soil, without a trace of organic structure, was all that could be discovered. This highly curious and satisfactory gradation of change, it is evident could be traced only under the peculiar conditions of this climate, where rain never falls so as to wash away the particles of shells in their last stage of decomposition. I was much interested by finding embedded, together with pieces of sea-weed in the mass of shells, in the 85 foot bed, a bit of *cotton-thread*, plaited rush, and the head of a stalk of Indian corn. This fact, coupled with another, which will be mentioned, proves I think the amount of 85-feet elevation since man inhabited this part of Peru. On the coast of Patagonia and La Plata, where perhaps the movements have been slower, there is evidence, as we have seen, that several mammalia have become extinct during a smaller change of level. At Valparaiso, where there exist abundant proofs of recent elevation to a greater altitude than in this part of Peru, I can show that the greatest possible change during the last 220 years, has not exceeded the small measure of 15 feet.

On the mainland in front of San Lorenzo, near Bellavista, there is an extensive and level plain, at the height of about 100 feet. The section on the coast shows that the lower part consists of alternating layers of sand and impure clay, together with some gravel; and the surface, to the depth of from 3 to 6 feet, of a reddish loam, containing a few scattered sea-shells, and numerous small fragments of coarse red earthenware. At first I was inclined to believe that this superficial bed must have been deposited beneath the sea; but I afterwards found in one spot, that it covered an artificial floor of round stones. The conclusion which then seemed most probable was, that at a period when the land stood at a less height, there was a plain very similar to the one now surrounding Callao, which being protected by a shingle beach, is raised but very little above the level of the sea. On this plain, with its clay beds, I imagine the Indians manufactured their earthen vessels; and that, during some violent earthquake, the sea broke over the beach and converted the plain into a temporary lake, as happened in 1713* around Callao. The water would then deposit mud containing fragments of pottery from the kilns, and shells from the sea. This bed with fossil earthenware occurring at about the same altitude with the terrace on San Lorenzo, confirms the supposed amount of elevation within the human period.

* Frezier's Voyage.

CHAPTER XIX

Islands volcanic – Number of craters – Leafless bushes – Colony at Charles Island – James Island – Salt-lake in crater – Character of vegetation – Ornithology, curious finches – Great tortoises, habits of, paths to the wells – Marine lizard feeds on sea-weed – Terrestrial species, burrowing habits, herbivorous – Importance of reptiles in the Archipelago – Few and minute insects – American type of organization – Species confined to certain islands – Tameness of birds – Falkland Islands – Fear of man an acquired instinct

GALAPAGOS ARCHIPELAGO

SEPTEMBER 15TH – The *Beagle* arrived at the southernmost of the Galapagos Islands. This archipelago consists of ten principal islands, of which five much exceed the others in size. They are situated under the equatorial line, and between 500 and 600 miles to the westward of the coast of America. The constitution of the whole is volcanic. With the exception of some ejected fragments of granite, which have been most curiously glazed and altered by the heat, every part consists of lava, or of sandstone resulting from the attrition of such materials. The higher islands, (which attain an elevation of 3,000, and even 4,000 feet) generally have one or more principal craters towards their centre, and on their flanks smaller orifices. I have no exact data from which to calculate, but I do not hesitate to affirm, that there must be, in all the islands of the archipelago, at least 2,000 craters. These are of two kinds; one, as in ordinary cases, consisting of scoriæ and lava, the other of finely stratified volcanic sandstone. The latter in most instances have a form beautifully symmetrical: their origin is due to the ejection of mud – that is, fine volcanic ashes and water – without any lava.

Considering that these islands are placed directly under the equator, the climate is far from being excessively hot; a circumstance which, perhaps, is chiefly owing to the singularly low temperature of the surrounding sea. Excepting during one short season, very little rain falls, and even then it is not regular: but the clouds generally hang low. From these circumstances the lower parts of the islands are extremely arid, whilst the summits, at an elevation of 1,000 feet or more, possess a tolerably luxuriant vegetation. This is especially the case on the windward side, which first receives and condenses the moisture from the atmosphere.

In the morning (17th) we landed on Chatham Island, which, like the others, rises with a tame and rounded outline, interrupted only here and there by scattered hillocks – the remains of former craters. Nothing could be less inviting than the first appearance. A broken field of black basaltic lava is every where covered by a stunted brushwood, which shows little signs of life. The dry and parched surface, having been heated by the noonday sun, gave the air a close and sultry feeling, like that from a stove: we fancied even the bushes smelt unpleasantly. Although I diligently tried to collect as many plants as possible, I succeeded in getting only ten kinds; and such wretched-looking little weeds would have better become an arctic, than an equatorial Flora.

The thin woods, which cover the lower parts of all the islands, excepting where the lava has recently flowed, appear from a short distance quite leafless, like the deciduous trees of the northern hemisphere in winter. It was some time before I discovered, that not only almost every plant was in full leaf, but that the greater number were now in flower. After the period of heavy rains, the islands are said to appear for a short time partially green. The only other country, in which I have seen a vegetation with a character at all approaching to this, is at the volcanic island of Fernando Noronha, placed in many respects under similar conditions.

The natural history of this archipelago is very remarkable: it seems to be a little world within itself; the greater number of its inhabitants, both vegetable and animal, being found nowhere else. As I shall refer to this subject again, I will only here remark, as forming a striking character on first landing, that the birds are strangers to man. So tame and unsuspecting were they, that they did not even understand what was meant by stones being thrown at them; and quite regardless of us, they approached so close that any number might have been killed with a stick.

The *Beagle* sailed round Chatham Island, and anchored in several bays. One night I slept on shore, on a part of the island where some black cones – the former chimneys of the subterranean heated fluids – were extraordinarily numerous. From one small eminence, I counted sixty of these truncated hillocks, which were all surmounted by a more or less perfect crater. The greater number consisted merely of a ring of red scoriæ, or slags, cemented together: and their height above the plain of lava, was not more than from 50 to 100 feet. From their regular form, they gave the country a *workshop* appearance, which strongly reminded me of those parts of Staffordshire where the great iron-foundries are most numerous.

The age of the various beds of lava was distinctly marked by the comparative growth, or entire absence, of vegetation. Nothing can be imagined more rough and horrid than the surface of the more modern streams. These have been aptly compared to the sea petrified in its most boisterous moments: no sea, however, would present such irregular undulations, or would be traversed by such deep chasms. All the craters are in an extinct condition; and although the age of the different streams of lava could be so clearly distinguished, it is probable they have remained so for many centuries. There is no account in any of the old voyagers of any volcano on this island having been seen in activity; yet since the time of Dampier (1684), there must have been some increase in the quantity of vegetation, otherwise so accurate a person would not have expressed himself thus: 'Four or five of the easternmost islands are rocky, barren, and hilly, producing neither tree, herb, nor grass, but a few dildoe (cactus) trees, except by the sea-side.'* This description is at present applicable only to the western islands, where the volcanic forces are in frequent activity.

The day, on which I visited the little craters, was glowing hot, and the scrambling over the rough surface, and through the

* Dampier's Voyage, vol. i, p. 101.

intricate thickets, was very fatiguing; but I was well repaid by the Cyclopian scene. In my walk I met two large tortoises, each of which must have weighed at least 200 pounds. One was eating a piece of cactus, and when I approached, it looked at me, and then quietly walked away: the other gave a deep hiss and drew in its head. These huge reptiles, surrounded by the black lava, the leafless shrubs, and large cacti, appeared to my fancy like some antediluvian animals.

SEPTEMBER 23RD – The *Beagle* proceeded to Charles Island. This archipelago has long been frequented, first by the Bucaniers, and latterly by whalers, but it is only within the last six years, that a small colony has been established on it. The inhabitants are between 200 and 300 in number: they nearly all consist of people of colour, who have been banished for political crimes from the Republic of the Equator (Quito is the capital of this state) to which these islands belong. The settlement is placed about four and a half miles inland, and at an elevation probably of 1,000 feet. In the first part of the road we passed through leafless thickets, as in Chatham Island. Higher up, the wood gradually became greener; and immediately we had crossed the ridge of the island, our bodies were cooled by the fine southerly trade wind, and our senses refreshed by the sight of a green and thriving vegetation. The houses are irregularly scattered over a flat space of ground, which is cultivated with sweet potatoes and bananas. It will not easily be imagined how pleasant the sight of black mud was to us, after having been so long accustomed to the parched soil of Peru and Chile.

The inhabitants, although complaining of poverty, gain, without much trouble, the means of subsistence from the fertile soil. In the woods there are many wild pigs and goats, but the main article of animal food is derived from the tortoise. Their numbers in this island have of course been greatly reduced, but the people yet reckon on two days' hunting supplying food for the rest of the week. It is said that formerly single vessels have taken away as many as 700 of these animals, and that the ship's company of a frigate some years since brought down 200 to the beach in one day.

We stayed at this island four days, during which time I collected many plants and birds. One morning I ascended the highest hill, which has an altitude of nearly 1,800 feet. The summit consists of a broken-down crater, thickly clothed with coarse grass and brushwood. Even in this one island, I counted thirty-nine hills, each of which was terminated by a more or less perfect circular depression.

SEPTEMBER 29TH – We doubled the south-west extremity of Albemarle Island, and the next day were nearly becalmed between it and Narborough Island. Both are covered with immense streams of black naked lava; which, having either flowed over the rims of the great caldrons, or having burst forth from the smaller orifices on the flanks, have in their descent spread over miles of the sea-coast. On both of these islands eruptions are known occasionally to take place; and in Albemarle we saw a small jet of smoke curling from the summit of one of the more lofty craters. In the evening we anchored in Bank's Cove, in Albemarle Island.

When morning came, we found that the harbour in which we were at anchor was formed by a broken-down crater, composed of volcanic sandstone. After breakfast I went out walking. To the southward of this first crater, there was another of similar composition, and beautifully symmetrical. It was elliptic in form; the longer axis being less than a mile, and its depth about 500 feet. The bottom was occupied by a shallow lake, and in its centre a tiny crater formed an islet. The day was overpoweringly hot, and the lake looked clear and blue. I hurried down the cindery slope, and choked with dust eagerly tasted the water – but to my sorrow I found it salt as brine.

The rocks on the coast abounded with great black lizards, between 3 and 4 feet long; and on the hills, another species was equally common. We saw several of the latter, some clumsily running out of our way, and others shuffling into their burrows. I shall presently describe in more detail the habits of both these reptiles.

OCTOBER 3RD – We sailed round the northern end of Albemarle Island. Nearly the whole of this side is covered with recent streams of dark-coloured lavas, and is studded with craters. I should think it would be difficult to find in any other part of the world, an island situated within the tropics, and of such considerable size (namely 75 miles long), so sterile and incapable of supporting life.

On the 8th we reached James Island.* Captain FitzRoy put Mr Bynoe, myself, and three others on shore, leaving with us a tent and provisions, to wait there till the vessel returned from watering. This was an admirable plan for the collections, as we had an entire week of hard work. We found here a party of Spaniards, who had been sent from Charles Island to dry fish, and to salt tortoise-meat.

At the distance of about 6 miles, and at the height of nearly 2,000 feet, the Spaniards had erected a hovel in which two men lived, who were employed in catching tortoises, whilst the others were fishing on the coast. I paid this party two visits, and slept there one night. In the same manner as in the other islands, the lower region is covered by nearly leafless bushes: but here many of them grow to the size of trees. I measured several which were 2 feet in diameter, and some even 2 feet 9 inches. The upper region being kept damp, from the moisture of the condensed clouds, supports a green and flourishing vegetation. So damp was the ground, that there were large beds of a coarse carex, in which great numbers of a very small water-rail lived and bred. While staying in this upper region, we lived entirely upon tortoise-meat. The breastplate roasted (as the Gauchos do *carne con cuero*), with the flesh attached to it, is very good; and the young tortoises make excellent soup; but otherwise the meat to my taste is very indifferent.

During another day we accompanied a party of the Spaniards in their whale-boat to a salina, or lake from which salt is procured. After landing, we had a very rough walk over a rugged field of recent lava, which has almost surrounded a sandstone crater, at the bottom of which the salt-lake is situated. The water was only three or four inches deep, and rested on a layer of beautifully crystallized white salt. The lake was quite circular, and fringed with a border of brightly green succulent plants: the precipitous walls of the crater were also clothed with wood, so that the scene

---

* Both Charles and James Islands take their names from the Stuarts. See Cowley's Voyage in 1684.

was both picturesque and curious. A few years since, the sailors belonging to a sealing-vessel murdered their captain in this quiet spot; and we saw his skull lying among the bushes.

During the greater part of our week on shore, the sky was cloudless, and if the trade wind failed for an hour, the heat became very oppressive. On two days, the thermometer within the tent stood for some hours at 93°; but in the open air, in the wind and sun, at only 85°. The sand was extremely hot; the thermometer placed in some of a brown colour immediately rose to 137°, and how much higher it would have risen, I do not know, for it was not graduated above that number. The *black* sand felt much hotter, so that even in thick boots it was disagreeable, on this account, to walk over it.

I will now offer a few general observations on the natural history of these islands. I endeavoured to make as nearly a perfect collection in every branch as time permitted. The plants have not yet been examined, but Professor Henslow, who has kindly undertaken the description of them, informs me that there are probably many new species, and perhaps even some new genera. They all have an extremely weedy character, and it would scarcely have been supposed, that they had grown at an inconsiderable elevation directly under the equator. In the lower and sterile parts, the bush, which from its minute brown leaves chiefly gives the leafless appearance to the brushwood, is one of the Euphorbiaceæ. In the same region an acacia and a cactus (*Opuntia Galapageia**), with large oval compressed articulations. springing from a cylindrical stem, are in some parts common. These are the only trees which in that part afford any shade. Near the summits of the different islands, the vegetation has a very different character; ferns and coarse grasses are abundant; and the commonest tree is one of the Compositæ. Tree-ferns are not present. One of the most singular characters of the Flora, considering the position of this archipelago. is the absence of every member of the palm family Cocos Island, on the other hand, which is the nearest point of land, takes its name from the great number of cocoa-nut trees on it. From the presence of the Opuntias and some other

* Magazine of Zoology and Botany, vol. i, p. 466.

plants, the vegetation partakes more of the character of that of America than of any other country.

Of mammalia a large kind of mouse forms a well-marked species. From its large thin ears, and other characters, it approaches in form a section of the genus, which is confined to the sterile regions of South America. There is also a rat which Mr Waterhouse believes is probably distinct from the English kind; but I cannot help suspecting that it is only the same altered by the peculiar conditions of its new country.

In my collections from these islands, Mr Gould considers that there are twenty-six different species of land birds. With the exception of one, all probably are undescribed kinds, which inhabit this archipelago, and no other part of the world. Among the waders and waterfowl it is more difficult, without detailed comparison, to say what are new. But a water-sail which lives near the summits of the mountains, is undescribed, as perhaps is a Totanus and a heron. The only kind of gull which is found among these islands, is also new; when the wandering habits of this genus are considered, this is a very remarkable circumstance. The species most closely allied to it, comes from the Strait of Magellan. Of the other aquatic birds, the species appear the same with well-known American birds.

The general character of the plumage of these birds is extremely plain, and like the Flora possesses little beauty. Although the species are thus peculiar to the archipelago, yet nearly all in their general structure, habits, colour of feathers, and even tone of voice, are strictly American. The following brief list will give an idea of their kinds. First: A buzzard, having many of the characters of Polyborus or Caracara; and in its habits not to be distinguished from that peculiar South American genus; second: Two owls; third: Three species of tyrant-flycatchers – a form strictly American. One of these appears identical with a common kind (*Muscicapa coronata?* Lath.), which has a very wide range, from La Plata throughout Brazil to Mexico; fourth: A sylvicola, an American form, and especially common in the northern division of the continent; fifth: Three species of mocking-birds, a genus common to both Americas; sixth: A finch, with a stiff tail and a long claw to its hinder toe, closely allied to a North American genus;

seventh: A swallow belonging to the American division of that genus; eighth: A dove, like, but distinct from, the Chilian species; ninth: A group of finches, of which Mr Gould considers there are thirteen species; and these he has distributed into four new sub-genera. These birds are the most singular of any in the archipelago. They all agree in many points; namely, in a peculiar structure of their bill, short tails, general form, and in their plumage. The females are gray or brown, but the old cocks jet-black. All the species, excepting two, feed in flocks on the ground, and have very similar habits. It is very remarkable that a nearly perfect gradation of structure in this one group can be traced in the form of the beak, from one exceeding in dimensions that of the largest gros-beak, to another differing but little from that of a warbler. Of the aquatic birds I have already remarked that some are peculiar to these islands, and some common to North and South America.

We will now turn to the order of reptiles, which forms, perhaps, the most striking feature in the zoology of these islands. The species are not numerous, but the number of individuals of each kind, is extraordinarily great. There is one kind both of the turtle and tortoise; of lizards four; and of snakes about the same number.

I will first describe the habits of the tortoise (*Testudo Indicus*) which has been so frequently alluded to. These animals are found, I believe, in all the islands of the Archipelago; certainly in the greater number. They frequent in preference the high damp parts, but likewise inhabit the lower and arid districts. I have already mentioned* proofs, from the numbers which have been taken in a single day, how very numerous they must be. Some individuals grow to an immense size: Mr Lawson, an Englishman, who had at the time of our visit charge of the colony, told us that he had seen several so large, that it required six or eight men to lift them from the ground; and that some had afforded as much as 200 pounds of meat. The old males are the largest, the females rarely

---

* Dampier says, 'The land-turtles are here so numerous, that five or six hundred men might subsist on them for several months without any other sort of provisions. They are so extraordinarily large and fat, and so sweet, that no pullet eats more pleasantly.' – Vol. i, p. 110.

growing to so great a size. The male can readily be distinguished from the female by the greater length of its tail. The tortoises which live on those islands where there is no water, or in the lower and arid parts of the others, chiefly feed on the succulent cactus. Those which frequent the higher and damp regions, eat the leaves of various trees, a kind of berry (called guayavita) which is acid and austere, and likewise a pale green filamentous lichen, that hangs in tresses from the boughs of the trees.

The tortoise is very fond of water, drinking large quantities, and wallowing in the mud. The larger islands alone possess springs, and these are always situated towards the central parts, and at a considerable elevation. The tortoises, therefore, which frequent the lower districts, when thirsty, are obliged to travel from a long distance. Hence broad and well-beaten paths radiate off in every direction from the wells even down to the sea-coast; and the Spaniards by following them up, first discovered the watering-places. When I landed at Chatham Island, I could not imagine what animal travelled so methodically along the well-chosen tracks. Near the springs it was a curious spectacle to behold many of these great monsters; one set eagerly travelling onwards with outstretched necks, and another set returning, after having drunk their fill. When the tortoise arrives at the spring, quite regardless of any spectator, it buries its head in the water above its eyes, and greedily swallows great mouthfuls, at the rate of about ten in a minute. The inhabitants say each animal stays three or four days in the neighbourhood of the water, and then returns to the lower country; but they differed in their accounts respecting the frequency of these visits. The animal probably regulates them according to the nature of the food which it has consumed. It is, however, certain, that tortoises can subsist even on those islands where there is no other water, than what falls during a few rainy days in the year.

I believe it is well ascertained, that the bladder of the frog acts as a reservoir for the moisture necessary to its existence: such seems to be the case with the tortoise. For some time after a visit to the springs, the urinary bladder of these animals is distended with fluid, which is said gradually to decrease in volume, and to become less pure. The inhabitants, when walking in the lower district, and overcome with thirst, often take advantage of this circumstance, by killing a tortoise, and if the bladder is full, drinking its contents. In one I saw killed, the fluid was quite limpid, and had only a *very slightly* bitter taste. The inhabitants, however, always drink first the water in the pericardium, which is described as being best.

The tortoises, when moving towards any definite point, travel by night and day, and arrive at their journey's end much sooner than would be expected. The inhabitants, from observations on marked individuals, consider that they can move a distance of about 8 miles in two or three days. One large tortoise, which I watched, I found walked at the rate of 60 yards in ten minutes, that is 360 in the hour, or 4 miles a day; allowing also a little time for it to eat on the road.

During the breeding season, when the male and female are together, the male utters a hoarse roar or bellowing, which it is said, can be heard at the distance of more than 100 yards. The female never uses her voice, and the male only at such times; so that when the people hear this noise, they know the two are together. They were at this time (October) laying their eggs. The female, where the soil is sandy, deposits them together, and covers them up with sand; but where the ground is rocky she drops them indiscriminately in any hollow. Mr Bynoe found seven placed in a line in a fissure. The egg is white and spherical; one which I measured was seven inches and three-eighths in circumference. The young animals, as soon as they are hatched, fall a prey in great numbers to the buzzard, with the habits of the Caracara. The old ones seem generally to die from accidents, as from falling down precipices. At least several of the inhabitants told me, they had never found one dead without some such apparent cause.

The inhabitants believe that these animals are absolutely deaf; certainly they do not overhear a person walking close behind them. I was always amused, when overtaking one of these great monsters as it was quietly pacing along, to see how suddenly, the instant I passed, it would draw in its head and legs, and uttering a deep hiss fall to the ground with a heavy sound, as if struck dead. I frequently got on their backs, and then, upon giving a few raps on the hinder part of the shell, they would rise up and walk away; but I found it very difficult to keep my balance.

The flesh of this animal is largely employed, both fresh and salted; and a beautifully clear oil is prepared from the fat. When a tortoise is caught, the man makes a slit in the skin near its tail, so as to see inside its body, whether the fat under the dorsal plate is thick. If it is not, the animal is liberated; and it is said to recover soon from this strange operation. In order to secure the tortoises, it is not sufficient to turn them like turtle, for they are often able to regain their upright position.

It was confidently asserted, that the tortoises coming from different islands in the archipelago were slightly different in form; and that in certain islands they attained a larger average size than in others. Mr Lawson maintained that he could at once tell from which island any one was brought. Unfortunately, the specimens which came home in the *Beagle* were too small to institute any certain comparison. This tortoise, which goes by the name of *Testudo Indicus*, is at present found in many parts of the world. It is the opinion of Mr Bell, and some others who have studied reptiles, that it is not improbable that they all originally came from this archipelago. When it is known how long these islands have been frequented by the bucaniers, and that they constantly took away numbers of these animals alive, it seems very probable that they should have distributed them in different parts of the world. If this tortoise does not originally come from these islands, it is a remarkable anomaly; inasmuch as nearly all the other land inhabitants seem to have had their birthplace here.

Of lizards there are four or five species; two probably belong to the South American genus Leiocephalus, and two to Amblyrhyncus. This remarkable genus was characterized by Mr Bell,[*] from a stuffed specimen sent from Mexico, but which I conceive there can be little doubt originally came through some whaling ship from these islands. The two species agree pretty closely in general appearance; but one is aquatic and the other terrestrial in its habits. Mr Bell thus concludes his description of *Amb. cristatus*: 'On a comparison of this animal with the true Iguanas, the most striking

[*] Zoological Journal, July, 1835.

and important discrepancy is in the form of the head. Instead of the long, pointed, narrow muzzle of those species, we have here a short, obtusely truncated head, not so long as it is broad, the mouth consequently only capable of being opened to a very short space. These circumstances, with the shortness and equality of the toes, and the strength and curvature of the claws, evidently indicate some striking peculiarity in its food and general habits, on which, however, in the absence of all certain information, I shall abstain from offering any conjecture.' The following account of these two lizards, will, I think, show with what judgment Mr Bell foresaw a variation in habit, accompanying change in structure.

First for the aquatic kind (*Amb. cristatus*). This lizard is extremely common on all the islands throughout the Archipelago. It lives exclusively on the rocky sea-beaches, and is never found, at least I never saw one, even 10 yards inshore. It is a hideous-looking creature, of a dirty black colour, stupid and sluggish in its movements. The usual length of a full-grown one is about a yard, but there are some even 4 feet long: I have seen a large one which weighed 20 pounds. On the island of Albemarle they seem to grow to a greater size than on any other. These lizards were occasionally seen some 100 yards from the shore swimming about; and Captain Collnett, in his *Voyage*, says, 'they go out to sea in shoals to fish'. With respect to the object, I believe he is mistaken; but the fact stated on such good authority cannot be doubted. When in the water the animal swims with perfect ease and quickness, by a serpentine movement of its body and flattened tail – the legs, during this time, being motionless and closely collapsed on its sides. A seaman on board sank one, with a heavy weight attached to it, thinking thus to kill it directly; but when an hour afterwards he drew up the line, the lizard was quite active. Their limbs and strong claws are admirably adapted for crawling over the rugged and fissured masses of lava, which every where form the coast. In such situations, a group of six or seven of these hideous reptiles may oftentimes be seen on the black rocks, a few feet above the surf, basking in the sun with outstretched legs.

I opened the stomach of several, and in each case found it largely distended with minced sea-weed, of that kind which grows in thin foliaceous expansions of a bright green or dull red colour. I do not recollect having observed this sea-weed in any quantity on the tidal rocks; and I have reason to believe it grows at the bottom of the sea, at some little distance from the coast. If such is the case, the object of these animals occasionally going out to sea is explained. The stomach contained nothing but the seaweed. Mr Bynoe, however, found a piece of a crab in one; but this might have got in accidentally, in the same manner as I have seen a caterpillar, in the midst of some lichen, in the paunch of a tortoise. The intestines were large, as in other herbivorous animals.

The nature of this lizard's food, as well as the structure of its tail, and the certain fact of its having been seen voluntarily swimming out at sea, absolutely prove its aquatic habits; yet there is in this respect one strange anomaly; namely, that when frightened it will not enter the water. From this cause, it is easy to drive these lizards down to any little point overhanging the sea, where they will sooner allow a person to catch hold of their tail than jump into the water. They do not seem to have any notion of biting; but when much frightened they squirt a drop of fluid from each nostril. One day I carried one to a deep pool left by the retiring tide, and threw it in several times as far as I was able. It invariably returned in a direct line to the spot where I stood. It swam near the bottom, with a very graceful and rapid movement, and occasionally aided itself over the uneven ground with its feet. As soon as it arrived near the margin, but still being under water, it either tried to conceal itself in the tufts of sea-weed, or it entered some crevice. As soon as it thought the danger was past, it crawled out on the dry rocks, and shuffled away as quickly as it could. I several times caught this same lizard, by driving it down to a point, and though possessed of such perfect powers of diving and swimming, nothing would induce it to enter the water; and so often as I threw it in, it returned in the manner above described. Perhaps this singular piece of apparent stupidity may be accounted for by the circumstance, that this reptile has no enemy whatever on shore, whereas at sea it must often fall a prey to the numerous sharks. Hence, probably urged by a fixed and hereditary instinct that the shore is its place of safety, whatever the emergency may be, it there takes refuge.

During our visit (in October) I saw extremely few small individuals of this species, and none I should think under a year old. From this circumstance it seems probable that the breeding season had not commenced. I asked several of the inhabitants if they knew where it laid its eggs: they said, that although well acquainted with the eggs of the other kind, they had not the least knowledge of the manner in which this species is propagated; a fact, considering how common an animal this lizard is, not a little extraordinary.

We will now turn to the terrestrial species (*Amb. subcristatus* of Gray).* This species, differently from the last, is confined to the central islands of the Archipelago, namely to Albemarle, James, Barrington, and Indefatigable. To the southward, in Charles, Hood, and Chatham islands, and to the northward, in Towers, Bindloes, and Abington, I neither saw nor heard of any. It would appear as if this species had been created in the centre of the Archipelago, and thence had been dispersed only to a certain distance.

In the central islands they inhabit both the higher and damp, as well as the lower and sterile parts; but in the latter they are much the most numerous. I cannot give a more forcible proof of their numbers, than by stating, that when we were left at James Island, we could not for some time find a spot free from their burrows, on which to pitch our tent. These lizards, like their brothers the sea-kind, are ugly animals; and from their low facial angle have a singularly stupid appearance. In size perhaps they are a little inferior to the latter, but several of them weighed between 10 and 15 pounds each. The colour of their belly, front legs, and head (excepting the crown which is nearly white), is a dirty yellowish-orange: the back is a brownish-red, which in the younger specimens is darker. In their movements they are lazy and half torpid. When not frightened, they slowly crawl along with their tails and

---

* Briefly characterized by Mr Gray in the Zoological Miscellany, from a specimen badly stuffed; from which cause one of its most important characters (the rounded tail, compared to the flattened one of the aquatic kind) was overlooked. Captain FitzRoy has presented some fine specimens of both species to the British Museum. I cannot omit here returning my thanks to Mr Gray, for the kind manner in which he has afforded me every facility as often as I have visited the British Museum.

bellies dragging on the ground. They often stop, and doze for a minute with closed eyes, and hind legs spread out on the parched soil.

They inhabit burrows; which they sometimes excavate between fragments of lava, but more generally on level patches of the soft volcanic sandstone. The holes do not appear to be very deep, and they enter the ground at a small angle; so that when walking over these lizard *warrens*, the soil is constantly giving way, much to the annoyance of the tired walker. This animal when excavating its burrow, alternately works the opposite sides of its body. One front leg for a short time scratches up the soil, and throws it towards the hind foot, which is well placed so as to heave it beyond the mouth of the hole. This side of the body being tired, the other takes up the task, and so on alternately. I watched one for a long time, till half its body was buried; I then walked up and pulled it by the tail; at this it was greatly astonished, and soon shuffled up to see what was the matter; and then stared me in the face, as much as to say, 'What made you pull my tail?'

They feed by day, and do not wander far from their burrows; and if frightened they rush to them with a most awkward gait. Except when running down hill, they cannot move very fast; which appears chiefly owing to the lateral position of their legs.

They are not at all timorous: when attentively watching anyone, they curl their tails, and raising themselves on their front legs, nod their heads vertically, with a quick movement, and try to look very fierce: but in reality they are not at all so; if one just stamps the ground, down go their tails, and off they shuffle as quickly as they can. I have frequently observed small muscivorous lizards, when watching any thing, nod their heads in precisely the same manner; but I do not at all know for what purpose. If this Amblyrhyncus is held, and plagued with a stick, it will bite it very severely; but I caught many by the tail, and they never tried to bite me. If two are placed on the ground and held together, they will fight and bite each other till blood is drawn.

The individuals (and they are the greater number) which inhabit the lower country, can scarcely taste a drop of water throughout the year; but they consume much of the succulent cactus, the branches of which are occasionally broken off by the wind. I have sometimes thrown a piece to two or three when together; and it was amusing enough to see each trying to seize and carry it away in its mouth, like so many hungry dogs with a bone. They eat very deliberately, but do not chew their food. The little birds are aware how harmless these creatures are: I have seen one of the thick-billed finches picking at one end of a piece of cactus (which is in request among all the animals of the lower region), whilst a lizard was eating at the other; and afterwards the little bird with the utmost indifference hopped on the back of the reptile.

I opened the stomachs of several, and found them full of vegetable fibres, and leaves of different trees, especially of a species of acacia. In the upper region they live chiefly on the acid and astringent berries of the guayavita, under which trees I have seen these lizards and the huge tortoises feeding together. To obtain the acacia-leaves, they crawl up the low stunted trees; and it is not uncommon to see one or a pair quietly browsing, whilst seated on a branch several feet above the ground.

The meat of these animals when cooked is white, and by those whose stomachs rise above all prejudices, it is relished as very good food. Humboldt has remarked that in inter-tropical South America, all lizards which inhabit *dry* regions are esteemed delicacies for the table. The inhabitants say, that those inhabiting the damp region drink water, but that the others do not travel up for it from the sterile country like the tortoises. At the time of our visit, the females had within their bodies numerous large elongated eggs. These they lay in their burrows, and the inhabitants seek them for food.

These two species of Amblyrhyncus agree, as I have already stated, in general structure, and in many of their habits. Neither have that rapid movement, so characteristic of true Lacerta and Iguana. They are both herbivorous, although the kind of vegetation consumed in each case is so very different. Mr Bell has given the name to the genus from the shortness of the snout: indeed, the form of the mouth may almost be compared to that of the tortoise. One is tempted to suppose this is an adaptation to their herbivorous appetites. It is very interesting thus to find a

well-characterized genus, having its aquatic and terrestrial species, belonging to so confined a portion of the world. The former
species is by far the most remarkable, because it is the only
existing Saurian, which can properly be said to be a maritime
animal. I should perhaps have mentioned earlier, that in the whole
archipelago, there is only one rill of fresh water that reaches the
coast; yet these reptiles frequent the sea-beaches, and no other
parts in all the islands. Moreover, there is no existing lizard, as far
as I am aware, excepting this Amblyrhyncus, that feeds exclusively on aquatic productions. If, however, we refer to epochs long
past, we shall find such habits common to several gigantic animals
of the Saurian race.

To conclude with the order of reptiles. Of snakes there are
several species, but all harmless. Of toads and frogs there are
none. I was surprised at this, considering how well the temperate
and damp woods in the elevated parts appeared adapted for their
habits. It recalled to my mind the singular statement made by
Bory St Vincent,* namely, that none of this family are to be found
on the volcanic islands in the great oceans. There certainly appears
to be some foundation for this observation; which is the more
remarkable, when compared with the case of lizards, which are
generally among the earliest colonists of the smallest islet. It may
be asked, whether this is not owing to the different facilities of
transport through salt water, of the eggs of the latter protected by
a calcareous coat, and of the slimy spawn of the former?

As I at first observed, these islands are not so remarkable for the
number of species of reptiles, as for that of individuals; when we
remember the well-beaten paths made by the many hundred great
tortoises – the warrens of the terrestrial Amblyrhyncus – and the
groups of the aquatic species basking on the coast-rocks – we
must admit that there is no other quarter of the world, where this
order replaces the herbivorous mammalia in so extraordinary a
manner. It is worthy of observation by the geologist (who will
probably refer back in his mind to the secondary periods, when
the Saurians were developed with dimensions, which at the
present day can be compared only to the cetaceous mammalia),

* Voyage aux quatre Iles d'Afrique.

that this archipelago, instead of possessing a humid climate and rank vegetation, cannot be considered otherwise than extremely arid, and for an equatorial region, remarkably temperate.

To finish with the zoology: I took great pains in collecting the insects, but I was surprised to find, even in the high and damp region, how exceedingly few they were in number. The forests of Tierra del Fuego are certainly much more barren; but with that exception I never collected in so poor a country. In the lower and sterile land I took seven species of Heteromera, and a few other insects; but in the fine thriving woods towards the centre of the islands, although I perseveringly swept under the bushes during all kinds of weather, I obtained only a few minute Diptera and Hymenoptera. Owing to this scarcity of insects, nearly all the birds live in the lower country; and the part which any one would have thought much the most favourable for them, is frequented only by a few of the small tyrant-flycatchers. I do not believe a single bird, excepting the water-rail, is confined to the damp region. Mr Waterhouse informs me that nearly all the insects belong to European forms, and that they do not by any means possess an equatorial character. I did not take a single one of large size, or of bright colours. This last observation applies equally to the birds and flowers. It is worthy of remark, that the only land-bird with bright colours, is that species of tyrant-flycatcher, which seems to be a wanderer from the continent. Of shells, there are a considerable number of land kinds, all of which, I believe are confined to this archipelago. Even of marine species, a large proportion were not known, before the collection made by Mr Cuming on these islands was brought to England.

I will not here attempt to come to any definite conclusions, as the species have not been accurately examined; but we may infer, that, with the exception of a few wanderers, the organic beings found on this archipelago are peculiar to it; and yet that their general form strongly partakes of an American character. It would be impossible for any one accustomed to the birds of Chile and La Plata to be placed on these islands, and not to feel convinced that he was, as far as the organic world was concerned, on American ground. This similarity in type, between distant islands and continents, while the species are distinct, has scarcely been sufficiently noticed. The circumstance would be explained, according to the views of some authors, by saying that the creative power had acted according to the same law over a wide area.

It has been mentioned, that the inhabitants can distinguish the tortoises, according to the islands whence they are brought. I was also informed that many of the islands possess trees and plants which do not occur on the others. For instance the berry-bearing tree, called Guayavita, which is common on James Island, certainly is not found on Charles Island, though appearing equally well fitted for it. Unfortunately, I was not aware of these facts till my collection was nearly completed: it never occurred to me, that the productions of islands only a few miles apart, and placed under the same physical conditions, would be dissimilar. I therefore did not attempt to make a series of specimens from the separate islands. It is the fate of every voyager, when he has just discovered what object in any place is more particularly worthy of his attention, to be hurried from it. In the case of the mocking-bird, I ascertained (and have brought home the specimens) that one species (*Orpheus trifasciatus*, Gould) is exclusively found in Charles Island; a second (*O. parvulus*) on Albemarle Island; and a third (*O. melanotus*) common to James and Chatham Islands. The two last species are closely allied, but the first would be considered by every naturalist as quite distinct. I examined many specimens in the different islands, and in each the respective kind was *alone* present. These birds agree in general plumage, structure, and habits; so that the different species replace each other in the economy of the different islands. These species are not character-ized by the markings on the plumage alone, but likewise by the size and form of the bill, and other differences. I have stated, that in the thirteen species of ground-finches, a nearly perfect gra-dation may be traced, from a beak extraordinarily thick, to one so fine, that it may be compared to that of a warbler. I very much suspect, that certain members of the series are confined to differ-ent islands; therefore, if the collection had been made on any *one* island, it would not have presented so perfect a gradation. It is clear, that if several islands have each their peculiar species of the same genera, when these are placed together, they will have a wide range of character. But there is not space in this work, to enter on this curious subject.

Before concluding my account of the zoology of these islands, I must describe more in detail the tameness of the birds. This disposition is common to all the terrestrial species; namely, to the mocking-birds, the finches, sylvicolæ, tyrant-flycatchers, doves, and hawks. There is not one which will not approach sufficiently near to be killed with a switch, and sometimes, as I have myself tried, with a cap or hat. A gun is here almost superfluous; for with the muzzle of one I pushed a hawk off the branch of a tree. One day a mocking-bird alighted on the edge of a pitcher (made of the shell of a tortoise), which I held in my hand whilst lying down. It began very quietly to sip the water, and allowed me to lift it with the vessel from the ground. I often tried, and very nearly succeeded, in catching these birds by their legs. Formerly the birds appear to have been even tamer than at present. Cowley* (in the year 1684) says that the 'Turtle-doves were so tame that they would often alight upon our hats and arms, so as that we could take them alive: they not fearing man, until such time as some of our company did fire at them, whereby they were rendered more shy.' Dampier† (in the same year) also says that a man in a morning's walk might kill six or seven dozen of these birds. At present, although certainly very tame, they do not alight on people's arms; nor do they suffer themselves to be killed in such numbers. It is surprising that the change has not been greater; for these islands during the last 150 years, have been frequently visited by bucaniers and whalers; and the sailors, wandering through the woods in search of tortoises, always take delight in knocking down the little birds.

These birds, although much persecuted, do not become wild in a short time: in Charles Island, which had then been colonized about six years, I saw a boy sitting by a well with a switch in his hand, with which he killed the doves and finches as they came to drink. He had already procured a little heap of them for his dinner; and he said he had constantly been in the habit of waiting there for

---

* Cowley's Voyage, p. 10, in Dampier's Collection of Voyages.
† Dampier's Voyage, vol. i, p. 103.

the same purpose. We must conclude that the birds, not having as yet learnt that man is a more dangerous animal than the tortoise, or the amblyrhyncus, disregard us, in the same manner as magpies in England do the cows and horses grazing in the fields.

The Falkland Islands offer a second instance of this disposition among its birds. The extraordinary tameness of the dark-coloured Furnarius has been remarked by Pernety, Lesson, and other voyagers. It is not, however, peculiar to that bird: the Caracara, snipe, upland and lowland goose, thrush, Emberiza, and even some true hawks, are all more or less tame. Both hawks and foxes are present; and as the birds are so tame, we may infer that the absence of all rapacious animals at the Galapagos, is not the cause of their tameness there. The geese at the Falklands, by the precaution they take in building on the islets, show that they are aware of their danger from the foxes; but they are not by this rendered wild towards man. This tameness of the birds, especially the waterfowl, is strongly contrasted with the habits of the same species in Tierra del Fuego, where for ages past they have been persecuted by the wild inhabitants. In the Falklands, the sportsman may sometimes kill more of the upland geese in one day, than he is able to carry home; whereas in Tierra del Fuego, it is nearly as difficult to kill one, as it is in England of the common wild species.

In the time of Pernety* (1763), all the birds appear to have been much tamer than at present. Pernety states that the Furnarius would almost perch on his finger; and that with a wand he killed ten in half an hour. At that period, the birds must have been about as tame as they now are at the Galapagos. They appear to have learnt caution more quickly at the Falklands than at the latter place, and they have had proportionate means of experience; for besides frequent visits from vessels, the islands have been at intervals colonized during the whole period.

Even formerly, when all the birds were so tame, by Pernety's account it was impossible to kill the black-necked swan. It is rather an interesting fact, that this is a bird of passage, and therefore brings with it the wisdom learnt in foreign countries.

* Pernety, Voyage aux Iles Malouines, vol. ii, p. 20.

I have not met with any account of the *land* birds being so tame, in any other quarter of the world, as at the Galapagos and Falkland Islands. And it may be observed that of the few archipelagoes of any size, which when discovered were uninhabited by man, these two are among the most important. From the foregoing statements we may, I think, conclude; first, that the wildness of birds with regard to man, is a particular instinct directed against *him*, and not dependent on any general degree of caution arising from other sources of danger; secondly, that it is not acquired by them in a short time, even when much persecuted; but that in the course of successive generations it becomes hereditary. With domesticated animals we are accustomed to see instincts becoming hereditary; but with those in a state of nature, it is more rare to discover instances of such acquired knowledge. In regard to the wildness of birds towards men, there is no other way of accounting for it. Few young birds in England have been injured by man, yet all are afraid of him: many individuals, on the other hand, both at the Galapagos and at the Falklands, have been injured, but yet have not learned that salutary dread. We may infer from these facts, what havoc the introduction of any new beast of prey must cause in a country, before the instincts of the aborigines become adapted to the stranger's craft or power.

CHAPTER XX

Tahiti – Aspect – Vegetation on the slope of the mountains –
View of Eimeo – Excursion in the interior – Profound ravines –
Succession of waterfalls – Number of wild useful plants –
Temperance of inhabitants – Their moral state – Parliament
convened – New Zealand – Bay of islands – Hippahs – Absence
of all government – Excursion to Waimate – Missionary establishment – English weeds now run wild – Waiomio – Funeral
service – Sail from New Zealand

TAHITI AND NEW ZEALAND

OCTOBER 20TH – The survey of the Galapagos Archipelago
being concluded, a course was steered towards Tahiti; and we
commenced our long passage of 3,200 miles. In the course of a
few days we sailed out of the gloomy and clouded region, which
extends during the winter far from the coast of South America.
We then enjoyed bright and clear weather, while running
pleasantly along at the rate of 150 or 160 miles a day before a
steady trade wind. The temperature in this more central part of
the Pacific, is higher than near the American shore. The thermometer in the poop cabin, both by night and day, ranged between
80° and 83°, which to my feelings was quite delightful; but with
one degree higher, the effect became oppressive. We passed
through the Dangerous or Low Archipelago, and saw several of
those most curious rings of land, just rising above the edge of the
water, which have been called Lagoon Islands. A long and
brilliantly white beach is capped by a margin of green vegetation;
and this strip appears on both hands rapidly to narrow away in the
distance, and then sinks beneath the horizon. From the mast-head
a wide expanse of smooth water can be seen within the annular

margin of land. These low islands bear no proportion to the vast ocean out of which they abruptly rise; and it seems wonderful, that such weak intruders are not overwhelmed, by the all-powerful and never-tiring waves of that great sea, miscalled the Pacific.

NOVEMBER 15TH – At daylight, Tahiti, an island which must for ever remain as classical to the voyager in the South Sea, was in view. At this distance the appearance was not very inviting. The luxuriant vegetation of the lower parts was not discernible, and as the clouds rolled past, the wildest and most precipitous peaks showed themselves towards the centre of the island. As soon as we came to an anchor in Matavai Bay, we were surrounded by canoes. This was our Sunday, but the Monday of Tahiti: if the case had been reversed, we should not have received a single visit; for the injunction not to launch a canoe on the sabbath is rigidly obeyed. After dinner we landed to enjoy all the delights of the first impressions produced by a new country, and that country the charming Tahiti. A crowd of men, women, and children, was collected on the memorable point Venus, ready to receive us with laughing, merry faces. They marshalled us towards the house of Mr Wilson, the missionary of the district, who met us on the road, and gave us a very friendly reception. After sitting a short time in his house, we separated to walk about, but returned there in the evening.

The land capable of cultivation is scarcely in any part more than a fringe of low alluvial soil, accumulated round the base of the mountains, and protected from the waves of the sea by a coral reef, which encircles at a distance the entire line of coast. The reef is broken in several parts so that ships can pass through, and the lake of smooth water within thus affords a safe harbour, as well as a channel for the native canoes. The low land which comes down to the beach of coral sand, is covered by the most beautiful productions of the inter-tropical regions. In the midst of bananas, orange, cocoa-nut, and breadfruit trees, spots are cleared where yams, sweet potatoes, sugar-cane, and pineapples, are cultivated. Even the brushwood is a fruit-tree, namely, the guava, which from its abundance is as noxious as a weed. In Brazil I have often admired the contrast of varied beauty in the banana, palm, and orange-tree: here we have in addition the bread-fruit, conspicuous from its large, glossy, and deeply digitated leaf. It is admirable to behold groves of a tree, sending forth its branches with the force of an English oak, loaded with large and most nutritious fruit. However little on most occasions utility explains the delight received from any fine prospect, in this case it cannot fail to enter as an element in the feeling. The little winding paths, cool from the surrounding shade, led to the scattered houses; and the owners of these every where gave us a cheerful and most hospitable reception.

I was pleased with nothing so much as with the inhabitants. There is a mildness in the expression of their countenances, which at once banishes the idea of a savage; and an intelligence, which shows they are advancing in civilization. Their dress is as yet incongruous; no settled costume having taken the place of the ancient one. But even in its present state, it is far from being so ridiculous as it has been described by travellers of a few years' standing. Those who can afford it wear a white shirt, and sometimes a jacket, with a wrapper of coloured cotton round their middles; thus making a short petticoat, like the chilipa of the Gauchos. This dress appears so general with the chiefs, that it will probably become the settled fashion. No one, even to the queen, wears shoes or stockings; and only the chiefs have a straw hat on their heads. The common people, when working, keep the upper part of their bodies uncovered; and it is then that the Tahitians are seen to advantage. They are very tall, broad-shouldered, athletic, and with well-proportioned limbs. It has been somewhere re-marked, that it requires little habit to make a darker tint of the skin more pleasing and natural, even to the eye of an European, than his own colour. To see a white man bathing by the side of a Tahitian, was like comparing a plant bleached by the gardener's art, with one growing in the open fields. Most of the men are tattooed; and the ornaments follow the curvature of the body so gracefully, that they have a very pleasing and elegant effect. One common figure, varying only in its detail, branches somewhat like a tuft of palm-leaves* from the line of the backbone, and curls

* The similarity is not closer than between the capital of a Corinthian column and a tuft of acanthus.

round each side. The simile may be a fanciful one, but I thought the body of a man thus ornamented, was like the trunk of a noble tree embraced by a delicate creeper.

Many of the older people had their feet covered with small figures, placed in order so as to resemble a sock. This fashion, however, is partly gone by, and has been succeeded by others. Here, although each man must for ever abide by the whim which reigned in his early days, yet fashion is far from immutable. An old man has thus his age for ever stamped on his body, and he cannot assume the airs of a young dandy. The women are also tattooed in the same manner as the men, and very commonly on their fingers. An unbecoming fashion in one respect is now almost universal: it is that of cutting the hair, or rather shaving it, from the upper part of the head, in a circular form, so as to leave only an outer ring of hair. The missionaries have tried to persuade the people to change this habit: but it is the fashion, and that is sufficient answer at Tahiti as well as at Paris. I was much disappointed in the personal appearance of the women; they are far inferior in every respect to the men. The custom of wearing a flower in the back of the head, or through a small hole in each ear, is pretty; the flower is generally either white or scarlet, and like the Camelia Japonica. They wear also a sort of crown of woven cocoa-nut leaves, as a shade to their eyes. The women appear to be in greater want of some becoming costume, even than the men.

Nearly all understand a little English; that is, they know the names of common things, and by the aid of this, together with signs, a lame sort of conversation could be carried on. In returning in the evening to the boat, we stopped to witness a very pretty scene; numbers of children were playing on the beach, and had lighted bonfires, which illuminated the placid sea and surrounding trees. Others, in circles, were singing Tahitian verses. We seated ourselves on the sand, and joined their party. The songs were impromptu, and I believe related to our arrival: one little girl sang a line, which the rest took up in parts, forming a very pretty chorus. The whole scene made us unequivocally aware that we were seated on the shores of an island in the South Sea.

* * *

NOVEMBER 18TH – In the morning I came on shore early, bringing with me some provisions in a bag, and two blankets for myself and servant. These were lashed to each end of a pole, and thus carried by my Tahitian companions: from custom these men are able to walk for a whole day, with as much as 50 pounds at each end. I told my guides to provide themselves with food and clothing: but for the latter, they said their skins were sufficient, and for the former, that there was plenty of food in the mountains. The line of march was the valley of Tia-auru, in which the river flows that enters the sea by Point Venus. This is one of the principal streams in the island, and its source lies at the base of the loftiest pinnacles, which attain the elevation of about 7,000 feet. The whole island may be considered as one group of mountains, so that the only way to penetrate the interior is to follow up the valleys. Our road, at first, lay through the wood which bordered each side of the river; and the glimpses of the lofty central peaks, seen as through an avenue, with here and there a waving cocoa-nut tree on one side, were extremely picturesque. The valley soon began to narrow, and the sides to grow lofty and more precipitous. After having walked between three and four hours, we found the width of the ravine scarcely exceeded that of the bed of the stream. On each hand the walls were nearly vertical; yet from the soft nature of the volcanic strata, trees and a rank vegetation sprung from every projecting ledge. These precipices must have been some 1,000 feet high: and the whole formed a mountain gorge, far more magnificent than any thing which I had ever before beheld. Until the mid-day sun stood vertically over the ravine, the air had felt cool and damp, but now it became very sultry. Shaded by a ledge of rock, beneath a façade of columnar lava, we ate our dinner. My guides had already procured a dish of small fish and fresh-water prawns. They carried with them a small net stretched on a hoop; and where the water was deep and in eddies, they dived, and like otters, by their eyesight followed the fish into holes and corners, and thus secured them.

The Tahitians have the dexterity of amphibious animals in the water. An anecdote mentioned by Ellis shows how much they feel at home in that element. When a horse was landing for

Pomarre in 1817, the slings broke, and it fell into the water: immediately the natives jumped overboard, and by their cries and vain efforts at assistance, almost drowned the animal. As soon, however, as it reached the shore, the whole population took to flight, and tried to hide themselves from the man-carrying-pig, as they christened the horse.

A little higher up, the river divided itself into three little streams. The two northern ones were impracticable, owing to a succession of waterfalls, which descended from the jagged summit of the highest mountain; the other to all appearance was equally inaccessible, but we managed to ascend it by a most extraordinary road. The sides of the valley were here nearly precipitous; but, as frequently happens with stratified rocks, small ledges projected, which were thickly covered by wild bananas, liliaceous plants, and other luxuriant productions of the tropics. The Tahitians, by climbing amongst these ledges, searching for fruit, had discovered a track by which the whole precipice could be scaled. The first ascent from the valley was very dangerous: for it was necessary to pass the face of a naked rock, by the aid of ropes, which we brought with us. How any person discovered that this formidable spot was the only point where the side of the mountain was practicable, I cannot imagine. We then cautiously walked along one of the ledges, till we came to the stream already alluded to. This ledge formed a flat spot, above which a beautiful cascade, of some 100 feet, poured down its waters, and beneath it another high one emptied itself into the main stream. From this cool and shady recess, we made a circuit to avoid the overhanging cascade. As before, we followed little projecting ledges, the apparent danger being partly hidden by the thickness of the vegetation. In passing from one of the ledges to another, there was a vertical wall of rock. One of the Tahitians, a fine active man, placed the trunk of a tree against this, climbed up it, and then by the aid of crevices reached the summit. He fixed the ropes to a projecting point, and lowered them for us, then hauled up a dog which accompanied us, and lastly our luggage. Beneath the ledge on which the dead tree was placed the precipice must have been 500 or 600 feet deep; and if the abyss had not been partly concealed by the overhanging ferns and lilies, my head would have turned giddy, and nothing should have induced me to have attempted it. We continued to ascend sometimes along ledges, and sometimes along knife-edged ridges, having on each hand profound ravines. In the Cordillera, I have seen mountains on a far grander scale, but for abruptness, no part of them at all comparable to this. In the evening we reached a flat little spot on the banks of the same stream, which I have mentioned as descending by a chain of waterfalls. Here we bivouacked for the night. On each side of the ravine there were great beds of the Feyé, or mountain-banana, covered with ripe fruit. Many of these plants were from 20 to 25 feet high, and from 3 to 4 in circumference. By the aid of strips of bark for twine, the stems of bamboos for rafters, and the large leaf of the banana for a thatch, the Tahitians in a few minutes built an excellent house; and with the withered leaves made a soft bed.

They then proceeded to make a fire, and cook our evening meal. A light was procured by rubbing a blunt-pointed stick in a groove made in another (as if with the intention of deepening it), until by friction the dust became ignited. A peculiarly white and very light wood (the *Hibiscus tiliaceus*) is alone used for this purpose: it is the same which serves for poles to carry any burden, and for the floating outrigger to steady the canoe. The fire was produced in a few seconds: but, to a person who does not understand the art, it requires the greatest exertion; as I found, before at last, to my great pride, I succeeded in igniting the dust. The Gaucho in the Pampas uses a different method: taking an elastic stick about 18 inches long, he presses one end on his breast, and the other (which is pointed) in a hole in a piece of wood, and then rapidly turns the curved part, like a carpenter's centre-bit. The Tahitians having made a small fire of sticks, placed a score of stones, of about the size of cricket-balls, on the burning wood. In about ten minutes' time the sticks were consumed and the stones hot. They had previously folded up in small parcels of leaves, pieces of beef, fish, ripe and unripe bananas, and the tops of the wild arum. These green parcels were laid in a layer between two layers of the hot stones, and the whole then covered up with earth, so that no smoke or steam could escape. In about a quarter of an hour, the whole was most deliciously cooked. The choice green parcels were now laid on a cloth of banana-leaves, and with a cocoa-nut shell we drank the cool water of the running stream; and thus we enjoyed our rustic meal.

I could not look on the surrounding plants without admiration. On every side were forests of banana; the fruit of which, though serving for food in various ways, lay in heaps decaying on the ground. In front of us there was an extensive brake of wild sugar-cane; and the stream was shaded by the dark green knotted stem of the ava — so famous in former days for its powerful intoxicating effects. I chewed a piece, and found that it had an acrid and unpleasant taste, which would have induced any one at once to have pronounced it poisonous. Thanks be to the missionaries, this plant now thrives only in these deep ravines, innocuous to every one. Close by I saw the wild arum, the roots of which, when well baked, are good to eat, and the young leaves better than spinach. There was the wild yam, and a liliaceous plant called Ti, which grows in abundance, and has a soft brown root, in shape and size like a huge log of wood. This served us for dessert, for it is as sweet as treacle, and with a pleasant taste. There were, moreover, several other wild fruits, and useful vegetables. The little stream, besides its cool water, produced eels and cray-fish. I did indeed admire this scene, when I compared it with an uncultivated one in the temperate zones. I felt the force of the observation, that man, at least savage man, with his reasoning powers only partly developed, is the child of the tropics.

As the evening drew to a close, I strolled beneath the gloomy shade of the bananas up the course of the stream. My walk was soon brought to a close, by coming to a waterfall between 200 and 300 feet high; and again above this there was another. I mention all these waterfalls in this one brook, to give a geneal idea of the inclination of the land. In the little recess where the water fell, it did not appear that a breath of wind had ever entered. The leaves of the banana, damp with spray, possessed an unbroken edge, instead of being split, as generally is the case, into a thousand shreds. From our position, almost suspended on the mountain-side, there were glimpses into the depths of the neighbouring valleys; and the lofty points of the central mountains, towering up within 60° of the zenith, hid half the evening sky. Thus seated, it was a sublime spectacle to watch the shades of night gradually obscuring the last and highest pinnacles.

Before we laid ourselves down to sleep, the elder Tahitian fell on his knees, and with closed eyes repeated a long prayer in his native tongue. He prayed as a Christian should do, with fitting reverence, and without the fear of ridicule or any ostentation of piety. At our meals neither of the men would taste food, without saying beforehand a short grace. Those travellers, who think that a Tahitian prays only when the eyes of the missionary are fixed on him, should have slept with us that night on the mountain-side. Before morning it rained very heavily; but the good thatch of banana-leaves kept us dry.

NOVEMBER 19TH — At daylight my friends, after their morning prayer, prepared an excellent breakfast in the same manner as in the evening. They themselves certainly partook of it largely; indeed I never saw any men eat nearly so much. I should suppose such capacious stomachs must be the result of a large part of their diet consisting of fruit and vegetables, which contain, in a given bulk, a comparatively small portion of nutriment. Unwittingly, I was the means of my companions breaking (as I afterwards learned) one of their own laws and resolutions. I took with me a flask of spirits, which they could not resolve to refuse; but as often as they drank a little, they put their fingers before their mouths, and uttered the word 'Missionary'. About two years ago, although the use of the ava was prevented, drunkenness from the introduction of spirits became very prevalent. The missionaries prevailed on a few good men, who saw their country rapidly going to ruin, to join with them in a Temperance Society. From good sense or shame all the chiefs and the queen were at last persuaded to join it. Immediately a law was passed that no spirits should be allowed to be introduced into the island, and that he who sold and he who bought the forbidden article, should be punished by a fine. With remarkable justice, a certain period was allowed for stock in hand to be sold, before the law came into effect. But when it did, a general search was made in which even the houses of the missionaries were not exempted, and all the ava (as the natives call all ardent spirits) was poured on the ground. When one reflects on the effect of intemperance on the aborigines of the two Americas, I think it will be acknowledged, that every well-wisher of Tahiti owes no common debt of gratitude to the missionaries. As long as the little island of St Helena remained under the government of the East India Company, spirits, owing to the great injury they had produced, were not allowed to be imported; but wine was supplied from the Cape of Good Hope. It is rather a striking, and not very gratifying fact, that in the same year that spirits were allowed to be sold on that island, their use was banished from Tahiti by the free will of the people.

After breakfast we proceeded on our journey. As my object was merely to see a little of the interior scenery, we returned by another track, which descended into the main valley lower down. For some distance we wound, by a most intricate path, along the side of the mountain which formed the valley. In the less precipitous parts we passed through extensive groves of the wild banana. The Tahitians, with their naked, tattooed bodies, their heads ornamented with flowers, and seen in the dark shade of the woods, would have formed a fine picture of man, inhabiting some primeval forest. In our descent we followed the line of ridges; these were exceedingly narrow, and for considerable lengths steep as a ladder; but all clothed with vegetation. The extreme care necessary in poising each step rendered the walk fatiguing. I am never weary of expressing my astonishment at these ravines and precipices: the mountains may almost be described, as rent by so many crevices. When viewing the surrounding country from the knife-edged ridges, the point of support was so small, that the effect was nearly the same, I should think, as from a balloon. In this descent we had occasion to use the ropes only once, at the point where we entered the main valley. We slept under the same ledge of rock, where, the day before, we had dined: the night was fine, but from the depth and narrowness of the gorge, profoundly dark.

Before actually seeing this country, I had difficulty in understanding two facts mentioned by Ellis; namely, that after the murderous battles of former times, the survivors on the conquered side retired into the mountains, where a handful of men could resist a multitude. Certainly half a dozen men, at the spot where the Tahitian reared the old tree, could easily have repulsed thousands. Secondly, that after the introduction of Christianity, there were wild men who lived in the mountains, and whose retreats were unknown to the more civilized inhabitants.

NOVEMBER 20TH – In the morning we started early, and reached Matavai at noon. On the road we met a large party of noble athletic men, going for wild bananas. I found that the ship, on account of the difficulty in watering, had moved to the harbour of Papawa, to which place I immediately walked. This is a very pretty spot. The cove is surrounded by reefs, and the water as smooth as that in a lake. The cultivated ground, with all its beautiful productions, and the cottages, comes close down to the water's edge.

From the varying accounts which I had read before reaching these islands, I was very anxious to form, from my own observation, a judgment of their moral state – although such judgment would necessarily be very imperfect. A first impression at all times very much depends on one's previously-acquired ideas. My notions were drawn from Ellis's *Polynesian Researches* – an admirable and most interesting work, but naturally looking at every thing under a favourable point of view; from Beechey's *Voyage*; and from that of Kotzebue, which is strongly adverse to the whole missionary system. He who compares these three accounts, will, I think, form a tolerably accurate conception of the present state of Tahiti. One of my impressions, which I took from the two last authorities, was decidedly incorrect; viz., that the Tahitians had become a gloomy race, and lived in fear of the missionaries. Of the latter feeling I saw no trace, unless, indeed, fear and respect be confounded under one name. Instead of discontent being a common feeling, it would be difficult in Europe to pick out of a crowd half so many merry and happy faces. The prohibition of the flute and dancing is inveighed against as wrong and foolish; the more than presbyterian manner of keeping the sabbath, is looked at in a similar light. On these points I will not pretend to offer any opinion in opposition to men who have resided as many years as I was days on the island.

On the whole it appears to me, that the morality and religion of the inhabitants is highly creditable. There are many who attack, even more acrimoniously than Kotzebue, both the missionaries, their system, and the effects produced by it. Such reasoners never compare the present state with that of the island only twenty years ago; nor even with that of Europe at this day; but they compare it with the high standard of Gospel perfection. They expect the missionaries to effect that, which the Apostles themselves failed to do. In as much as the condition of the people falls short of this high order, blame is attached to the missionary, instead of credit for that which he has effected. They forget, or will not remember, that human sacrifices, and the power of an idolatrous priesthood – a system of profligacy unparalleled in the world, and infanticide a consequence on that system – bloody wars, where the conquerors spared neither women nor children – that all these have been abolished; and that dishonesty, intemperance, and licentiousness have been greatly reduced by the introduction of Christianity. In a voyager to forget these things is base ingratitude; for should he chance to be at the point of shipwreck on some unknown coast, he will most devoutly pray that the lesson of the missionary may be found to have extended thus far.

In point of morality the virtue of the women, it has been often said, is most open to exception. But before they are blamed too severely, it will be well distinctly to call to mind the scenes described by Captain Cook and Mr Banks, in which the grand-mothers and mothers of the present race played a part. Those who are most severe, should consider how much of the morality of the women in Europe is owing to the system early impressed by mothers on their daughters, and how much in each individual case to the precepts of religion. But it is useless to argue against such reasoners: I believe that disappointed in not finding the field of licentiousness quite so open as formerly, they will not give credit to a morality which they do not wish to practise, or to a religion which they undervalue, if not despise.

SUNDAY 22ND – The harbour of Papiete, which may be considered as the capital of the island, is about 7 miles distant from Matavai, to which point the *Beagle* had returned. The queen resides there, and it is the seat of government, and the chief resort of shipping. Captain FitzRoy took a party there to hear divine service, first in the Tahitian language, and afterwards in our own. Mr Pritchard, the leading missionary in the island, performed the service, which was a most interesting spectacle. The chapel consisted of a large airy framework of wood; and it was filled to excess by tidy, clean people, of all ages and both sexes. I was rather disappointed in the apparent degree of attention; but I believe my expectations were raised too high. At all events the appearance was quite equal to that in a country church in England. The singing of the hymns was decidedly very pleasing; but the language from the pulpit, although fluently delivered, did not sound well. A constant repetition of words, like '*tata ta, mata mai*', rendered it monotonous. After English service, a party returned on foot to Matavai. It was a pleasant walk, sometimes along the sea-beach and sometimes under the shade of the many beautiful trees.

About two years ago, a small vessel under English colours was plundered by the inhabitants of the Low Islands, which were then under the dominion of the Queen of Tahiti. It was believed that the perpetrators were instigated to this act by some indiscreet laws issued by her majesty. The British government demanded compensation; which was acceded to, and a sum of nearly $3,000 was agreed to be paid on the first of last September. The commodore at Lima ordered Captain FitzRoy, to inquire concerning this debt, and to demand satisfaction if it were not paid. Captain FitzRoy accordingly requested an interview with the queen: and a parliament was held to consider the question; at which all the principal chiefs of the island and the queen were assembled. I will not attempt to describe what took place, after the interesting account given by Captain FitzRoy. The money it appeared had not been paid. Perhaps the alleged reasons for the failure were rather equivocating: but otherwise I cannot sufficiently express our general surprise, at the extreme good sense, the reasoning powers, moderation, candour, and prompt resolution, which were displayed on all sides. I believe every one of us left the meeting with a very different opinion of the Tahitians, from that which we entertained when entering. The chiefs and people resolved to subscribe and complete the sum which was wanting: Captain FitzRoy urged that it was hard that their private property should be sacrificed for the crimes of distant islanders. They replied, that they were grateful for his consideration, but that

Pomarre was their Queen, and they were determined to help her in this her difficulty. This resolution and its prompt execution (for a book was opened early the next morning), made a perfect conclusion to this very remarkable scene of loyalty and good feeling.

After the main discussion was ended, several of the chiefs took the opportunity of asking Captain FitzRoy many intelligent questions, concerning international customs and laws. These related to the treatment of ships and foreigners. On some points, as soon as the decision was made, the law was issued verbally on the spot. This Tahitian parliament lasted for several hours; and when it was over Captain FitzRoy invited the queen to pay the *Beagle* a visit.

NOVEMBER 26TH—In the evening, with a gentle land-breeze, a course was steered for New Zealand, and as the sun set we took a farewell look at the mountains of Tahiti – the island to which every voyager has offered up his tribute of admiration.

DECEMBER 19TH – In the evening we saw New Zealand in the distance. We may now consider ourselves as having nearly crossed the Pacific ocean. It is necessary to sail over this great sea to understand its immensity. Moving quickly onwards for weeks together we meet with nothing, but the same blue, profoundly deep, ocean. Even within the Archipelagoes, the islands are mere specks, and far distant one from the other. Accustomed to look at maps, drawn on a small scale, where dots, shading, and names are crowded together, we do not judge rightly how infinitely small the proportion of dry land is to the water of this great sea. The meridian of the Antipodes likewise has now been passed; and every league, thanks to our good fortune, which we travel onwards, is one league nearer to England. These Antipodes call to mind old recollections of childish doubt and wonder. Only the other day, I looked forward to this airy barrier, as a definite point in our voyage homewards; but now I find it, and all such resting-places for the imagination, are like shadows which a man moving onwards cannot catch. A gale of wind, which lasted for some days, has lately given us time and inclination to measure the future stages in our long voyage, and to wish most earnestly for its termination.

DECEMBER 21ST – Early in the morning we entered the Bay of Islands, and being becalmed for some hours near the mouth, we did not reach the anchorage till the middle of the day.

* * *

In the evening I went with Captain FitzRoy, and Mr Baker, one of the missionaries, to pay a visit to Kororadika. This is the largest village, and will one day, no doubt increase till it becomes the chief town: besides a considerable native population, there are many English residents. These latter are men of the most worthless character: and among them are many runaway convicts from New South Wales. There are many spirit-shops; and the whole population is addicted to drunkenness and all kinds of vice. As this is the capital, a person would be inclined to form his opinion of the New Zealanders from what he here saw; but in this case his estimate of their character would be too low. This little village is the very stronghold of vice. Although many tribes in other parts have embraced Christianity, here the greater part yet remain in heathenism. In such places the missionaries are held in little esteem: but they complain far more of the conduct of their countrymen, than of that of the natives. It is strange, but I have heard these worthy men say, that the only protection which they need, and on which they rely, is from the native chiefs against Englishmen.

We wandered about the village, and saw and conversed with many of the people, both men, women, and children. Looking at the New Zealander, one naturally compares him with the Tahitian; both belonging to the same family of mankind. The comparison, however, tells heavily against the New Zealander. He may, perhaps, be superior in energy, but in every other respect, his character is of a much lower order. One glance at their respective expressions, brings conviction to the mind, that one is a savage, the other a civilized man. It would be vain to seek in the whole of New Zealand, a person with the face and mien of the old Tahitian chief, Utamme. No doubt the extraordinary manner in which tattooing is here practised, gives a disagreeable expression to their countenances. The complicated but symmetrical figures covering the whole face, puzzle and mislead an unaccustomed eye: it is moreover probable, that the deep incisions, by destroying the play of the superficial muscles, give an air of rigid inflexibility. But besides this, there is a twinkling in the eye, which cannot indicate any thing but cunning and ferocity. Their figures are tall and bulky; but in elegance are not comparable with those of the working classes in Tahiti.

Both their persons and houses are filthily dirty and offensive: the idea of washing either their bodies or their clothes never seems to enter their heads. I saw a chief, who was wearing a shirt black and matted with filth; and when asked how it came to be so dirty, he replied, with surprise, 'Do not you see it is an old one?' Some of the men have shirts; but the common dress is one or two large blankets, generally black with dirt, which are thrown over their shoulders in a very inconvenient and awkward fashion. A few of the principal chiefs have decent suits of English clothes; but these are only worn on great occasions.

Considering the number of foreigners residing in New Zealand, and the amount of commerce carried on there, the state of government of the country is most remarkable. It is, however, incorrect to use the term government, where absolutely no such thing exists. The land is divided, by well-determined boundaries, between various tribes, independent of each other. The individuals in each tribe consist of freemen, and slaves taken in war; and the land is common to all the free born; that is, each may occupy and till any part that is vacant. In a sale, therefore, of land, every such person must receive part-payment. Among the freemen, there will always be some one, who from riches, from talents, or from descent from some noted character, will take the lead; and in this respect he may be considered as the chief. But if the united tribe should be asked, who was their chief, no one would be acknowledged. Without doubt, in many cases, individuals have obtained great influence; but as far as I can understand the system, their power is not legitimate. Even the authority of a master over his slave, or a parent over his child, appears to be regulated by no kind of ordinary custom. Proper laws of course are quite unknown: certain lines of action are generally considered right, and others wrong: if such customs are infringed, the injured person and his tribe, if they have power, seek retribution; if not, they treasure up the recollection of the injury till the day of revenge arrives. If the state in which the Fuegians live should be fixed at zero in the scale of government, I am afraid New Zealand would rank but a few degrees higher; while Tahiti, even when first discovered, would have occupied a respectable position.

DECEMBER 23RD – At a place called Waimate, about 15 miles from the Bay of Islands, and midway between the eastern and western coasts, the missionaries have purchased some land for agricultural purposes. I had been introduced to the Rev. W. Williams, who, upon my expressing the wish, invited me to pay him a visit there. Mr Bushby, the British Resident, offered to take me in his boat by a creek, where I should see a pretty waterfall, and by which means my walk would be shortened. He likewise procured for me a guide. Upon asking a neighbouring chief to recommend a man, the chief himself offered to go; but his ignorance of the value of money was so complete, that at first he asked how many pounds I would give him; but, afterwards was well contented with two dollars. When I showed the chief a very small bundle, which I wanted carried, it became absolutely necessary to take a slave for that purpose. These feelings of pride are beginning to wear away; but formerly a leading man would sooner have died than undergone the indignity of carrying the smallest burden. My companion was a light active man, dressed in a dirty blanket, and with his face completely tattooed. He had formerly been a great warrior. He appeared to be on very cordial terms with Mr Bushby; but at various times they had quarrelled violently. Mr Bushby remarked that a little quiet irony would frequently silence any one of these natives in their most blustering moments. This chief has come and harangued Mr Bushby in a hectoring manner, saying, 'A great chief, a great man, a friend of mine, has come to pay me a visit – you must give him something good to eat, some fine presents, &c.' Mr Bushby has allowed him to finish his discourse, and then has quietly replied by some such answer as, 'What else shall your slave do for you?' The man would then instantly, with a very comical expression cease his braggadocio.

Some time ago, Mr Bushby suffered a far more serious attack. A chief and a party of men tried to break into his house in the middle of the night, and not finding this so easy, commenced a brisk firing with their muskets. Mr Bushby was slightly wounded; but the party was at length driven away. Shortly afterwards it was discovered who was the aggressor; and a general meeting of the chiefs was convened to consider the case. It was considered by the New Zealanders as very atrocious, inasmuch as it was a night attack, and that Mrs Bushby was lying ill in the house: this latter circumstance, much to their honour, being considered in all cases as a protection. The chiefs agreed to confiscate the land of the aggressor to the King of England. The whole proceeding, however, in thus trying and punishing a chief was entirely without precedent. The aggressor, moreover, lost cast in the estimation of his equals; and this was considered by the British as of more consequence, than the confiscation of his land.

As the boat was shoving off, a second chief stepped into her, who only wanted the amusement of the passage up and down the creek. I never saw a more horrid and ferocious expression, than this man had. It immediately struck me, I had somewhere seen his likeness: it will be found in Retzsch's outlines to Schiller's ballad of Fridolin, where two men are pushing Robert into the burning iron furnace. It is the man who has his arm on Robert's breast. Physiognomy here spoke the truth; this chief had been a notorious murderer, and was to boot an arrant coward. At the point where the boat landed, Mr Bushby accompanied me a few hundred yards on the road: I could not help admiring the cool impudence of the hoary old villain, whom we left lying in the boat, when he shouted to Mr Bushby, 'Do not you stay long, I shall be tired of waiting here.'

We now commenced our walk. The road lay along a well-beaten path, bordered on each side by the tall fern, which covers the whole country. After travelling some miles, we came to a little country village, where a few hovels were collected together, and some patches of ground cultivated for potato crops. The introduction of the potato, has been the most essential benefit to the island; it is now much more used, than any native vegetable. New Zealand is favoured by one great natural advantage; namely, that the inhabitants can never perish from famine. The whole country abounds with fern; and the roots of this plant, if not very palatable, yet contain much nutriment. A native can always subsist on these, and on the shells which are abundant on all parts of the sea-coast. The villages are chiefly conspicuous, by the platforms which are raised on four posts 10 or 12 feet above the ground, and on which the produce of the fields is kept secure from all accidents.

On coming near one of the huts, I was much amused by seeing in due form the ceremony of rubbing, or as it should more properly be called, pressing noses. The women, on our first approach, began uttering something in a most dolorous voice; they then squatted themselves down and held up their faces; my companions standing over them, placed the bridge of their own noses at right angles to theirs, and commenced pressing. This lasted rather longer than a cordial shake of the hand would with us; and as we vary the force of the grasp of the hand in shaking, so do they in pressing. During the process they uttered comfortable little grunts, very much in the same manner as two pigs do, when rubbing against each other. I noticed, that the slave would press noses with any one he met, indifferently either before or after his master the chief. Although among savages the chief has absolute power of life and death over his slave, yet there is an entire absence of ceremony between them. Mr Burchell has remarked the same thing in Southern Africa with respect to the rude Bachapins. Where civilization has arrived at a certain point, as among the Tahitians, complex formalities are soon instituted between the different grades of society. For instance, in the above island, formerly all were obliged to uncover themselves as low as the waist in presence of the king.

The ceremony of pressing noses having been completed with all present, we seated ourselves in a circle in the front of one of the houses, and rested there half an hour. All the native hovels which I have seen, have nearly the same form and dimensions, and all agree in being filthily dirty. They resemble a cow-shed with one end open, but having a partition a little way within, with a square hole in it, which thus cuts off a part, and makes a small gloomy chamber. In this the inhabitants keep all their property, and when the weather is cold they sleep there. They eat, however, and pass their time in the open part in front.

My guides having finished their pipes, we continued our walk. The path led through the same undulating country, the whole uniformly clothed as before with fern. On our right hand, we had a serpentine river, the banks of which were fringed with trees, and here and there on the hill-sides there were clumps of wood. The whole scene, in spite of its green colour, bore rather a desolate aspect. The sight of so much fern impresses the mind with an idea of sterility. This, however, is not the case; for wherever the fern grows thick and breast-high, the land by tillage becomes productive. Some of the residents, with much probability think that all this extensive open country was originally covered with forests, and that it has been cleared by the aid of fire. It is said that by digging in the barest spots, lumps of the kind of resin which flows from the kauri pine, are frequently found. The natives had an evident motive in thus clearing the country; for in such parts the fern, formerly so staple an article of food, flourishes best. The almost entire absence of associated grasses, which forms so remarkable a feature in the vegetation of this island, may perhaps be accounted for, by the open parts being the work of man, while nature had designed the country for forest land.

The soil is volcanic; in several parts we passed over slaggy and vesicular lavas, and the form of a crater could clearly be distinguished in several of the neighbouring hills. Although the scenery is nowhere beautiful, and only occasionally pretty, I enjoyed my walk. I should have enjoyed it more, if my companion, the chief, had not possessed extraordinary conversational powers. I only knew three words; 'good', 'bad', and 'yes': and with these I answered all his remarks, without of course having understood one word he said. This, however, was quite sufficient: I was a good listener, an agreeable person, and he never ceased talking to me.

At length we reached Waimate. After having passed over so many miles of an uninhabited useless country, the sudden appearance of an English farm-house, and its well-dressed fields, placed there as if by an enchanter's wand, was exceedingly pleasing. Mr Williams not being at home, I received in Mr Davies's house a cordial and pleasant welcome. After drinking tea with his family party, we took a stroll about the farm. At Waimàte there are three large houses, where the missionary gentlemen Messrs Williams, Davies, and Clarke, reside; and near them are the huts of the native labourers. On an adjoining slope fine crops of barley and wheat in full ear were standing; and, in another part, fields of potatoes and clover. But I cannot attempt to describe all I saw; there were large gardens, with every fruit and vegetable which England produces; and many belonging to a warmer clime. I may instance, asparagus, kidney beans, cucumbers, rhubarb, apples, pears, figs, peaches, apricots, grapes, olives, gooseberries, currants, hops, gorze for fences, and English oaks; also many different kinds of flowers. Around the farm-yard there were stables, a thrashing-barn with its winnowing machine, a blacksmith's forge, and on the ground plough-shares and other tools: in the middle was that happy mixture of pigs and poultry, which may be seen so comfortably lying together in every English farm-yard. At the distance of a few hundred yards, where the water of a little rill was dammed up into a pool, a large and substantial water-mill had been erected.

All this is very surprising, when it is considered, that five years ago, nothing but the fern flourished here. Moreover, native workmanship, taught by the missionaries, has effected this change: the lesson of the missionary is the enchanter's wand. The house has been built, the windows framed, the fields ploughed, and even the trees grafted, by the New Zealander. At the mill, a New Zealander may be seen powdered white with flour, like his brother miller in England. When I looked at this whole scene, I thought it admirable. It was not merely that England was vividly brought before my mind; yet, as the evening drew to a close, the domestic sounds, the fields of corn, the distant country with its trees now appearing like pasture-land, all might well be mistaken for some part of it. Nor was it the triumphant feeling at seeing what Englishmen could effect, but it was something of far more consequence; the object for which this labour had been bestowed – the moral effect on the aborigines of this fine country.

The missionary system here appears to me different from that

of Tahiti; much more attention is there paid to religious instruction, and to the direct improvement of the mind; here, more to the
arts of civilization. I do not doubt that in both cases, the same
object is kept in view. Judging from the success alone, I should
rather lean to the Tahiti side; probably, however, each system is
best adapted to the country where it is followed. The mind of a
Tahitian is certainly one of a higher order; and on the other hand,
the New Zealander, not being able to pluck from the tree that
shades his house the bread-fruit and banana, would naturally turn
his attention with more readiness to the arts. When comparing the
state of New Zealand with that of Tahiti, it should always be
remembered, that from the respective forms of government of
the two countries, the missionaries here have had to labour at a
task, many times more difficult. The reviewer of Mr Earle's
travels in the *Quarterly Journal*, by pointing out a more advantageous line of conduct for the missionaries, evidently considers
that too much attention has been paid to religious instruction, in
proportion to other subjects. This opinion being so very different
from the one at which I arrived, any third person hearing the two
sides, would probably conclude, that the missionaries had been
the best judges, and had chosen the right path.

Several young men were employed about the farm, who had
been brought up by the missionaries; having been redeemed by
them from slavery. They were dressed in a shirt, jacket and
trousers, and had a respectable appearance. Judging from one
trifling anecdote, I should think they must be honest. When
walking in the fields, a young labourer came up to Mr Davies, and
gave him a knife and gimlet, saying he had found them on the
road, and did not know to whom they belonged! These young
men and boys appeared very merry and good-humoured. In the
evening I saw a party of them at cricket: when I thought of the
austerity of which the missionaries have been accused, I was
amused by observing one of their own sons taking an active part
in the game. A more decided and pleasing change was manifested
in the young women, who acted as servants within the houses.
Their clean, tidy, and healthy appearance, like that of dairy-maids
in England, formed a wonderful contrast with the women of the
filthy hovels in Kororadika. The wives of the missionaries tried to

persuade them not to be tattooed; but a famous operator having arrived from the south, they said, 'We really must just have a few lines on our lips; else when we grow old our lips will shrivel, and we shall be so very ugly.' Tattooing is not nearly so much practised as formerly; but as it is a badge of distinction between the chief and the slave, it will not probably very soon be disused. So soon does any train of ideas become habitual, that the missionaries told me, that even in their eyes, a plain face looked mean, and not like that of a New Zealand gentleman.

Late in the evening I went to Mr Williams's house, where I passed the night. I found there a very large party of children, collected together for Christmas-day, and all sitting round a table at tea. I never saw a nicer or more merry group: and to think, that this was in the centre of the land of cannibalism, murder, and all atrocious crimes! The cordiality and happiness so plainly pictured in the faces of the little circle, appeared equally felt by the older persons of the mission.

* * *

CHRISTMAS-DAY – In a few more days, the fourth year of our absence from England will be completed. Our first Christmas-day was spent at Plymouth; the second at St Martin's Cove, near Cape Horn; the third at Port Desire, in Patagonia; the fourth at anchor in a harbour in the Peninsula of Tres Montes; this fifth here; and the next, I trust in providence, will be in England. We attended divine service in the chapel of Pahia; part of the service was read in English, and part in the New Zealand language.

As far as I was able to understand, the greater number of people in this northern part of the island profess Christianity. It is curious, that the religion even of those who do not profess it, has been modified and is now partly Christian, partly heathen. Moreover, so excellent is the Christian faith, that the outward conduct even of the unbelievers is said to have been decidedly improved by the spread of its doctrines. It is beyond doubt, however, that much immorality still exists; that there are many who would not hesitate to kill a slave for a trifling offence; and that polygamy is still common – indeed, I believe, general.

We did not hear of any recent act of cannibalism; but Mr Stokes found burnt human bones, strewed round an old fireplace, on a small island near the anchorage: these remains of some quiet banquet might, indeed, have been lying there for several years. Notwithstanding the above facts, it is probable that the moral state of the people will rapidly improve. Mr Bushby mentioned one pleasing anecdote as a proof of the sincerity of some, at least, of those who profess Christianity. One of his young men left him, who had been accustomed to read prayers to the rest of the servants. Some weeks afterwards, happening to pass late in the evening by an outhouse, he saw and heard one of his men reading the bible with difficulty, by the light of the fire, to the others. After this, the party knelt and prayed: in their prayers they mentioned Mr Bushby and his family, and the missionaries, each separately in his respective district.

DECEMBER 26TH – Mr Bushby offered to take Mr Sulivan and myself in his boat, some miles up the river to Cawa-Cawa; and proposed afterwards to walk on to the village of Waiomio, where there are some curious rocks. Following one of the arms of the bay, we enjoyed a pleasant row, and passed through pretty scenery, until we came to a village, beyond which the boat could not proceed. From this place a chief and a party of men volunteered to walk with us to Waiomio, a distance of 4 miles. The chief was at this time rather notorious, from having lately hung one of his wives and a slave, for adultery. When one of the missionaries remonstrated with him, he seemed surprised, and said he thought he was exactly following the English method. Old Shongi, who happened to be in England during the Queen's trial, expressed great disapprobation at the whole proceeding: he said he had five wives, and he would rather cut off all their heads, than be so much troubled about one. Leaving this village, we crossed over to another, seated on a hill-side at a little distance. The daughter of a chief, who was still a heathen, had died here five days before. The hovel in which she had expired had been burnt to the ground: her body being enclosed between two small canoes was placed upright on the ground, and protected by an enclosure bearing wooden images of their gods, and the whole was painted bright red, so as to be conspicuous from afar. Her gown was fastened to the coffin, and her hair being cut off was cast at its foot. The relatives of the family had torn the flesh of their arms, bodies, and faces, so that they were covered with clotted blood; and the old women looked most filthy, disgusting objects. On the following day some of the officers visited this place, and found the women still howling and cutting themselves.

We continued our walk, and soon reached Waiomio. Here there are some singular masses of limestone, resembling ruined castles. These rocks have long served for burial-places, and in consequence are held sacred. One of the young men cried out, 'Let us all be brave,' and ran on ahead; but when within 100 yards, the whole party thought better of it, and stopped short. With perfect indifference, however, they allowed us to examine the whole place. At this village we rested some hours, during which time there was a long discussion with Mr Bushby, concerning the right of sale of certain lands. One old man, who appeared a perfect genealogist, illustrated the successive possessors by bits of stick driven into the ground. Before leaving the houses, a little basket-ful of roasted sweet-potatoes was given to each of our party; and we all, according to the custom, carried them away to eat on the road. I noticed that among the women employed in cooking, there was a man-slave: it must be an humiliating thing for a man in this warlike country to be employed in doing that which is considered as the lowest woman's work. Slaves are not allowed to go to war; but this perhaps can hardly be considered as a hardship. I heard of one poor wretch who, during hostilities, ran away to the opposite party; being met by two men, he was immediately seized; but they not agreeing to whom he should belong, each stood over him with a stone hatchet, and seemed determined that the other at least should not take him away alive. The poor man, almost dead with fright, was only saved by the address of a chief's wife. We afterwards enjoyed a pleasant walk back to the boat, but did not reach the ship till late in the evening.

DECEMBER 30TH – In the afternoon we stood out of the Bay of Islands on our course to Sydney. I believe we were all glad to leave New Zealand. It is not a pleasant place. Amongst the natives there is absent that charming simplicity which is found at Tahiti; and the greater part of the English are the very refuse of society. Neither is the country itself attractive. I look back but to one bright spot, and that is Waimate, with its Christian inhabitants.

CHAPTER XXI

AUSTRALIA

JANUARY 12TH, 1836 – Early in the morning, a light air carried us
towards the entrance of Port Jackson. Instead of beholding a
verdant country scattered over with fine houses: a straight line of
yellowish cliff brought to our minds the coast of Patagonia. A
solitary lighthouse, built of white stone, alone told us we were
near a great and populous city. Having entered the harbour, it
appeared fine and spacious; but the level country, showing on the
cliff-formed shores bare and horizontal strata of sandstone, was
covered by woods of thin scrubby trees, that bespoke useless
sterility. Proceeding further inland, the country improved;
beautiful villas and nice cottages were here and there scattered
along the beach. In the distance stone houses, two and three
stories high, and windmills, standing on the edge of a bank,
pointed out to us the neighbourhood of the capital of Australia.

At last we anchored within Sydney Cove. We found the little
basin occupied by many large ships, and surrounded by warehouses. In the evening I walked through the town, and returned
full of admiration at the whole scene. It is a most magnificent

testimony to the power of the British nation. Here, in a less promising country, scores of years have effected many times more, than the same number of centuries have done in South America. My first feeling was to congratulate myself that I was born an Englishman. Upon seeing more of the town afterwards, perhaps my admiration fell a little; but yet it is a fine town; the streets are regular, broad, clean, and kept in excellent order; the houses are of a good size, and the shops well furnished. It may be faithfully compared to the large suburbs, which stretch out from London and a few other great towns in England: but not even near London or Birmingham is there an appearance of such rapid growth. The number of large houses just finished and others building was truly surprising; nevertheless, every one complained of the high rents and difficulty in procuring a house. In the streets, gigs, phaetons, and carriages with livery servants, were driving about; and of the latter, many were extremely well equipped. Coming from South America, where in the towns every man of property is known, no one thing surprised me more than not being able to ascertain readily to whom this or that carriage belonged.

Many of the older residents say, that formerly they knew every face in the colony, but now that in a morning's ride it is a chance if they know one. Sydney has a population of 23,000, and is rapidly increasing: it must contain much wealth. It appears that a man of business can hardly fail to make a large fortune. I saw on all sides fine houses – one built from the profits of steam-vessels – another from building, and so on. An auctioneer, who was a convict, it is said, intends to return home, and will take with him £100,000. Another has an income so large that scarcely any body ventures to guess at it – the least sum assigned being 15,000 a year. But the two crowning facts are – first, that the public revenue has increased 60,000*l.* during this last year; and secondly, that less than an acre of land within the town of Sydney sold for 8,000*l.* sterling.

I hired a man and two horses to take me to Bathurst; a village about 120 miles in the interior, and the centre of a great pastoral district. By this means I hoped to get a general idea of the appearance of the country. On the morning of the 16th (January) I set out on my excursion. The first stage took us to Paramatta, a small country-town, the second to Sydney in importance. The roads were excellent, and made upon the MacAdam principle: whinstone having been brought for the purpose from the distance of several miles. The road appeared much frequented by all sorts of carriages; and I met two stage-coaches. In all these respects there was a close resemblance to England; perhaps the number of alehouses was here in excess. The iron gangs, or parties of convicts, who have here committed some trifling offence, appeared the least like England; they were working in chains, under the charge of sentries with loaded arms. The power, which the government possesses, by means of forced labour, of at once opening good roads throughout the country, has been, I believe, one main cause of the early prosperity of this colony.

I slept at night at a very comfortable inn at Emu ferry, 35 miles from Sydney, and near the ascent of the Blue Mountains. This line of road is the most frequented, and has been longest inhabited of any in the colony. The whole land is enclosed with high railings, for the farmers have not succeeded in rearing hedges. There are many substantial houses and good cottages scattered about; but although considerable pieces of land are under cultivation, the greater part yet remains as when first discovered. Making allowances for the cleared parts, the country here resembled all that I saw during the ten succeeding days.

The extreme uniformity of the vegetation is the most remarkable feature in the landscape of the greater part of New South Wales. Every where we have an open woodland; the ground being partially covered with a very thin pasture. The trees nearly all belong to one family; and mostly have the surface of their leaves placed in a vertical, instead of as in Europe, a nearly horizontal position: the foliage is scanty, and of a peculiar, pale green tint, without any gloss. Hence the woods appear light and shadowless: this, although a loss of comfort to the traveller under the scorching rays of summer, is of importance to the farmer, as it allows grass to grow where it otherwise could not. The leaves are not shed periodically: this character appears common to the entire southern hemisphere, namely, South America, Australia, and the Cape of Good Hope. The inhabitants of this hemisphere and of the inter-tropical regions, thus lose perhaps one of the most glorious, though to our eyes common, spectacles in the world – the first bursting into full foliage of the leafless tree. They may, however, say that we pay dearly for our spectacle, by having the land covered with mere naked skeletons for so many months. This is too true; but our senses thus acquire a keen relish for the exquisite green of the spring, which the eyes of those living within the tropics, sated during the long year with the gorgeous productions of those glowing climates, can never experience. The greater number of the trees, with the exception of some of the blue gums, do not attain a large size; but they grow tall and tolerably straight, and stand well apart. The bark of some falls annually, or hangs dead in long shreds, which swing about with the wind; and hence the woods appear desolate and untidy. Nowhere is there an appearance of verdure, but rather that of arid sterility. I cannot imagine a more complete contrast in every respect than between the forests of Valdivia, or Chiloe, and the woods of Australia.

Although this colony flourishes so remarkably, the appearance of infertility is to a certain degree real. The soil without doubt is good, but there is so great a deficiency both of rain and running water, that it cannot produce much. The agricultural crops, and often those in gardens, are estimated to fail once in three years; and this has even happened on successive years. Hence the colony cannot supply itself with the bread and vegetables, which its inhabitants consume. It is essentially pastoral, and chiefly so for sheep, and not the larger quadrupeds. The alluvial land near Emu ferry was some of the best cultivated which I saw; and certainly the scenery on the banks of the Nepean, bounded to the west by the Blue Mountains, was pleasing even to the eye of a person thinking of England.

At sunset, a party of a score of the black aborigines passed by, each carrying, in their accustomed manner, a bundle of spears and other weapons. By giving a leading young man a shilling, they were easily detained, and threw their spears for my amusement. They were all partly clothed, and several could speak a little English; their countenances were goodhumoured and pleasant; and they appeared far from being such utterly degraded beings as they are usually represented. In their own arts they are admirable: a cap being fixed at 30 yards distance, they transfixed it with a spear, delivered by the throwing stick, with the rapidity of an arrow from the bow of a practised archer. In tracking animals or men they show most wonderful sagacity; and I heard of several of their remarks which manifested considerable acuteness. They will not, however, cultivate the ground, or build houses and remain stationary, or even take the trouble of tending a flock of sheep when given to them. On the whole they appear to me to stand some few degrees higher in the scale of civilization than the Fuegians.

It is very curious thus to see in the midst of a civilized people, a set of harmless savages wandering about without knowing where they shall sleep at night, and gaining their livelihood by hunting in the woods. As the white man has travelled onwards, he has spread over the country belonging to several tribes. These, although thus enclosed by one common people, keep up their ancient distinctions, and sometimes go to war with each other. In an engagement which took place lately, the two parties most singularly chose the centre of the village of Bathurst, for the field of battle. This was of service to the defeated side, for the runaway warriors took refuge in the barracks.

The number of aborigines is rapidly decreasing. In my whole ride, with the exception of some boys brought up in the houses, I saw only one other party; these were rather more numerous than the first, and not so well clothed. This decrease, no doubt, must be partly owing to the introduction of spirits, to European diseases (even the milder ones of which, as the measles,* prove very destructive), and to the gradual extinction of the wild animals. It is said that numbers of their children invariably perish in very early infancy from the effects of their wandering life. As the difficulty of procuring food increases, so must their wandering habits; and hence the population, without any apparent deaths

---

* It is remarkable how the same disease is modified in different climates. At the little island of St Helena, the introduction of scarlet fever is dreaded as a plague. In some countries, foreigners and natives are as differently affected by certain contagious disorders, as if they had been different animals; of which fact some instances have occurred in Chile; and, according to Humboldt, in Mexico. (Polit. Essay on Kingdom of New Spain, vol. iv.)

from famine, is repressed in a manner extremely sudden compared to what happens in civilized countries, where the father may add to his labour, without destroying his offspring.

Besides these several evident causes of destruction, there appears to be some more mysterious agency generally at work. Wherever the European has trod, death seems to pursue the aboriginal. We may look to the wide extent of the Americas, Polynesia, the Cape of Good Hope, and Australia, and we shall find the same result. Nor is it the white man alone, that thus acts the destroyer; the Polynesian of Malay extraction has in parts of the East Indian archipelago, thus driven before him the dark-coloured native. The varieties of man seem to act on each other; in the same way as different species of animals – the stronger always extirpating the weaker. It was melancholy at New Zealand to hear the fine energetic natives saying, they knew the land was doomed to pass from their children. Every one has heard of the inexplicable reduction of the population in the beautiful and healthy island of Tahiti since the date of Captain Cook's voyages: although in that case we might have expected it would have been otherwise; for infanticide, which formerly prevailed to so extraordinary a degree, has ceased, and the murderous wars have become less frequent.

* * *

JANUARY 18TH – Very early in the morning, I walked about three miles to see Govett's Leap: a view of a similar but even perhaps more stupendous character than that near the Weatherboard. So early in the day the gulf was filled with a thin blue haze, which, although destroying the general effect, added to the apparent depth at which the forest was stretched below the country on which we were standing. Soon after leaving the Blackheath, we descended from the sandstone platform by the pass of Mount Victoria. To effect this pass, an enormous quantity of stone has been cut through; the design, and its manner of execution, would have been worthy of any line of road in England – even that of Holyhead. We now entered upon a country less elevated by nearly 1,000 feet, and consisting of granite. With the change of rock, the vegetation improved; the trees were both finer, and stood further apart, and the pasture between them was a little greener, and more plentiful.

At Hassan's Walls, I left the high road, and made a short detour to a farm called Walerawang; to the superintendent of which, I had a letter of introduction from the owner in Sydney. Mr Browne had the kindness to ask me to stay the ensuing day, which I had much pleasure in doing. This place offers an example of one of the large farming, or rather sheep-grazing, establishments of the colony. Cattle and horses are, however, in this case, rather more numerous than usual, owing to some of the valleys being swampy, and producing a coarser pasture. The sheep were 15,000 in number, of which the greater part were feeding under the care of different shepherds, on unoccupied ground, at the distance of more than 100 miles, and beyond the limits of the colony. Mr Browne had just finished, this day, the last of the shearing of 7,000 sheep; the rest being sheared in another place. I believe the profit of the average produce of wool from 15,000 sheep, would be more than 5,000l. sterling. Two or three flat pieces of ground near the house were cleared and cultivated with corn, which the harvest men were now reaping: but no more wheat is sown than sufficient for the annual support of the labourers employed on the establishment. The usual number of assigned convict servants here is about forty, but at the present time there were rather more. Although the farm was well stocked with every requisite, there was an apparent absence of comfort; and not even a single woman resided here. The sunset of a fine day will generally cast an air of happy contentment on any scene; but here, at this retired farm-house, the brightest tints on the surrounding woods could not make me forget that forty hardened, profligate men, were ceasing from their daily labours, like the slaves from Africa, yet without their just claim for compassion.

Early on the next morning, Mr Archer, the joint superintend-ent, had the kindness to take me out Kangaroo-hunting. We continued riding the greater part of the day, but had very bad sport, not seeing a kangaroo, or even a wild dog. The greyhounds pursued a kangaroo rat into a hollow tree, out of which we dragged it: it is an animal as big as a rabbit, but with the figure of a kangaroo. A few years since, this country abounded with wild animals; but now the emu is banished to a long distance, and the kangaroo is become scarce; to both, the English greyhound is utterly destructive. It may be long before these animals are altogether exterminated, but their doom is fixed. The natives are always anxious to borrow the dogs from the farm-houses: the use of them, the offal when an animal is killed, and milk from the cows, are the peace-offerings of the settlers, who push further and further towards the interior. The thoughtless aboriginal, blinded by these trifling advantages, is delighted at the approach of the white man, who seems predestined to inherit the country of his children.

Although having bad sport, we enjoyed a pleasant ride. The woodland is generally so open that a person on horseback can gallop through it. It is traversed by a few flat-bottomed valleys, which are green and free from trees: in such spots the scenery was like that of a park, and pretty. In the whole country I scarcely saw a place without the marks of fire; whether these had been more or less recent – whether the stumps were more or less black, was the greatest change which varied the uniformity, so wearisome to the traveller's eye. In these woods there are not many birds; I saw, however, some large flocks of the white cockatoos feeding in a corn-field, and a few most beautiful parrots; crows like our jackdaws were not uncommon, and another bird something like the magpie. The English have not been very particular in giving names to the productions of Australia; trees of one genus (Casuarina) are called oaks for no one reason that I can discover, without it is that there is no one point of resemblance. Some quadrupeds are called tigers and hyenas, simply because they are carnivorous, and so on in many other cases.

In the dusk of the evening I took a stroll along a chain of ponds, which in this dry country represented the course of a river, and had the good fortune to see several of the famous Platypus, or *Ornithorhyncus paradoxus*. They were diving and playing about the surface of the water, but showed so little of their bodies that they might easily have been mistaken for water-rats. Mr Browne shot one: certainly it is a most extraordinary animal; the stuffed specimens do not at all give a good idea of the recent appearance of its head and beak; the latter becoming hard and contracted.

A little time before this I had been lying on a sunny bank, and was reflecting on the strange character of the animals of this country as compared with the rest of the world. An unbeliever in every thing beyond his own reason might exclaim, 'Two distinct Creators must have been at work; their object, however, has been the same, and certainly the end in each case is complete.' While thus thinking, I observed the hollow conical pitfall of the lion-ant: first a fly fell down the treacherous slope and immediately disappeared; then came a large but unwary ant; its struggles to escape being very violent, those curious little jets of sand, described by Kirby* as being flirted by the insect's tail, were promptly directed against the expected victim. But the ant enjoyed a better fate than the fly, and escaped the fatal jaws which lay concealed at the base of the conical hollow. There can be no doubt but that this predacious larva belongs to the same genus with the European kind, though to a different species. Now what would the sceptic say to this? Would any two workmen ever have hit upon so beautiful, so simple, and yet so artificial a contrivance? It cannot be thought so: one Hand has surely worked throughout the universe.

* * * *

Before arriving here the three things which interested me most were – the state of society amongst the higher classes, the condition of the convicts, and the degree of attraction sufficient to induce persons to emigrate. Of course, after so very short a visit one's opinion is worth scarcely any thing; but it is as difficult not to form some opinion, as it is to form a correct judgment. On the whole, from what I heard, more than from what I saw, I was disappointed in the state of society. The whole community is rancorously divided into parties on almost every subject. Among those, who from their station in life ought to be the best, many

---

* Kirby's Entomology, vol. i, p. 425. The Australian pitfall is only about half the size of the one made by the European species.

live in such open profligacy, that respectable people cannot associate with them. There is much jealousy between the children of the rich emancipist and the free settlers; the former being pleased to consider honest men as interlopers. The whole population, poor and rich, are bent on acquiring wealth; amongst the higher orders wool and sheep-grazing form the constant subject of conversation. The very low ebb of literature is strongly marked by the emptiness of the booksellers' shops; for they are inferior even to those in the smaller country-towns of England.

There are many serious drawbacks to the comforts of families; the chief of which, perhaps, is being surrounded by convict servants. How thoroughly odious to every feeling to be waited on by a man, who the day before, perhaps, was flogged, from your representation, for some trifling misdemeanor. The female servants are of course much worse; hence children learn the vilest expressions, and it is fortunate if not equally vile ideas.

On the other hand, the capital of a person without any trouble on his part, produces him treble interest to what it will in England; and with care he is sure to grow rich. The luxuries of life are in abundance and very little dearer, and most articles of food cheaper, than in England. The climate is splendid and quite healthy; but to my mind its charms are lost by the uninviting aspect of the country. Settlers possess a great advantage in finding their sons of service, when very young. At the age of from 16 to 20 they frequently take charge of distant farming stations; this, however, must happen at the expense of their boys associating entirely with convict servants. I am not aware that the tone of society has assumed any peculiar character; but with such habits, and without intellectual pursuits, it can hardly fail to deteriorate. My opinion is such, that nothing but rather severe necessity should compel me to emigrate.

The rapid prosperity and future prospects of this colony are to me, not understanding these subjects, very puzzling. The two main exports are wool and whale-oil; and to both of these productions there is a limit. The country is totally unfit for canals; therefore there is a line not very distant, beyond which the land carriage of wool will not repay the expense of shearing and tending sheep. Pasture every where is so thin, that settlers have already pushed far into the interior: moreover the country further inland becomes extremely poor. I have before said that agriculture can never succeed on a very extended scale; therefore so far as I can see, Australia must ultimately depend upon being the centre of commerce for the southern hemisphere, and perhaps on her future manufactories. Possessing coal, she always has the moving power at hand. From the habitable country extending along the coast, and from her English extraction she is sure to be a maritime nation. I formerly imagined that Australia would rise to be as grand and powerful a country as North America; but now it appears to me such future grandeur is rather problematical.

With respect to the state of the convicts, I had still fewer opportunities of judging than on other points. The first question is, whether their condition is at all one of punishment: no one will maintain that it is a very severe one. This, however, I suppose is of little consequence as long as it continues to be an object of dread to criminals at home. The corporeal wants of the convicts are tolerably well supplied; their prospect of future liberty and comfort is not distant, and after good conduct certain. A 'ticket of leave', which, as long as a man keeps clear of suspicion as well as of crime, makes him free within a certain district, is given upon good conduct after years proportional to the length of the sentence. For life, eight years is the time of probation; for seven years, four, &c. Yet with all this, and overlooking the previous imprisonment and wretched passage out, I believe the years of assignment are passed away with discontent and unhappiness. As an intelligent man remarked to me, the convicts know no pleasure beyond sensuality, and in this they are not gratified. The enormous bribe which government possesses in offering free pardons, together with the deep horror of the secluded penal settlements, destroys confidence between the convicts, and so prevents crime. As to a sense of shame, such a feeling does not appear to be known, and of this I witnessed some very singular proofs. Though it is a curious fact, I was universally told, that the character of the convict population is one of arrant cowardice: not unfrequently some become desperate and quite indifferent of life, yet a plan requiring cool or continued courage is seldom put into execution. The worst feature in the whole case is, that although there exists what may be called a legal reform, and comparatively little which the law can touch is committed, yet that any moral reform should take place appears to be quite out of the question. I was assured by well-informed people, that a man who should try to improve, could not while living with other assigned servants: his life would be one of intolerable misery and persecution. Nor must the contamination of the convict ships and prisons both here and in England be forgotten. On the whole, as a place of punishment the object is scarcely gained; as a real system of reform it has failed, as perhaps would every other plan: but as a means of making men outwardly honest – of converting vaga-bonds most useless in one hemisphere into active citizens of another, and thus giving birth to a new and splendid country – a grand centre of civilization – it has succeeded to a degree perhaps unparalleled in history.

VAN DIEMEN'S LAND

JANUARY 30TH – The *Beagle* sailed for Hobart Town in Van Diemen's Land. On the 5th of February, after a six days' passage, of which the first part was fine, and the latter very cold and squally, we entered the mouth of Storm Bay: the weather justified this awful name. The bay should rather be called an estuary, for it receives at its head the waters of the Derwent. Near the mouth there are some extensive basaltic platforms; but higher up, the land becomes mountainous and is covered by a light wood. The lower parts of the hills which skirt the bay are cleared; and the bright yellow fields of corn, and dark green ones of potatoes appeared very luxuriant. Late in the evening we anchored in the snug cove, on the shores of which stands the capital of Tasmania, as Van Diemen's Land is now called. The first aspect of the place was very inferior to that of Sydney; the latter might be called a city, this only a town.

In the morning I walked on shore. The streets are fine and broad; but the houses rather scattered: the shops appeared good. The town stands at the base of Mount Wellington, a mountain, 3,100 feet high, but of very little picturesque beauty: from this source, however, it receives a good supply of water. Round the cove there are some fine warehouses, and on one side a small fort. Coming from the Spanish settlements, where such magnificent care has generally been paid to the fortifications, the means of defence in these colonies appeared very contemptible. Comparing the town to Sydney, I was chiefly struck with the comparative fewness of the large houses, either built or building. This circumstance must indicate that fewer people are gaining large fortunes. The growth, however, of small houses has been most abundant; and the vast number of little red brick dwellings, scattered on the hill behind the town, sadly destroys its picturesque appearance. Hobart Town, from the census of this year, contained 13,826 inhabitants, and the whole of Tasmania 36,505.

All the aborigines have been removed to an island in Bass's Straits, so that Van Diemen's Land enjoys the great advantage of being free from a native population. This most cruel step seems to have been quite unavoidable, as the only means of stopping a fearful succession of robberies, burnings, and murders, committed by the blacks; but which sooner or later must have ended in their utter destruction. I fear there is no doubt that this train of evil and its consequences, originated in the infamous conduct of some of our countrymen. Thirty years is a short period, in which to have banished the last aboriginal from his native island, – and that island nearly as large as Ireland. I do not know a more striking instance of the comparative rate of increase of a civilized over a savage people.

* * *

The *Beagle* stayed here ten days, and in this time I made several pleasant little excursions, chiefly with the object of examining the geological structure of the immediate neighbourhood. The main points of interest consist, first in the presence of certain basaltic rocks which evidently have flowed as lava; secondly, in some great unstratified masses of greenstone; thirdly, in proofs of an exceedingly small rise of the land; fourthly, in some ancient fossiliferous strata, probably of the age of the Silurian system of Europe; and lastly, in a solitary and superficial patch of yellowish limestone or travertin, which contains numerous impressions of leaves of trees and plants, not now existing. It is not improbable that this one small quarry, includes the only remaining record of the vegetation of Van Diemen's Land during one former epoch.

* * *

FEBRUARY 17TH – The *Beagle* sailed from Tasmania, and, on the 6th of the ensuing month, reached King George's Sound, situated near the SW corner of Australia. We staid there eight days; and I do not remember, since leaving England, having passed a more dull, uninteresting time. The country, viewed from an eminence, appears a woody plain, with here and there rounded and partly bare hills of granite protruding. One day I went out with a party, in hopes of seeing a kangaroo hunt, and walked over a good many miles of country. Every where we found the soil sandy, and very poor; it either supported a coarse vegetation of thin, low brushwood and wiry grass, or a forest of stunted trees. The scenery resembled the elevated sandstone platform of the Blue Mountains; the Casuarina (a tree somewhat resembling a Scotch fir) is, however, here in greater number, as the Eucalyptus is in rather less. In the open parts there were many grass-trees; a plant which, in appearance, has some affinity with the palm; but, instead of being surmounted by a crown of noble fronds, it can boast merely of a tuft of coarse grass. The general bright green colour of the brushwood and other plants, viewed from a distance, seemed to bespeak fertility. A single walk, however, will quite dispel such an illusion; and he who thinks with me, will never wish to walk again in so uninviting a country.

One day I accompanied Captain FitzRoy to Bald Head; the place mentioned by so many navigators, where some imagined they saw coral, and others petrified trees, standing in the position in which they grew. According to our view, the rock was formed by the wind heaping up calcareous sand, during which process, branches and roots of trees, and land-shells were enclosed; the mass being afterward consolidated by the percolation of rain-water. When the wood had decayed, lime was washed into the cylindrical cavities, and became hard, sometimes even like that in a stalactite. The weather is now wearing away the softer rock, and in consequence the casts of roots and branches project above the surface: their resemblance to the stumps of a dead shrubbery was so exact, that, before touching them, we were sometimes at a loss to know which were composed of wood, and which of calcareous matter.

A large tribe of natives, called the White Cockatoo men, happened to pay the town a visit while we were there. These men, as well as those of the tribe belonging to King George's Sound, being tempted by the offer of some tubs of rice and sugar, were persuaded to hold a 'corrobery', or great dancing-party. As soon as it grew dark, small fires were lighted, and the men commenced their toilet, which consisted in painting themselves white in spots and lines. As soon as all was ready, large fires were kept blazing, round which the women and children were collected as spectators; the Cockatoo and King George's men formed two distinct parties, and danced generally in answer to each other. The dancing consisted in the whole set running either sideways or in Indian file, into an open space, and stamping the ground with great force as they marched together. Their heavy footsteps were accompanied by a kind of grunt, and, by beating their clubs and weapons, and various other gesticulations, such as extending their arms, and wriggling their bodies. It was a most rude, barbarous scene, and, to our ideas, without any sort of meaning; but we observed that the women and children watched the whole proceeding with the greatest pleasure. Perhaps these dances originally represented some scenes, such as wars and victories; there was one called the Emu dance, in which each man extended his arm in a bent manner, so as to imitate the neck of that bird. In another dance, one man took off the movements of a kangaroo grazing in the woods, whilst a second crawled up, and pretended to spear him. When both tribes mingled in the dance, the ground trembled with the heaviness of their steps, and the air resounded with their wild cries. Every one appeared in high spirits, and the group of nearly naked figures, viewed by the light of the blazing fires, all moving in hideous harmony, formed a perfect representation of a festival amongst the lowest barbarians. In Tierra del Fuego, we have beheld many curious scenes in savage life, but never, I think, one where the natives were in such high spirits, and so perfectly at their ease. After the dancing was over, the whole party formed a great circle on the ground, and the boiled rice and sugar was distributed, to the delight of all.

After several tedious delays from clouded weather, on the 14th of March, we gladly stood out of King George's Sound on our course to Keeling Island. Farewell, Australia! you are a rising infant and doubtless some day will reign a great princess in the south: but you are too great and ambitious for affection, yet not great enough for respect. I leave your shores without sorrow or regret.

CHAPTER XXII

CORAL FORMATIONS

APRIL 1ST – We arrived in view of the Keeling or Cocos Islands, situated in the Indian ocean, and about 600 miles distant from the coast of Sumatra. This is one of the lagoon islands of coral formation, similar to those we passed in the Dangerous Archipelago. An excellent idea of the general appearance of these extraordinary rings of land, which rise out of the depths of the ocean, may be obtained from the characteristic sketch of Whitsunday Island, in Beechey's *Voyage*.

* * *

The next morning after anchoring, I went on shore on Direction Island. The strip of dry land is only a few hundred yards wide; on the lagoon side we have a white calcareous beach, the radiation from which in such a climate is very oppressive; and on the outer coast, a solid broad flat of coral rock, which serves to break the violence of the open sea. Excepting near the lagoon where there is some sand, the land is entirely composed of rounded fragments of coral. In such a loose, dry, stony soil, the climate of the inter-tropical regions alone could produce a vigorous vegetation. On some of the smaller islets, nothing could be more elegant, than the manner in which the young and full-grown cocoa-nut trees, without destroying each other's symmetry, were mingled into one wood. A beach of glittering white sand formed a border to these fairy spots.

I will now give a sketch of the natural history of these islands, which, from its very paucity, possesses a peculiar interest. The cocoa-nut tree, at the first glance, seems to compose the whole wood; there are, however, five or six other kinds. One of these grows to a very large size, but, from the extreme softness of its wood, is useless; another sort affords excellent timber for ship-building. Besides the trees, the number of plants is exceedingly limited, and consists of insignificant weeds. In my collection, which includes, I believe, nearly the perfect Flora, there are twenty species, without reckoning a moss, lichen, and fungus. To this number two trees must be added; one of which was not in flower, and the other I only heard of. The latter is a solitary tree of its kind in the whole group, and grows near the beach, where, without doubt, the one seed was thrown up by the waves. I do not include in the above list, the sugar-cane, banana, some other vegetables, fruit-trees, and imported grasses. As these islands consist entirely of coral, and at one time probably existed as a mere water-washed reef, all the productions now living here, must have been transported by the waves of the sea. In accordance to this, the Flora has quite the character of a refuge for the destitute: Professor Henslow informs me, that of the twenty species, nineteen belong to different genera, and these again to no less than sixteen orders!

In Holman's* Travels, an account is given on the authority of Mr A. S. Keating, who resided twelve months on these islands, of the various seeds, and other bodies, which have been known to have been washed on shore:

Seeds and plants from Sumatra and Java have been driven up by the surf on the windward side of the islands. Among them have been found the Kimiri, native of Sumatra and the peninsula of Malacca; the cocoa-nut of

---

* Holman's Travels, vol. iv, p. 378.

Balci, known by its shape and size; the Dadass, which is planted by the Malays with the pepper-vine, the latter intwining round its trunk, and supporting itself by the prickles on its stem; the soap-tree; the castor-oil plant; trunks of the sago palm; and various kinds of seeds unknown to the Malays who settled on the islands. These are all supposed to have been driven on shore by the NW monsoon to the coast of New Holland, and thence to these islands by the SE tradewind. Large masses of Java teak, and yellow wood, have also been found, besides immense trees, of red and white cedar, and the blue gum-wood of New Holland, in a perfectly sound condition. All the hardy seeds, such as creepers, retain their germinating power, but the softer kinds, among which is the mangostin, are destroyed in the passage. Fishing-canoes, apparently from Java, have at times been washed on shore.

It is interesting thus to discover how numerous the seeds are, which, coming from several countries, are drifted over the wide ocean. Professor Henslow tells me, he believes that nearly all the plants which I brought from this island, are common littoral species in the East Indian archipelago. From the direction, however, of the winds and currents, it seems scarcely possible that they can have come here in a direct line. If, as suggested with much probability by Mr Keating, they have first been carried towards the coast of New Holland, and thence drifted back again, together with the productions of that country, the seeds, before germinating, must have travelled between 1,800 and 2,400 miles.

Chamisso,[*] when describing the Radack Archipelago, situated in the central part of the Western Pacific, states that, 'The sea brings to these islands the seeds and fruits of many trees, most of which have yet not grown here. The greater part of these seeds appear to have not yet lost the capability of growing.' It is also said that trunks of northern firs are washed on shore, which must have been floated from an immense distance. These facts are highly interesting. It cannot be doubted, if there were land-birds to pick up the seeds when first cast on shore, and a soil more adapted for their growth than the loose blocks of coral, that such islands, although so isolated, would soon possess a more abundant Flora.

The list of land-animals is even poorer than that of plants. Some of the islets are inhabited by rats; and their origin is known to be

---

[*] Kotzebue's First Voyage, vol. iii, p. 155.

duc to a ship from the Mauritius, which was wrecked here. These rats have rather a different appearance from the English kind; they are smaller and much more brightly coloured. There are no true land-birds; for a snipe and a rail (*Rallus phillippensis*), though living entirely among the dry herbage, belong to the order of Waders. Birds of this order are said to occur on several of the low islands in the Pacific. At Ascension a rail (*Porphyrio?*) was shot near the summit of the mountain; and it was evidently a solitary straggler. From these circumstances, I believe, the waders are the first colonists of any island, after the innumerable web-footed species. I may add, that whenever I have noticed birds, which were not pelagic, very far out at sea, they always belonged to this order; and hence they would naturally become the earliest colonist of any distant point.

Of reptiles, I saw only one small lizard. Of insects, I took pains to collect every kind. Exclusive of spiders, which were numerous, there were thirteen species.* Of these, one only was a beetle. A small species of ant, swarmed by thousands under the loose dry blocks of coral, and was the only true insect which was abundant. Although the productions of the land are thus scanty; if we look to the waters of the surrounding sea, the number of organic beings is indeed infinite.

Chamisso† has described the natural history of Romanzoff, a lagoon island in the Radack Archipelago. The number and kind of productions there is very nearly the same with those here. One small lizard was seen: wading birds (Numenius and Scolopax) were numerous, and very tame. Of plants, he states there were nineteen species (including one fern); and some of them are the same species with those I collected here, although on an island situated in a different ocean.

These strips of land are raised only to that height, to which the surf can throw fragments, and the wind heap up sand. Their protection is due to the outward and lateral increase of the reef,

* The thirteen species belong to the following orders. *Coleoptera*, a species of minute Elater; *Orthoptera*, a Gryllus and Blatta; *Hemiptera*, one; *Homoptera*, two; *Neuroptera*, a Chrysopa; *Hymenoptera*, two ants; *Lepidoptera Nocturna*, a Diopæa, and a Pterophorus (?). *Diptera*, two.
† Kotzebue's First Voyage, vol. iii, p. 222.

which thus breaks the sea. The aspect and constitution of these islets at once call up the idea, that the land and the ocean are here struggling for mastery: although terra firma has obtained a footing, the denizens of the other element think their claim at least equal. In every part one meets hermit-crabs of more than one species,* carrying on their backs the houses they have stolen from the neighbouring beach. Overhead, the trees are occupied by numbers of gannets, frigate-birds, and terns. From the many nests and smell of the atmosphere, this might be called a sea-rookery. The gannets, sitting on their rude nests, look at an intruder with a stupid, yet angry air. The noddies, as their name expresses, are silly little creatures. But there is one charming bird; it is a small and snow-white tern, which smoothly hovers at the distance of an arm's length from your head; its large black eye scanning with quiet curiosity your expression. Little imagination is required to fancy, that so light and delicate a body must be tenanted by some wandering fairy spirit.

* * *

APRIL 6TH – I accompanied Captain FitzRoy to an island at the head of the lagoon: the channel was exceedingly intricate, winding through fields of delicately branched corals. We saw several turtle, and two boats were then employed in catching them. The method is rather curious: the water is so clear and shallow, that although at first a turtle quickly dives out of sight, yet in a canoe, or boat under sail, the pursuers after no very long chase come up to it. A man standing ready in the bows, at this moment dashes through the water upon the turtle's back; then clinging with both hands by the shell of its neck, he is carried away till the animal becomes exhausted and is secured. It was quite an interesting chase to see the two boats thus doubling about, and the men dashing into the water trying to seize their prey.

---

* The large claws or pincers of some of these crabs, are most beautifully adapted, when drawn back, to form an operculum to the shell, which is nearly as perfect as the proper one that belonged to the original molluscous animal. I was assured, and as far as my observation went it was confirmed, that there are certain kinds of these hermits, which always use certain kinds only of old shells.

When we arrived at the head of the lagoon, we crossed the narrow islet and found a great surf breaking on the windward coast. I can hardly explain the cause, but there is to my mind a considerable degree of grandeur in the view of the outer shores of these lagoon islands. There is a simplicity in the barrier-like beach, the margin of green bushes and tall cocoa-nuts, the solid flat of coral rock, strewed here and there with great fragments, and the line of furious breakers, all rounding away towards either hand. The ocean throwing its waters over the broad reef appears an invincible, all-powerful enemy, yet we see it resisted and even conquered by means which at first seem most weak and inefficient.

It is not that the ocean spares the rock of coral; the great fragments scattered over the reef, and accumulated on the beach, whence the tall cocoa-nut springs, plainly bespeak the unrelenting power of its waves. Nor are there any periods of repose granted. The long swell, caused by the gentle but steady action of the trade wind always blowing in one direction over a wide area, causes breakers, which even exceed in violence those of our temperate regions, and which never cease to rage. It is impossible to behold these waves without feeling a conviction that an island, though built of the hardest rock, let it be porphyry, granite, or quartz, would ultimately yield and be demolished by such irresistible forces. Yet these low, insignificant coral islets stand and are victorious: for here another power, as antagonist to the former, takes part in the contest. The organic forces separate the atoms of carbonate of lime one by one from the foaming breakers, and unite them into a symmetrical structure. Let the hurricane tear up its thousand huge fragments; yet what will this tell against the accumulated labour of myriads of architects at work night and day, month after month. Thus do we see the soft and gelatinous body of a polypus, through the agency of the vital laws, conquering the great mechanical power of the waves of an ocean, which neither the art of man, nor the inanimate works of nature could successfully resist.

We did not return on board till late in the evening, as we staid some time in the lagoon collecting specimens of the giant Chama, and looking at the coral fields. Near the head of the lagoon I was much surprised to find a wide area, considerably more than a mile square, covered with a forest of branching coral, which though standing upright was all dead and rotten. At first I was quite at a loss to understand the cause; afterwards it occurred to me that it was owing to the following rather curious combination of circumstances. It should, however, first be stated, that corals are never able to survive even a short exposure in the air to the sun's rays, so that their upward limit of growth is determined by that of lowest water at spring tides. It appears from some old charts, that the long island to windward was formerly separated by wide channels into several islets; this fact is likewise indicated by the less age of the trees in certain portions. Under this former condition of the reef, a strong breeze, by throwing more water over the barrier, would tend to raise the level of the lagoon. Now it acts in a directly contrary manner; for the water, not only is not increased by currents from the outside, but is blown outwards by the force of the wind. Hence, it is observed, that the tides near the head of the lagoon do not rise so high during strong breezes as on ordinary occasions. This difference of level, although no doubt very small, has I believe caused the death of those coral groves, which under the former condition of things had attained the utmost possible limit of upward growth.

A few miles north of Keeling there is another small lagoon island, the centre of which is nearly filled up. Captain Ross found in the conglomerate of the outer coast a well rounded fragment of greenstone, rather larger than a man's head; he and the men with him were so much surprised at this, that they brought it away and preserved it as a curiosity. The occurrence of this one stone, where every other particle of matter is calcareous, certainly is very puzzling. The island has scarcely ever been visited, nor is it probable that a ship had been wrecked there. From the absence of any better explanation, I came to the conclusion that it must have come there entangled in the roots of some large tree: when, however, I considered the great distance from the nearest land, the combination of chances against a stone thus being entangled, the tree washed into the sea, floated so far, then landed safely, and the stone finally so embedded as to allow of its discovery, I was almost ashamed of imagining a means of transport so improbable.

It was therefore with great interest that I found Chamisso,* the justly distinguished naturalist who accompanied Kotzebue, stating that the inhabitants of the Radack Archipelago, a group of lagoon islands in the midst of the Pacific, obtained stones for sharpening their instruments by searching the roots of trees which are cast up on the beach. It will be evident that this must have happened several times, since laws have been established that such stones belong to the chief, and a punishment is inflicted on any one who attempts to defraud him of this right. When the isolated position of these small islands in the midst of a vast ocean – their great distance from any land excepting that of coral formation, a fact well attested by the value which the inhabitants, who are such bold navigators, attach to a stone of any kind,† – and the slowness of the currents of the open sea are all considered, the occurrence of pebbles thus transported does appear wonderful. Stones may often be thus transported; and if the island on which they are stranded is constructed of any other substance besides coral, they would scarcely attract attention, and their origin at least would never have been guessed. Moreover this agency may long escape discovery from the probability of trees, especially those loaded with stones, floating beneath the surface. In the channels of Tierra del Fuego large quantities of drift timber are cast upon the beach, yet it is extremely rare to meet a tree swimming on the water. It is easy to conceive that waterlogged wood might be transported, when floating close to the bottom, and occasionally even just touching it. The knowledge of any result which (with sufficient time allowed) can be produced by causes, though appearing infinitely improbable, is valuable to the geologist, for he by his creed deals with centuries and thousands of years as others do with minutes. If a few isolated stones are discovered in a mass of

---

* Kotzebue's first voyage, vol. iii, p. 155. It is said, 'The sea throws up on the reefs of Radack the trunks of northern firs (!) and trees of the torrid zone (palms, bamboos). It provides the inhabitants not only with timber for boats, but it also brings them in wrecks of European ships, the iron which they want.' – 'They receive, in a similar manner, another treasure, hard stones fit for whetting. They are sought for in the roots and hollows of the trees which the sea throws up.'

† Some natives carried by Kotzebue to Kamtschatka collected stones among other valuable articles to take back to their country.

fine sedimentary strata, it cannot, after the above facts, be considered as very improbable that they may have been drifted there
by the floating timber of a former epoch.

* * *

I was a good deal surprised by finding two species of coral of the
genus Millepora, possessed of the property of stinging. The stony
branches or plates when taken fresh from the water have a harsh
feel and are not slimy, although possessing a strong and disagreeable odour. The stinging property seems to vary within certain
limits in different specimens: when a piece was pressed or rubbed
on the tender skin of the face or arm, a pricking sensation was
generally caused, which came on after the interval of a second,
and lasted only for a short time. One day, however, by merely
touching my face with one of the branches the pain was instan-
taneous; it increased as usual after a few seconds, and remaining
sharp for some minutes, was perceptible for half an hour afterwards. The sensation was as bad as that from a nettle, but more
like that caused by the Portuguese man-of-war (*Physalia*). Little
red spots were produced on the tender skin of the arm, which
appeared as if they would have formed watery pustules, but did
not. The circumstance of this stinging property is not new,
though it has scarcely been sufficiently remarked on. M. Quoy*
mentions it, and I have heard of stinging corals in the West Indies.
In the East Indian sea a stinging seaweed also is found.

There was another and quite distinct kind of coral, which was
remarkable from the change of colour, which it underwent
shortly after death; when alive it was of a honey-yellow, but some
hours after being taken out of water, it became as black as ink. I
may just mention, as partly connected with the above subjects,
that there are here two species of fish, of the genus Sparus, which
exclusively feed on coral. Both are coloured of a splendid bluish-
green, one living invariably in the lagoon, and the other amongst
the outer breakers. Mr Liesk assured us that he had repeatedly
seen whole shoals grazing with their strong bony jaws on the tops

* Freycinet's Voyage, vol. i., p. 597.

of the coral branches.* I opened the intestines of several, and found them distended with a yellowish calcareous matter. These fish, together with the lithophagous shells and nereidous animals, which perforate every block of dead coral, must be very efficient agents in producing the finest kind of mud, and this, when derived from such materials, appears to be the same with chalk.

APRIL 12TH – In the morning, we stood out of the Lagoon. I am glad we have visited these islands: such formations surely rank high amongst the wonderful objects of this world. It is not a wonder, which at first strikes the eye of the body, but rather, after reflection, the eye of reason. We feel surprised, when travellers relate accounts of the vast extent of certain ancient ruins; but how utterly insignificant are the greatest of these, when compared to the pile of stone here accumulated by the work of various minute animals. Throughout the whole group of islands, every single atom,† even from the smallest particle to large fragments of rock, bears the stamp of having been subjected to the power of organic arrangement. Captain FitzRoy, at the distance of but little more than a mile from the shore, sounded with a line, 7,200 feet long, and found no bottom. This island is, therefore, a lofty submarine mountain, which has a greater inclination than even those of volcanic origin on the land. I will now give a sketch‡ of the general results at which I have arrived, respecting the origin of the various classes of reefs, which occur scattered over such large spaces of the intertropical seas.

The first consideration to attend to, is, that every observation leads to the conclusion that those lamelliform corals, which are the efficient agents in forming a reef, cannot live at any considerable depth. As far as I have personally seen, I judge of this from carefully examining the impressions on the soundings, which were taken by Captain FitzRoy at Keeling Island, close outside

---

* It has sometimes been thought (*vide* Quoy in Freycinet's Voyage), that coral-eating fish were poisonous; such certainly was not the case with these Spari.

† I exclude, of course, the soil which has been brought here in vessels from Malacca and Java, and the small fragments of pumice, drifted here, together with the seeds of East Indian plants. The one block of greenstone, moreover, on the Northern Lagoon must be excepted.

‡ This sketch was read before the Geological Society, May, 1837.

the breakers, and from some others which I obtained at the Mauritius. At a depth under 10 fathoms, the arming came up as clean as if it had been dropped on a carpet of thick turf; but as the depth increased, the particles of sand brought up became more and more numerous, until, at last, it was evident the bottom consisted of a smooth layer of calcareous sand, interrupted only at intervals by shelves, composed probably of dead coral rock. To carry on the analogy, the blades of grass grew thinner and thinner, till, at last, the soil was so sterile, that nothing sprung from it.

As long as no facts, beyond those relating to the structure of lagoon islands were known, so as to establish some more comprehensive theory, the belief that corals constructed their habitations, or, speaking more correctly, their skeletons, on the circular crests of submarine craters, was both ingenious and very plausible. Yet the sinuous margin of some, as in the Radack Islands of Kotzebue, one of which is 52 miles long, by 20 broad, and the narrowness of others, as in Bow Island (of which there is a chart on a large scale, forming part of the admirable labours of Captain Beechey), must have startled every one who considered this subject.

The very general surprise of all those who have beheld lagoon islands, has perhaps been one chief cause why other reefs, of an equally curious structure have been almost overlooked:* I allude to the encircling reefs. We will take, as an instance, Vanikoro, celebrated on account of the shipwreck of La Peyrouse. The reef there runs at the distance of nearly 2, and in some parts 3 miles from the shore, and is separated from it by a channel having a general depth between 30 and 40 fathoms, and, in one part, no less than 50, or 300 feet. Externally, the reef rises from an ocean profoundly deep. Can any thing be more singular than this structure? It is analogous to that of a lagoon, but with an island standing, like a picture in its frame, in the middle. A fringe of low alluvial land in these cases generally surrounds the base of the mountains; this, covered by the most beautiful productions of a

---

* Mr De la Beche, however, seems to have been fully aware of the difficulty. He says, 'there are certain situations, where coral reefs run, as it were, in a line with the coast, but separated from it by deep water, which would seem to require a different explanation.' *Geological Manual*, p. 142.

tropical land, backed by the abrupt mountains and fronted by a lake of smooth water, only separated from the dark waves of the ocean by a line of breakers, form the elements of the beautiful scenery of Tahiti – so well called the Queen of Islands. We cannot suppose these encircling reefs are based on an external crater, for the central mass sometimes consists of primary rock, or on any accumulation of sedimentary deposits, for the reefs follow indifferently the island itself, or its submarine prolongation. Of this latter case there is a grand instance in New Caledonia, where the reefs extend no less than 140 miles beyond the island.

The great Barrier which fronts the NE coast of Australia, forms a third class of reef. It is described by Flinders as having a length of nearly 1,000 miles, and as running parallel to the shore, at a distance of between 20 and 30 miles from it, and, in some parts, even of 50 and 70. The great arm of the sea thus included, has a usual depth of between 10 and 20 fathoms, but this increases towards one end to 40 and even 60. This probably is both the grandest and most extraordinary reef now existing in any part of the world.

It must be observed, that the reef itself in the three classes, namely, lagoon, encircling, and barrier, agrees in structure, even in the most minute details: but these I have not space here even to allude to. The difference entirely lies in the absence or presence of neighbouring land, and the relative position which the reefs bear to it. In the two last-mentioned classes, there is one difficulty in undertaking their origin, which must be pointed out. Since the time of Dampier it has been remarked, that high land and deep seas go together. Now when we see a number of mountainous islands coming abruptly down to the sea-shore, we must suppose the strata of which they are composed, are continued with nearly the same inclination beneath the water. But, in such cases, where the reef is distant several miles from the coast, it will be evident upon a little consideration, that a line drawn perpendicularly from its outer edge down to the solid rock on which the reef must be based, very far exceeds that small limit at which the efficient lamelliform corals exist.

In some parts of the sea, as we shall hereafter mention, reefs do occur which fringe rather than encircle islands – the distance from the shore being so small, where the inclination of the land is great, that there is no difficulty in understanding the growth of the coral. Even in these 'fringing' reefs, as I shall call them in contradistinction to the 'encircling', the reef is not attached quite close to the shore. This appears to be the result of two causes: namely, first, that the water immediately adjoining the beach is rendered turbid by the surf, and therefore injurious to all zoophytes; and, secondly, that the larger and efficient kinds only flourish on the outer edge amidst the breakers of the open sea. The shallow space between the skirting reef and the shore has, however, a very different character from the deep channel, similarly situated with respect to those of the encircling order.

Having thus specified the several kinds of reefs, which differ in their forms and relative position with regard to the neighbouring land, but which are most closely similar in all other respects (as I could show if I had space), it will, I think, be allowed that no explanation can be satisfactory which does not include the whole series. The theory which I would offer, is simply, that as the land with the attached reefs subsides very gradually from the action of subterranean causes, the coral-building polypi soon raise again their solid masses to the level of the water: but not so with the land; each inch lost is irreclaimably gone; as the whole gradually sinks, the water gains foot by foot on the shore, till the last and highest peak is finally submerged.

Before I explain this view more in detail, I must enter on a few considerations, which render such changes of level not improbable. Indeed, the simple fact of a large portion of the continent of South America, still rising under our eyes, and abounding with proofs of similar elevations on a grander scale during the recent period, takes away any excessive improbability of a movement similar in kind, but in an opposite direction. Mr Lyell, who first suggested the idea of a general subsidence with reference to coral reefs, has remarked that the existence of so small a portion of land in the Pacific, where so many causes both aqueous and igneous tend to its production, renders such sinking of the foundation probable. There is, however, another argument of much greater weight, which may be inferred from the inconsiderable depth at which corals grow. We see large extents of ocean, of more than

..o miles in one direction and several hundreds in another,
...attered over with islands, none of which rise to a greater height
than that to which waves can throw fragments, or the wind heap
up sand. Now if we leave out of the question subsidence, the
foundation on which these reefs are built, must in every case come
to the surface within that small limit (we may say 20 fathoms) at
which corals can live. This conclusion is so extremely improbable
that it may at once be rejected: for in what country can there be
found a broad and grand range of mountains of the same height
within 120 feet? But on the idea of subsidence, the case is at once
clear: as each point, one after the other according to its altitude,
was submerged, the coral grew upwards, and formed the many
islets now standing at one level.

Having endeavoured on general grounds not only to remove
any extreme degree of improbability in the belief of a general
subsidence, but likewise to show that it is almost necessary to
account for the existence of a vast number of reefs on one level, we
will now see how far the same idea will apply to the peculiar
configuration in the several classes. Let us imagine an island
merely fringed by reefs extending to a short distance from the
shore; in which case, as we have before remarked, there is no
difficulty in understanding their structure. Now let this island
subside by a series of movements of extreme slowness, the coral at
each interval growing up to the surface. Without the aid of
sections it is not very easy to follow out the result, but a little
reflection will show that a reef encircling the shore at a greater or
less distance, according to the amount of subsidence, would be
produced. If we suppose the sinking to continue, the encircled
island must, by the submergence of the central land but upward
growth of the ring of coral, be converted into a lagoon island. If
we take a section of some encircled island on a true scale, as for
instance Gambier, which has been so well described by Captain
Beechey, we shall not find the amount of movement very
great, which would be necessary to change a well-characterized
encircling reef, into as characteristic a lagoon island.

It will at once be evident that a coral reef, closely skirting the
shore of a continent, would, in like manner after each subsidence,
rise to the surface; the water, however, always encroaching on the

land. Would not a barrier reef necessarily be produced, similar to the one extending parallel to the coast of Australia? It is indeed but uncoiling one of those reefs which encircle at a distance so many islands.

Thus the three great classes of reef, lagoon, encircling, and barrier, are connected by one theory. It will perhaps be remarked, if this be true, there ought to exist every intermediate form between a closely encircled and a lagoon island. Such forms actually occur in various parts of the ocean: we have one, two, or more islands encircled in one reef; and of these some are of small proportional size to the area enclosed by the coral formation; so that a series of charts might be given, showing a gradation of character between the two classes. In New Caledonia, where the double line of reef projects 140 miles beyond the island, we may imagine we see this change in progress. At the northern extremity, reefs occur, some of which are of the encircling kind, and others almost with the character of true lagoon islands. The line of reef which fronts the whole west coast of this great island, has by some been called a barrier. It is 400 miles long; and may be said thus to form a link between an ordinary encircling reef and the great Australian barrier.

I should perhaps have entered before into the consideration of one apparent difficulty in the origin of lagoon islands. It may be said, granting the theory of subsidence, a mere circular disc of coral would be formed, and not a cup-shaped mass. In the first place, even in reefs closely fringing the land (as before remarked), the corals do not grow on the shore itself, but leave a shallow channel. Secondly, the strong and vigorous species which alone build a solid reef, are never found within the lagoon; they only flourish amidst the foam of the never-tiring breakers. Nevertheless, the more delicate corals, though checked by several causes, such as strong tides and deposits of sand, do constantly tend to fill up the lagoon; but the process must become slower and slower, as the water in the shallow expanse is rendered subject to accidental impurities. A curious instance of this happened at Keeling Island, where a heavy tropical storm of rain killed nearly all the fish. When the coral at last has filled up the lagoon to the height of lowest water at spring-tides, which is the extreme limit possible –

how, afterwards, is the work to be completed? There is no high land whence sediment can be poured down; and the dark-blue colour of the ocean bespeaks its purity. The wind, carrying calcareous dust from the outer coast, is the only agent which can finally convert the lagoon island into solid land, and how slow must this process be!

Subsidence of the land must always be most difficult to detect, excepting in countries long civilized – for the movement itself tends to conceal all evidence of it. Nevertheless, at Keeling Island, tolerably conclusive evidence of such movement could be observed. On every side of the lagoon, in which the water is as tranquil as in the most sheltered lake, old cocoa-nut trees were undermined and falling. Captain FitzRoy likewise pointed out to me on the beach the foundation-posts of a storehouse, which the inhabitants said had stood, seven years before, just above high-water mark, but now was daily washed by the tide. Upon asking the people whether they ever experienced earthquakes, they said, that lately the island had been shaken by a very bad one; and that they remembered two others during the last ten years. I no longer doubted concerning the cause which made the trees fall, and the storehouse to be washed by the daily tide.

At Vanikoro, the encircled island already mentioned, I gathered from Captain Dillon's account, that the alluvial land at the foot of the mountain was very small in quantity, the channel extremely deep, and the islets on the reef itself, which result from the gradual accumulation of fragments, singularly few in number; all of which, together with the wall-like structure of the reef both inside as well as outside, indicated to my mind, that, without doubt, the movements of subsidence had lately been rapid. At the end of the chapter, it is stated that this island is shaken by earthquakes of extreme violence.

I may here mention a circumstance, which to my mind had the same weight as positive evidence, though bearing on another part of the question. M. Quoy, when discussing in general terms the nature of coral reefs, gives a description which is applicable only to those which, *skirting* the shore, do not require a foundation at any greater depth than that from which the coral-building polypi can spring. I was at first astonished at this, as I knew he had crossed both the Pacific and Indian oceans, and must, as I thought, have seen the class of widely encircling reefs, which indicate a subsiding land. He subsequently mentions several islands as instances of his description of the general structure; by a singular chance, the whole can be shown, by his own words, in different parts of his account, to have been recently elevated. Therefore, that which appeared so adverse to the theory, became as strong in its confirmation.

Continental elevations, as observed in South America and other parts, seem to act over wide areas with a very uniform force; we may therefore suppose that continental subsidences act in a nearly similar manner. On this assumption, and taking on the one hand, lagoon islands, encircling and barrier reefs, as indications of subsidence; and on the other, raised shells and corals, together with mere skirting reefs, as our proof of elevation, we may test the truth of the theory – that their configuration has been determined by the kind of subterranean movement – by observing whether any uniform results can be obtained. I think it can be shown that such is the case in a very remarkable degree; and that certain laws may be inferred from the examination, of far more importance than the mere explanation of the origin of the circular or other kinds of reef.

If there had been space, I should have made a few general remarks, before entering into any detail. I may, however, just notice the remarkable absence of the reef-building polypi over certain wide areas within the tropical sea: for instance, on the whole west coast of America, and, as I believe, of Africa (?), and round the eastern islands in the Atlantic ocean. Although certain species of lamelliform zoophytes are found on the shores of the latter islands, and though calcareous matter is abundant to excess, yet reefs are never formed. It would appear that the effective species do not occur there; of which circumstance I apprehend no explanation can be given, any more than why it has been ordained that certain plants, as heaths, should be absent from the New World, although so common in the Old.

Without entering into any minute geographical details, I must observe, that the usual direction of the island groups in the central parts of the Pacific, is N W and S E. This must be noticed, because subterranean disturbances are known to follow the coast lines of the land. Commencing on the shores of America, there are abundant proofs that the greater part has been elevated within the recent period, but as coral reefs do not occur there, it is not immediately connected with our present subject. Immediately adjoining the continent there is an extent of ocean remarkably free from islands, and where of course there exists no possible indication of any change of level. We then come to a NW by W line dividing the open sea from one strewed with lagoon islands, and including the two beautiful groups of encircled islands the Society and Georgian Archipelagoes. This great band having a length of more than 4,000 miles by 600 broad must, according to our view, be an area of subsidence. We will at present for convenience sake pass over the space of ocean immediately adjoining it, and proceed to the chain of islands including the New Hebrides, Solomon, and New Ireland. Any one who examines the charts of the separate islands in the Pacific, engraved on a large scale, will be struck with the absence of all distant or encircling reefs round these groups: yet it is known that coral occurs abundantly close in shore. Here, then, according to the theory, there are no proofs of subsidence; and in conformity to this we find in the works of Forster, Lesson, Labillardière, Quoy, and Bennett, constant allusion to the masses of elevated coral. These islands form, therefore, a well-determined band of elevation: between it and the great area of subsidence first mentioned there is a broad space of sea irregularly scattered with islets of all classes; some with proofs of recent elevation and merely fringed by reefs; others encircled; and some lagoon islands. One of the latter is described by Captain Cook as a grand circle of breakers without a single spot of land; in this case we may believe that an ordinary lagoon island has been recently submerged. On the other hand, there are proofs of other lagoon islands having been lifted up several yards above the level of the sea, but which still retain a pool of salt water in their centres. These facts show an irregular action in the subterranean forces; and when we remember that the space lies directly between the well-marked area of elevation and the enormous one of subsidence, an alternate and irregular movement seems almost probable.

To the westward of the New Hebrides line of elevation we have New Caledonia, and the space included between it and the Australian barrier, which Flinders, on account of the number of reefs, proposed to call the Corallian Sea. It is bounded on two sides by the grandest and most extraordinary reefs in the world, and is likewise terminated to the northward by the coast of Louisiade – most dangerous on account of its distant reefs. This, then, according to our theory, is an area of subsidence. I may here remark, that as the Barrier is supposed to be produced by the subsidence of the coast of the mainland, it may be expected that any outlying islands would have formed lagoon islands. Now Bligh and others distinctly state that some of the islands there are precisely similar to the well-known lagoon islands in the Pacific; there are also encircled islands, so that the three classes supposed to be produced by the same movement are there found in juxtaposition; as likewise happens, but in a less evident manner, at New Caledonia and in the Society Archipelago.

The New Hebrides line of islands, may be observed to bend abruptly at New Britain, thence to run nearly east and west; and, lastly, to resume its former north-west direction in Sumatra and the peninsula of Malacca. The figure may be compared to the letter S laid obliquely, but the line is often double. We have shown that the southern part, as far north as New Ireland, abounds with proofs of elevation, so is it with the rest. Since the time of Bougainville every voyager adduces some fresh instance of such changes throughout a great part of the East Indian archipelago. I may specify New Guinea, Wageeoo, Ceram, Timor, Java, and Sumatra. Coral reefs are abundant in the greater part of these seas, but they merely skirt the shores. In the same manner as we have followed the curved line of elevation, so may we that of subsidence. At Keeling Island, I have already mentioned that there exist proofs of the latter movement: and it is a very interesting circumstance, that during the last earthquake, by which that island was affected, Sumatra, though distant nearly 600 miles, was violently shaken. Bearing in mind that there is evidence of recent elevation on the coast of the latter, one is strongly tempted to believe that as one end of the lever goes up, the other goes down: that as the East Indian archipelago rises, the bottom of the neighbouring sea sinks and carries with it Keeling Island, which would have been submerged long ago in the depths of the ocean, had it not been for the wonderful labours of the reef-building polypi.

As I have remarked, the islands in this great archipelago are only skirted with reefs; and it appears from the statements of those who have visited them, as well as from an examination of the charts, that lagoon islands are not found there. This in itself is remarkable, but it becomes far more so when it is known, that according to all accounts (and distinctly stated by Mr De la Beche*) they are likewise absent in the West Indian sea, where coral is most abundant: now every one is aware of the numerous proofs of recent elevation in most parts of that archipelago. Again, Ehrenberg has observed that lagoon islands do not occur in the Red Sea: in Lyell's *Geology*, and in the *Geographical Journal*, proofs are given of recent elevation on the shores of a large part of that sea. Excepting on the theory of the form of reefs being determined by the kind of movement to which they have been subjected; it is a most anomalous circumstance, and which has never been attempted to be solved, that the lagoon structure being universal and considered as characteristic in certain parts of the ocean, should be entirely absent in others of equal extent.

I may here also just recall to mind the cases of skirting reefs mentioned by M. Quoy (to which number several others might be added), where proofs of elevation occurred. Some general law must determine the marked difference between reefs merely skirting the shore, and others rising from a deep ocean in the form of distant rings. We have endeavoured to show that with a subsiding movement, the first and simple class must necessarily pass into the second, and more remarkable structure.

To proceed with our examination: to the westward of the prolongation of the line of subsidence, of which Keeling Island is the index, we have an area of elevation. For on the northern end of Ceylon and on the eastern shores of India, elevated shells and corals, such as now exist in the neighbouring sea, have been observed. Again in the middle of the Indian ocean, the Laccadive,

* Geological Manual, p. 141.

Maldive, and Chagos line of atolls or lagoons show a line of subsidence. The best characterized of these, namely, the Maldive islands, extend in length for 480 miles, with an average breadth of 60. These atolls agree in most respects with the lagoons of the Pacific; they differ, however, in several of them being crowded together – such little groups being separated from other groups by profoundly deep channels. Now if we look in a chart, at the prolongation of the reef towards the northern end of New Caledonia, and then complete the work of subsidence, so as to continue producing the same results; we should have the original reef broken up into many patches; each of which, from the vigorous growth of coral on the outside, would have a constant tendency to assume a rounded form. Every accidental break in the continuity of the first line would determine a fresh circle. In the case, therefore, of the Low or Dangerous Archipelago in the Pacific, I believe that the lagoon islands were moulded round the flanks of so many distinct islands; but in the Maldives, that one single mountainous island, bordered by reefs, and very nearly of the same actual figure and dimensions with New Caledonia, formerly occupied that part of the ocean.

Lastly, to the extreme westward, the coast of Africa is closely skirted by coral reefs, and according to facts stated in Captain Owen's voyage, has probably been uplifted within a recent period. The same remark applies to the nothern part of Madagascar, and, judging from the reefs likewise at the Seychelles, situated on the submarine prolongation of that great island. Between these two, NNE and SSW lines of elevation, some lagoon and widely encircled islands indicate a band of subsidence.

When we consider the absence both of widely encircling reefs and lagoon islands in the several archipelagoes and wide areas, where there are proofs of elevations; and on the other hand the converse case of the absence of such proof where reefs of those classes do occur; together with the juxtaposition of the different kinds produced by movements of the same order, and the symmetry of the whole, I think it will be difficult (even independently of the explanation it offers of the peculiar configuration of each class) to deny a great probability to this theory. Its importance, if true, is evident; because we get at one glance an insight into the system by which the surface of the land has been broken up, in a manner somewhat similar, but certainly far less perfect, to what a geologist would have done who had lived his 10,000 years, and kept a record of the passing changes. We see the law almost established, that linear areas of great extent undergo movements of an astonishing uniformity, and that the bands of elevation and subsidence alternate. Such phenomena at once impress the mind with the idea of a fluid most gradually propelled onwards, from beneath one part of the solid crust to another.

I cannot at present do more than allude to some of the results which may be deduced from these views. If we examine the points of eruption over the Pacific and Indian oceans, we shall find that all the *active volcanoes* occur within *the areas of elevation*. (The Asiatic band must be excepted; insasmuch as we are entirely in want of information of all kinds respecting it.) On the other hand, in the great spaces supposed to be now subsiding, between the Radack and Dangerous Archipelagoes, in the Corallian sea, and among the atolls which front the west coast of India, not one occurs. If we look at the changes of level as a consequence of the propulsion of fluid matter beneath the crust, as before suggested, then the area to which the force is directed might be expected to yield more readily than that whence it was gradually retiring. I am the more convinced that the above law is true, because, if we look to other parts of the world, proofs of recent elevation almost invariably occur, where there are active vents: I may instance the West Indies, the Cape de Verds, Canary Islands, southern Italy, Sicily, and other places. But in answer to this, those geologists, who, judging from the history of the isolated volcanic mounds of Europe, were inclined to believe that the level of the ground was constantly oscillating up and down, might maintain that on these same areas, the amount of subsidence had been equal to that of elevation, but that we possessed no means of knowing it. I conceive it is by eliminating this source of doubt, that the alternate bands of opposite movement, deduced from the configuration of the reefs, directly bear on this law. I need not do more than simply state, that we thus obtain (if the view is correct) a means of forming some judgment of the prevailing movements, during the formation of even the oldest series, where volcanic rocks occur interstratified with sedimentary deposits.

Any thing which throws light on the movements of the ground is well worthy of consideration; and the history of coral reefs may, in another manner, elucidate such changes in the older formations. As there is every reason to believe that the lamelliform corals grow only abundantly at a small depth, we may feel sure, where a great thickness of coral limestone occurs, that the reefs on which the zoophytes flourished, must have been sinking. Until we are enabled to judge by some means what were the prevailing movements at different epochs, it will scarcely ever be possible to speculate with any safety on the circumstances under which the complicated European formations, composed of such different materials and in such different states, were accumulated.

Nor can I quite pass over the probability of the above views illustrating those admirable laws first brought forward by Mr Lyell – of the geographical distribution of plants and animals, as consequent on geological changes. M. Lesson has remarked on the singular uniformity* of the Indio-Polynesian Flora throughout the immense area of the Pacific; the dispersion of forms having been directed against the course of the trade wind. If we believe that lagoon islands, those monuments raised by infinite numbers of minute architects, record the former existence of an archipelago or continent in the central part of Polynesia, whence the germs could be disseminated, the problem is rendered far more intelligible. Again, if the theory should hereafter be so far established, as to allow us to pronounce that certain districts fall within areas either of elevation or subsidence, it will directly bear upon that most mysterious question, whether the series of organized beings peculiar to some isolated points, are the last remnants of a former population, or the first creatures of a new one springing into existence.

Briefly to recapitulate. In the first place, reefs are formed around islands, or on the coast of the mainland, at that limited depth at which the efficient classes of zoophytes can live; and

---

* Perhaps this is stated rather too strongly; but M. Lesson, of course, had grounds for his assertion.

where the sea is shallow, irregular patches may likewise be produced. Afterwards from the effects of a series of small subsidences, encircling reefs, grand barriers, or lagoon islands, are mere modifications of one necessary result. Secondly, it can be shown on the above views, that the inter-tropical ocean, throughout more than a hemisphere, may be divided into linear and parallel bands, of which the alternate ones have undergone, within a recent period, the opposite movements of elevation and subsidence. Thirdly, that the points of eruption seem invariably to fall within areas subject to a propulsion from below. The traveller who is an eyewitness of some great and overwhelming earthquake, at one moment of time loses all former associations of the land being the type of solidity, so will the geologist, if he believe in these oscillations of level (the deeply seated origin of which is betrayed by their forms and vast dimensions), perhaps be more deeply impressed with the never-ceasing mutability of the crust of this our World.

CHAPTER XXIII

MAURITIUS TO ENGLAND

* * *

MAY 9TH – We sailed from Port Louis, on our way to the Cape of
Good Hope, and on the evening of the 31st anchored in Simon's
Bay. The little town offers but a cheerless aspect to a stranger's
eye. About a couple of hundred, square, whitewashed houses,
with scarcely a single tree in the neighbourhood, and very few
gardens, are scattered along the beach, at the foot of a lofty, steep,
bare wall, of horizontally stratified sandstone.

The next day I set out for Cape Town, which is 20 miles distant.
Both towns are situated within the headlands, but at the opposite
extremities of a range of mountains, which extending parallel to
the mainland, is joined to it by a low sandy flat. The road skirted
the base of these mountains: for the first 14 miles the country is
very desert, and with the exception of the pleasure which the sight
of an entirely new vegetation never fails to communicate, there
was very little of interest. The view however of the mountains on
the opposite side of the flat, brightened by the declining sun, was
fine. Within 7 miles of Cape Town, in the neighbourhood of
Wynberg, a great improvement was visible, and here the country
houses of the more wealthy residents of the capital are situated.

The numerous woods of young Scotch firs and stunted oak-trees form the chief attraction of this locality. There is, indeed, a great charm in shade and retirement, after the unconcealed bleakness of so open a country as this. The houses and plantations are backed by a grand wall of mountains, which gives the scene a degree of uncommon beauty. I arrived late in the evening in Cape Town, and had a good deal of difficulty in finding quarters. In the morning several ships from India had arrived at this great inn on the great highway of nations, and they had disgorged on shore a host of passengers, all longing to enjoy the delights of a temperate climate. There is only one good hotel, so that strangers generally live in boarding-houses; a very uncomfortable fashion to which I was obliged to conform, although I was fortunate in my quarters.

In the morning I walked to a neighbouring hill to look at the town. It is laid out with the rectangular precision of a Spanish city: the streets are in good order, and Macadamized, and some of them have rows of trees on each side; the houses are all whitewashed, and look clean. In several trifling particulars the town had a foreign air, but it is daily becoming more English. There is scarcely a resident, excepting amongst the lowest order, who does not speak some English. In this facility in becoming Anglefied, there appears to exist a wide difference between this colony and that of the Mauritius. It does not, however, arise from the popularity of the English; for the Dutch as well as the French, although they have profited to an immense degree by the English government, yet thoroughly dislike our whole nation.

All the fragments of the civilized world which we have visited in the southern hemisphere, appear to be flourishing: little embryo Englands are springing into life in many quarters. Although the Cape colony possesses only a moderately fertile country, it appears in a very prosperous condition. In one respect it suffers like New South Wales, namely, in the absence of water communication, and in the interior being separated from the coast by a high chain of mountains. This country does not possess coal; and there is no timber, excepting at a considerable distance. Hides, tallow, and wine are the chief exports, and latterly a considerable quantity of corn. The farmers are beginning also to pay attention to sheep-grazing – a hint taken from Australia. It is no small triumph to Van Diemen's Land, that live sheep have been exported from a colony of thirty-three years' standing, to this which was founded in 1651.

In Cape Town it is said that the present number of inhabitants is about 15,000, and in the whole colony, including coloured people, 200,000. Many different nations are here mingled together; the Europeans consist of Dutch, French, and English, and scattered people from other parts. The Malays, descendants of slaves brought from the East Indian archipelago, form a large body. They are a fine set of men, and can always be distinguished by a conical hat, like the roof of a circular thatched cottage, or by a red handkerchief on their heads. The number of negroes is not very great; and the Hottentots, the ill-treated aborigines of the country, are, I should think, in a still smaller proportion. The first object in Cape Town which strikes the eye of a stranger, is the number of bullock-waggons. Several times I saw eighteen, and I heard of twenty-four oxen being all yoked together in one team. Besides these, waggons with four, six, and eight horses in hand, go trotting about the streets. I have as yet not mentioned the well-known Table Mountain. This great mass of horizontally stratified sandstone rises quite close behind the town to a height of 3,500 feet: the upper part forms an absolute wall, often reaching into the region of the clouds. I should think so high a mountain, not forming part of an extensive platform, and yet being composed of horizontal strata, must be a rare phenomenon. It certainly gives the landscape a very peculiar, and from some points of view, a grand character.

*   *   *

JUNE 18TH – We put to sea, and, on the 29th, crossed the Tropic of Capricorn for the sixth and last time. On the 8th of July, we arrived off St Helena. This island, the forbidding aspect of which has been so often described, rises like a huge castle from the ocean. A great wall, built of successive streams of black lava, forms around the whole circuit a bold coast. Near the town, as if in aid of the natural defence, small forts and guns are every where built up, and mingled with the rugged rocks. The town extends up a flat and very narrow valley; the houses look respectable, and they are interspersed with a very few green trees. When approaching the anchorage there is one striking view: an irregular castle perched on the summit of a lofty hill, and surrounded by a few scattered fir-trees, boldly projects against the sky.

The next day I obtained lodgings within a stone's throw of Napoleon's tomb.* I confess, however, this had little attraction for me: but it was a capital central situation, whence I could make excursions in every direction. During the four days I staid here, from morning to night I wandered over the island, and examined its geological history. The house was situated at an elevation of about 2,000 feet; here the weather was cold and very boisterous, with constant showers of rain; and every now and then the whole scene was veiled in thick clouds.

Near the coast the rough lava is entirely destitute of vegetation: in the central and higher parts, a different series of rocks have from extreme decomposition produced a clayey soil, which, where not covered by vegetation, is stained in broad bands of many bright colours. At this season the land, moistened by constant showers, produces a singularly bright green pasture; this lower and lower down, gradually fades away, and at last disappears. In latitude 16°, and at the trifling elevation of 1,500 feet, it is surprising to behold a vegetation possessing a character decidedly English. The hills are crowned with irregular plantations of Scotch firs; and the sloping banks are thickly scattered over with thickets of gorze, covered with its bright yellow flowers. Weeping-willows are common along the course of the rivulets, and the hedges are made of the blackberry, producing its well-known fruit. When we consider that the number of plants now found on the island is 746, and that out of these, 52 alone are native species, the rest being imported, and many of them from England, we see a good reason for this English character in the vegetation. The numerous species which have been so recently

---

* After the volumes of eloquence which have poured forth on this subject, it is dangerous even to mention the tomb. A modern traveller in twelve lines, burdens the poor little island with the following titles – it is a grave, tomb, pyramid, cemetery, sepulchre, catacomb, sarcophagus, minaret, and mausoleum!

introduced can hardly have failed to have destroyed some of the native kinds. I believe there is no accurate account of the state of the vegetation at the period when the island was covered with trees; such would have formed a most curious comparison with its present sterile condition, and limited Flora. Many English plants appear to flourish here better than in their native country; some also from the opposite quarter of Australia succeed remarkably well. It is only on the highest and steepest ridges, where the native Flora is still predominant.

The English, or rather the Welsh character of the scenery, is preserved by the numerous cottages and small white houses; some buried at the bottom of the deepest valleys, and others stuck up on the crests of the lofty hills. Some of the views are very striking; I may instance that from near Sir W. Doveton's house, where the bold peak called Lott is seen over a dark wood of firs, the whole being backed by the red water-worn mountains of the southern shore.

On viewing the island from an eminence, the first circumstance which strikes one, is the very great number of roads, and forts: the labour bestowed on the public works, if one forgets its character as a prison, seem out of all proportion to its extent or value. There is so little level or useful land, that it seems surprising how so many people (about 5,000) can subsist here. The lower orders, or the emancipated slaves, are I believe extremely poor: they complain of want of work, a fact which is likewise shown by the very cheap labour. From the reduction in the number of public servants owing to the island having been given up by the East India Company, and the consequent emigration of many of the richer people, the poverty probably will increase. The chief food of the working class is rice with a little salt meat; as neither of these articles are the products of the island, but must be purchased with money, the low wages tell heavily on the poor people. The fine times, as my old guide called them, when 'Bony' was here, can never return again. Now that the people are blessed with freedom, a right which I believe they value fully, it seems probable that their numbers will quickly increase: if so, what is to become of the little state of St Helena?

My guide was an elderly man, who had been a goatherd when a

boy, and knew every step amongst the rocks. He was of a race many times mixed, and although with a dusky skin, he had not the disagreeable expression of a mulatto. He was a very civil, quiet old man, and such appears the character of the greater number among the lower classes. It was strange to my ears to hear a man, nearly white, and respectably dressed, talking with indifference of the times when he was a slave. With my companion who carried our dinner and a horn of water, which latter is quite necessary as all in the lower valleys is saline, I every day took long walks.

Beneath the limits of the elevated and central green circle, the wild valleys are quite desolate and untenanted. Here, to the geologist, there are scenes of high interest, which show the successive changes, and complicated disturbances which have in past times happened. According to my views, St Helena has existed as an island from a very remote epoch: some obscure proofs, however, of the elevation of the land are still extant. I believe that the central and highest peaks form parts of the rim of a great crater; the southern half of which has been removed by the waves of the sea. There is, moreover, an external margin of black volcanic rocks, which belong to an anterior condition of things. These have been dislocated and broken up by forces acting from below, so that the confusion in structure from these different causes is extreme. On the higher parts of the island considerable numbers of shells occur embedded in the soil, which have always been supposed to be of marine origin: and the fact has been adduced as a proof of the retreat of the sea. The shell turns out to be a Bulimus, or terrestrial species. It is however very remarkable, that it is not now found in a living state: a circumstance which in all probability may be attributed to the entire destruction of the woods, and consequent loss of food and shelter, which occurred during the early part of the last century.

The history of the changes, which the elevated plains of Longwood and Deadwood have undergone, as given in General Beatson's account of the island, is extremely curious. It is said the plain in former times was covered with wood, and was therefore called the Great Wood. So late as the year 1716 there were many trees upon it, but in 1724 the old trees had mostly fallen; and as goats and hogs were at that time suffered to range about, all the young trees had been devoured. It appears also from the official records, that the trees were unexpectedly, some years afterwards, succeeded by indigenous wire grass, which now spreads over its whole extent.* He then adds, 'These are curious facts, since they trace the changes which this remarkable spot of land has undergone, for now this formerly naked plain (after the trees had fallen) is covered with fine sward, and is become the finest piece of pasture on the island.' The extent of surface, which was probably covered by wood at a former period is estimated at no less than 2,000 acres; at the present day scarcely a tree can be found there. It is said, that in 1709 there were quantities of dead wood in Sandy Bay: this place is now so utterly desert, that nothing but so well-attested an account could make me believe that trees had ever existed there. The fact, that the goats and hogs destroyed all the young trees as they sprung up, and that in the course of time the old ones, which were safe from their attacks, perished from age, seems clearly made out. Goats were introduced in the year 1502; 86 years afterwards, in the time of Cavendish, it is known they were exceedingly numerous. More than a century afterwards, in 1731, when the evil was completed and found irretrievable, an order was issued that all stray animals should be destroyed.

When at Valparaiso, I heard it positively affirmed, that the Sandal-wood tree had been found on the island of Juan Fernandez in considerable numbers, but that all without exception were standing dead. At the time, I thought it was some mysterious case of the natural death of a species; but when it is remembered, that goats for very many years have abounded on that island, it seems most probable that the young trees were prevented growing, and that the old ones perished from age. It is a very interesting fact, to observe that the arrival of animals at St Helena in 1501 did not change the whole aspect of the island, until a period of 220 years had elapsed: for they were introduced in 1502, and in 1724 it is said 'the old trees had mostly fallen.' There can be no doubt, this change affected not only the Bulimus and probably some other

* Beatson's St Helena. Introductory chapter, p. iv.

land-shells (of which I obtained specimens from the same bed),
but likewise a multitude of insects.

St Helena, situated so remote from any continent, in the midst
of a great ocean, and possessing a unique Flora – this little world
within itself – excites our curiosity. Birds and insects,* as might
have been expected, are very few in number; indeed I believe all
the birds have been introduced within late years. Partridges and
pheasants are tolerably abundant: the island is much too English,
not to be subject to strict game-laws. I was told of a more unjust
sacrifice to such ordinances, than I ever heard of even in England.
The poor people formerly used to burn a plant, which grows on
the coast rocks, and export soda; but a peremptory order came

* Among these few insects, I was surprised to find a small Aphodius (*nov. spec.*)
and an Oryctes, both extremely common under dung. When the island was
discovered it certainly possessed no quadruped, excepting *perhaps* a mouse; it
becomes, therefore, a difficult point to ascertain, whether these stercovorous
insects have since been imported by accident, or if aborigines, on what food
they formerly subsisted. On the banks of the Plata, where, from the vast
number of cattle and horses, the fine plains of turf are richly manured, it is vain
to seek the many kinds of dung-feeding beetles, which occur so abundantly in
Europe. I observed only an Oryctes (the insects of this genus in Europe
generally feed on decayed vegetable matter) and two species of Phanæus,
common in such situations. On the opposite side of the Cordillera in Chiloe,
another species of this genus is exceedingly abundant, and it buries the dung of
cattle in large earthen balls beneath the ground. There is reason to believe that
the genus Phanæux, before the introduction of cattle acted as scavengers to man.
In Great Britain those beetles, which find support in the matter, which has
already contributed towards the life of other and larger animals, are so numerous, that I suppose there are at least 100 different kinds. Considering this, and
observing what a quantity of food is thus lost on the plains of La Plata, I
imagined I saw an instance where man had disturbed that chain, by which so
many animals are linked together in their native country. To this view,
however, Van Diemen's Land offers an exception (in the same manner as St
Helena does in a much lesser degree), for I found there four species of
Onthophagus, two of Aphodius, and one of a third genus, very abundant under
the dung of cows; yet these latter animals had then been introduced only
thirty-three years. Previously to that time, the Kangaroo and some other small
animals were the only quadrupeds; and their dung is of a very different quality
from that of their successors introduced by man. In England the greater number
of stercovorous beetles are confined in their appetites; that is, they do not
depend indifferently on any quadruped for the means of subsistence. The
change, therefore, in habits, which must have taken place in Van Diemen's
Land, is the more remarkable – I am indebted to the Rev. F. W. Hope, who, I
hope, will permit me to call him my master in Entomology, for information
respecting the foregoing, and other insects.

out prohibiting this practice, and giving as a reason, that the partridges would have nowhere to build!

*   *   *

I so much enjoyed my rambles among the rocks and mountains of St Helena, that I felt almost sorry on the morning of the 14th to descend to the town. Before noon I was on board, and the *Beagle* made sail for Ascension.

We reached the anchorage of the latter place on the evening of the 19th (July). Those who have beheld a volcanic island, situated within an arid climate, will be able at once to picture to themselves the aspect of Ascension. They will imagine smooth conical hills of a bright red colour, with their summits generally truncated, rising distinct out of a level surface of black rugged lava. A principal mound in the centre of the island, seems the father of the lesser cones. It is called Green Hill; its name is taken from the faintest tinge of that colour, which at this time of the year was barely perceptible from the anchorage. To complete this desolate scene, the black rocks on the coast are lashed by a wild and turbulent sea.

The settlement is near the beach; it consists of several houses and barracks placed irregularly, but well built of white freestone. The only inhabitants are marines, and some negroes liberated from slave-ships, who are paid and victualled by government. There is not a private person on the island. Many of the marines appeared well contented with their situation; they think it better to serve their one-and-twenty years on shore, let it be what it may, than in a ship: in which choice, if I were a marine, I would most heartily agree.

The next morning I ascended Green Hill, 2,840 feet high, and thence walked across the island to the windward point. A good cart-road leads from the coast settlement to the houses, gardens, and fields, placed near the summit of the central mountain. On the roadside there are milestones, and likewise cisterns, where each thirsty passer-by, can drink some good water. Similar care is displayed in each part of the establishment, and especially in the management of the springs, so that a single drop of water shall not be lost: indeed the whole island may be compared to a huge ship kept in first-rate order. I could not help, when admiring the active industry which had created such effects out of such means, at the same time regretting that it was wasted on so poor and trifling an end. M. Lesson has remarked with justice, that the English nation alone would have thought of making the island of Ascension a productive spot; any other people would have held it without any further views, as a mere fortress in the ocean.

*   *   *

Upon leaving Ascension the ship's head was directed towards the coast of South America, and on August 1st, we anchored at Bahia or San Salvador. We staid here four days, in which time I took several long walks. I was glad to find my enjoyment of tropical scenery had not decreased even in the slightest degree, from the want of novelty. The elements of the scenery are so simple, that they are worth mentioning, as a proof on what trifling circumstances exquisite natural beauty depends.

The country may be described as a level plain of about 300 feet in elevation, which in every part has been worn into flat-bottomed valleys. This structure is remarkable in a granitic land, but is nearly universal in all those softer formations, of which plains usually are composed. The whole surface is covered by various kinds of stately trees, interspersed with patches of culti-vated ground, out of which houses, convents, and chapels arise. It must be remembered that within the tropics, the wild luxuriance of nature is not lost even in the vicinity of large cities; for the natural vegetation of the hedges and hill-sides, overpowers in picturesque effect the artificial labour of man. Hence, there are only a few spots where the bright red soil affords a strong contrast with the universal clothing of green. From the edges of the plain there are distant glimpses either of the ocean, or of the great bay bordered by low wooded shores, and on the surface of which numerous boats and canoes show their white sails. Excepting from these points, the range of vision is very limited: following the level pathways, on each hand alternate peeps into the wooded valleys below can alone be obtained. Finally, I may add that the houses, and especially the sacred edifices, are built in a peculiar and rather fantastic style of architecture. They are all whitewashed; so that when illuminated by the brilliant sun of midday, and as seen against the pale blue sky of the horizon, they stand out more like shadows than substantial buildings.

Such are the elements of the scenery, but to paint the effect is a hopeless endeavour. Learned naturalists describe these scenes of the tropics by naming a multitude of objects, and mentioning some characteristic feature of each. To a learned traveller, this possibly may communicate some definite ideas: but who else from seeing a plant in an herbarium can imagine its appearance when growing in its native soil? Who from seeing choice plants in a hothouse can magnify some into the dimensions of forest trees, and crowd others into an entangled jungle? Who when examining in the cabinet of the entomologist the gay exotic butterflies, and singular cicadas, will associate with these objects, the ceaseless harsh music of the latter, and the lazy flight of the former – the sure accompaniments of the still, glowing, noonday of the tropics. It is, when the sun has attained its greatest height, that such views should be beheld: then the dense splendid foliage of the mango hides the ground with its darkest shade, whilst the upper branches are rendered from the profusion of light of the most brilliant green. In the temperate zones, as it appears to me, the case is different, the vegetation there is not so dark or so rich, and hence the rays of the declining sun, tinged of a red, purple, or yellow colour, add most to the beauties of the scenery of those climes.

When quietly walking along the shady pathways, and admiring each successive view, one wishes to find language to express one's ideas. Epithet after epithet is found too weak to convey to those, who have not visited the intertropical regions, the sensation of delight which the mind experiences. I have said the plants in a hothouse fail to communicate a just idea of the vegetation, yet I must recur to it. The land is one great wild, untidy, luxuriant hothouse, which nature made for her menagerie, but man has taken possession of it, and has studded it with gay houses and formal gardens. How great would be the desire in every admirer of nature to behold, if such were possible, another planet; yet to every one in Europe, it may be truly said, that at the distance of a few degrees from his native soil, the glories of another world are open to him. In my last walk, I stopped again and again to gaze on these beauties, and endeavoured to fix for ever in my mind an impression, which at the time I knew, sooner or later must fail. The form of the orange-tree, the cocoa-nut, the palm, the mango, the tree-fern, the banana, will remain clear and separate; but the thousand beauties which unite these into one perfect scene must fade away; yet they will leave, like a tale heard in childhood, a picture full of indistinct, but most beautiful figures.

* * *

On the 17th we took our final leave of the coast of South America, and on the last day of the month anchored at Porto Praya. We staid there only five days, and on the 5th of September steered for the Azores. On the 19th we anchored off the town of Angra, the capital of Terceira.

This island is moderately lofty, and has a rounded outline, with detached conical hills evidently of volcanic origin. The land is well cultivated, and is divided into a multitude of rectangular fields by stone walls, extending from the water's edge to high up on the central hills. There are few or no trees, and the yellow stubble land at this time of the year gave a burnt-up and unpleasant character to the scenery. Small hamlets and single whitewashed houses are scattered in all parts. In the evening a party went on shore: we found the city a very clean and tidy little place, containing about 10,000 inhabitants, which includes nearly the fourth part of the total number on the island. There are no good shops, and there is little appearance of activity, excepting the intolerable creaking of an occasional bullock-waggon. The churches are very respectable, and there were formerly a good many convents; but Dom Pedro destroyed several. He levelled three nunneries to the ground, and gave permission to the nuns to marry, which, except by some very old ones, was gladly received.

Angra was formerly the capital of the whole Archipelago, but it has now only one division of the islands under its government, and its glory has departed. The city is defended by a strong castle on Mount Brazil, and by a line of batteries encircling the base of this extinct volcano, which overlooks the town. Terceira was the first place that received Dom Pedro, and from this beginning he conquered the other islands, and finally Portugal. A loan was scraped together in this one island of no less than $400,000, of which sum not one farthing has ever been paid to these first supporters of the present right royal and honourable family.

The next day the Consul kindly lent me his horse, and furnished me with guides to proceed to a spot in the centre of the island, which was described as an active crater. Ascending in deep lanes, bordered on each side by high stone walls, for the first 3 miles we passed many houses and gardens. We then entered on a very irregular country, consisting of more recent streams of hummocky basaltic lava. The rocks are covered in some parts by a thick brushwood about 3 feet high, and in others by heath, fern, and short pasture: a few broken down old stone walls completed the resemblance with the mountains of Wales. I saw, moreover, some old English friends amongst the insects; and of birds, the starling, water-wagtail, chaffinch, and blackbird. There are no houses in this elevated and central part, and the ground is only used for the pasture of cattle and goats. On every side besides the ridges of more ancient lavas, there were cones of various dimensions, which still partly retained their crater-formed summits; and where broken down, showed a pile of cinders, such as those from an iron-foundry.

When we reached the so-called crater, I found it consisted of a slight depression, or rather of a short valley abutting against a higher range and without any exit. The bottom was traversed by several large fissures, out of which, in nearly a dozen places, small jets of steam issued as from the cracks in the boiler of a steam engine. The steam close to the irregular orifices was far too hot for the hand to endure it. It had but little smell, yet from everything made of iron being blackened, and from a peculiar rough sensation communicated to the skin, the vapour cannot be pure; I imagine it contains some muriatic acid. The effect on the surrounding trachytic lava was singular, the solid stone being entirely converted either into pure snow-white porcelain clay, or into a kind of the brightest red, or the two colours were marbled together. The steam has thus been emitted during many years; and it is said that flames once issued from the cracks. During rain, the water from each bank must flow into these cracks; and it is probable that this same water trickling down to the neighbourhood of some heated subterranean lava, causes the above effects. Throughout the island, the powers from below have been unusually active during the last year; several small earthquakes have been felt, and during a few days a jet of steam issued from a bold precipice (part of Mount Brazil) overhanging the sea, not far from the town of Angra.

I enjoyed my day's ride, though I did not find much worth seeing. It was pleasant to meet the peasantry; I do not recollect ever having beheld a set of handsomer young men, with more goodhumoured expressions. The greater number whom we met, were employed in the mountains gathering sticks for firewood. A whole family, from the father to the least boy, might be seen, each carrying his bundle on his head to sell in the town. Their burdens were very heavy; this hard labour and the ragged state of their clothes plainly bespoke poverty; yet I am told it is not that they want food, but there is an absence of all luxuries – a case parallel to that of Chiloe. Hence, although the whole land is not cultivated, numbers are emigrating to Brazil, where the contract to which they are bound differs but little from slavery. It seems a great pity that so fine a population should be compelled to leave a land of plenty, where every article of food – meat, vegetables, and fruit – is exceedingly cheap and most abundant: but the labourer finds his labour of proportionally little value.

Another day I set out early in the morning to visit the town of Praya situated towards the NE extremity of the island. The distance is about 15 miles; the road ran during the greater part of the way, not far from the coast. The country is all cultivated, and scattered over with houses and small villages. I noticed in several places, that the solid lava, which in part formed the road, was worn into ruts of the depth of 12 inches from the long traffic of the bullock-waggons. This circumstance has been noticed with surprise in the ancient pavement of Pompeii, for it does not occur in any of the present towns of Italy. The waggon-wheels here have a tire surmounted by singularly large knobs of iron; perhaps the old

Roman wheels were thus furnished. The country during our morning's ride was not interesting; excepting always when enlivened by the pleasant sight of the healthy peasantry. The harvest was lately over, and near the houses, the fine yellow heads of the Indian corn were tied in large bundles, to be dried, to the poplar-trees; and these, seen from a distance, appeared weighed down by some beautiful fruit – the very emblem of fertility.

One part of the road crossed a broad stream of lava, which from its rocky and black surface seemed to be of comparatively recent origin: indeed, the crater, whence it had flowed, could be distinguished. The industrious inhabitants have turned this space into vineyards; but for this purpose it was necessary to clear away the loose fragments, and to pile them up into a multitude of walls, which enclose little patches of ground a few yards square, thus covering the country with a network of black lines.

The town of Praya is a quiet, forlorn, little place: many years since a large city was here overwhelmed by an earthquake. It is asserted that the land then subsided, and a wall of a convent now bathed by the sea, is shown as a proof of it: the fact is probable, but the supposed proof not conclusive. I returned home by another road, which first led along the northern shore, and then crossed the central part of the island. This north-eastern extremity is particularly well cultivated, and produces a large quantity of fine wheat. The square open fields, and small villages with whitewashed churches, gave to the view, as seen from the heights, an aspect resembling the less picturesque parts of central England. We soon reached the region of clouds, which during our whole visit hung very low and concealed the tops of the mountains. For a couple of hours we crossed this central and elevated part, which is not inhabited and has a desolate appearance. When we descended from the clouds to the city, I heard the good news that astronomical observations had been obtained, and that we should go to sea the same evening.

On the 25th we called at the island of St Michael's for letters, and then steered a direct course for England. On October 2nd the *Beagle* anchored at Falmouth, where I left her, having lived on board the little vessel very nearly five years.

* * *

Our voyage having come to an end, I will take a short retrospect of the advantages and disadvantages, the pains and pleasures, of our five years' wandering. If a person should ask my advice, before undertaking a long voyage, my answer would depend upon his possessing a decided taste for some branch of knowledge, which could by such means be improved. No doubt it is a high satisfaction to behold various countries, and the many races of mankind, but the pleasures gained at the time do not counterbalance the evils. It is necessary to look forward to a harvest, however distant it may be, when some fruit will be reaped, some good effected.

Many of the losses which must be experienced are obvious; such as that of the society of all old friends, and of the sight of those places, with which every dearest remembrance is so intimately connected. These losses, however, are at the time partly relieved by the exhaustless delight of anticipating the long wished-for day of return. If, as poets say, life is a dream, I am sure in a voyage these are the visions which serve best to pass away the long night. Other losses, although not at first felt, tell heavily after a period; these are, the want of room, of seclusion, of rest; the jading feeling of constant hurry; the privation of small luxuries, the comforts of civilization and domestic society, and, lastly, even of music and the other pleasures of imagination. When such trifles are mentioned, it is evident that the real grievances (excepting from accidents) of a sea life are at an end. The short space of sixty years has made an astonishing difference in the facility of distant navigation. Even in the time of Cook, a man who left his comfortable fireside for such expeditions, underwent severe privations. A yacht now with every luxury of life might circumnavigate the globe. Besides the vast improvements in ships and naval resources, the whole western shores of America are thrown open, and Australia has become the metropolis of a rising continent. How different are the circumstances to a man shipwrecked at the present day in the Pacific, to what they were in the time of Cook! since his voyage a hemisphere has been added to the civilized world.

If a person suffer much from sea-sickness, let him weigh it heavily in the balance. I speak from experience: it is no trifling evil which may be cured in a week. If, on the other hand, he takes pleasure in naval tactics, he will assuredly have full scope for his taste. But it must be borne in mind, how large a proportion of the time, during a long voyage, is spent on the water, as compared with the days in harbour. And what are the boasted glories of the illimitable ocean? A tedious waste, a desert of water, as the Arabian calls it. No doubt there are some delightful scenes. A moonlight night, with the clear heavens and the dark glittering sea, and the white sails filled by the soft air of a gently-blowing trade wind; a dead calm, with the heaving surface polished like a mirror, and all still, except the occasional flapping of the sails. It is well once to behold a squall with its rising arch and coming fury, or the heavy gale of wind and mountainous waves. I confess, however, my imagination had painted something more grand, more terrific in the full-grown storm. It is an incomparably finer spectacle when beheld on shore, where the waving trees, the wild flight of the birds, the dark shadows and bright lights, the rushing of the torrents, all proclaim the strife of the unloosed elements. At sea the albatross and petrel fly as if the storm were their proper sphere, the water rises and sinks as if fulfilling its usual task, the ship alone and its inhabitants seem the objects of wrath. On a forlorn and weather-beaten coast, the scene is indeed different, but the feelings partake more of horror than of wild delight.

Let us now look at the brighter side of the past time. The pleasure derived from beholding the scenery and the general aspect of the various countries we have visited, has decidedly been the most constant and highest source of enjoyment. It is probable that the picturesque beauty of many parts of Europe exceeds any thing we have beheld. But there is a growing pleasure in comparing the character of scenery in different countries, which to a certain degree is distinct from merely admiring its beauty. It depends more on an acquaintance with the individual parts of each view. I am strongly induced to believe that, as in music, the person who understands every note will, if he also possesses a proper taste, more thoroughly enjoy the whole, so he who examines each part of a fine view, may also thoroughly comprehend the full and combined effect. Hence, a traveller should be a botanist, for in all views plants form the chief embellishment.

Group masses of naked rock even in the wildest forms, and they may for a time afford a sublime spectacle, but they will soon grow monotonous. Paint them with bright and varied colours, they will become fantastic; clothe them with vegetation, they must form at least a decent, if not a most beautiful picture.

When I said that the scenery of Europe was probably superior to any thing which we have beheld, I excepted, as a class by itself, that of the intertropical regions. The two classes cannot be compared together; but I have already often enlarged on the grandeur of these climates. As the force of impressions generally depends on preconceived ideas, I may add, that all mine were taken from the vivid descriptions in the *Personal Narrative* of Humboldt, which far exceed in merit any thing I have read on the subject. Yet with these high-wrought ideas, my feelings were far from partaking of any tinge of disappointment on first landing on the shores of Brazil.

Among the scenes which are deeply impressed on my mind, none exceed in sublimity the primeval forests undefaced by the hand of man; whether those of Brazil, where the powers of Life are predominant, or those of Tierra del Fuego, where Death and Decay prevail. Both are temples filled with the varied productions of the God of Nature: no one can stand in these solitudes unmoved, and not feel that there is more in man than the mere breath of his body. In calling up images of the past, I find the plains of Patagonia frequently cross before my eyes: yet these plains are pronounced by all most wretched and useless. They are characterized only by negative possessions; without habitations, without water, without trees, without mountains, they support merely a few dwarf plants. Why then, and the case is not peculiar to myself, have these arid wastes taken so firm possession of the memory? Why have not the still more level, the greener and more fertile Pampas, which are serviceable to mankind, produced an equal impression? I can scarcely analyze these feelings: but it must be partly owing to the free scope given to the imagination. The plains of Patagonia are boundless, for they are scarcely practicable, and hence unknown: they bear the stamp of having thus lasted for ages, and there appears no limit to their duration through future time. If, as the ancients supposed, the flat earth was surrounded by an impassable breadth of water, or by deserts heated to an intolerable excess, who would not look at these last boundaries to man's knowledge with deep but ill-defined sensations?

Lastly, of natural scenery, the views from lofty mountains, though certainly in one sense not beautiful, are very memorable. When looking down from the crest of the highest Cordillera, the mind undisturbed by minute details, was filled wth the stupendous dimensions of the surrounding masses.

Of individual objects, perhaps no one is more certain to create astonishment than the first sight in his native haunt of a real barbarian – of man in his lowest and most savage state. One's mind hurries back over past centuries, and then asks, could our progenitors have been such as these? Men, whose very signs and expressions are less intelligible to us than those of the domesticated animals; men, who do not possess the instinct of those animals, nor yet appear to boast of human reason, or at least of arts consequent on that reason. I do not believe it is possible to describe or paint the difference between savage and civilized man. It is the difference between a wild and tame animal: and part of the interest in beholding a savage, is the same which would lead every one to desire to see the lion in his desert, the tiger tearing his prey in the jungle, the rhinoceros on the wide plain, or the hippopotamus wallowing in the mud of some African river.

Among the other most remarkable spectacles which we have beheld, may be ranked the stars of the southern hemisphere – the water-spout – the glacier leading its blue stream of ice in a bold precipice overhanging the sea – a lagoon island raised by the coral-forming polypi – an active volcano – and the overwhelming effects of a violent earthquake. The three latter phenomena, perhaps, possess for me a peculiar interest, from their intimate connexion with the geological structure of the world. The earthquake must however, be to every one a most impressive event: the earth, considered from our earliest childhood as the type of solidity, has oscillated like a thin crust beneath our feet; and in seeing the most beautiful and laboured works of man in a moment overthrown, we feel the insignificance of his boasted power.

It has been said, that the love of the chase is an inherent delight in man – a relic of an instinctive passion. If so, I am sure the pleasure of living in the open air, with the sky for a roof, and the ground for a table, is part of the same feeling: it is the savage returning to his wild and native habits. I always look back to our boat cruises, and my land journeys, when through unfrequented countries, with a kind of extreme delight, which no scenes of civilization could have created. I do not doubt that every traveller must remember the glowing sense of happiness he experienced, from the simple consciousess of breathing in a foreign clime, where the civilized man has seldom or never trod.

There are several other souces of enjoyment in a long voyage, which are, perhaps, of a more reasonable nature. The map of the world ceases to be a blank; it becomes a picture full of the most varied and animated figures. Each part assumes its true dimensions: continents are not looked at in the light of islands, or those islands considered as mere specks, which are, in truth, larger than many kingdoms of Europe. Africa, or North and South America, are well-sounding names, and easily pronounced; but it is not till having sailed for some weeks along small portions of their coasts that one is thoroughly convinced how large a portion of our immense world these names imply.

From seeing the present state, it is impossible not to look forward with high expectation to the future progress of nearly an entire hemisphere. The march of improvement, consequent on the introduction of Christianity throughout the South Sea, probably stands by itself on the records of history. It is the more striking when we remember that only sixty years since, Cook, whose most excellent judgment none will dispute, could foresee no prospect of such change. Yet these changes have now been effected by the philanthropic spirit of the British nation.

In the same quarter of the globe Australia is rising, or indeed may be said to have risen, into a grand centre of civilization, which at some not very remote period, will rule as empress over the southern hemisphere. It is impossible for an Englishman to behold these distant colonies, without a high pride and satisfaction. To hoist the British flag, seems to draw with it as a certain consequence, wealth, prosperity, and civilization.

In conclusion, it appears to me that nothing can be more

improving to a young naturalist, than a journey in distant countries. It both sharpens, and partly likewise allays that want and craving, which, as Sir J. Herschel* remarks, a man experiences although every corporeal sense is fully satisfied. The excitement from the novelty of objects, and the chance of success, stimulate him to increased activity. Moreover as a number of isolated facts soon become uninteresting, the habit of comparison leads to generalization. On the other hand, as the traveller stays but a short space of time in each place, his descriptions must generally consist of mere sketches, instead of detailed observation. Hence arises, as I have found to my cost, a constant tendency to fill up the wide gaps of knowledge, by inaccurate and superficial hypotheses.

But I have too deeply enjoyed the voyage, not to recommend any naturalist, although he must not expect to be so fortunate in his companions as I have been, to take all chances, and to start, on travels by land if possible, if otherwise on a long voyage. He may feel assured, he will meet with no difficulties or dangers (excepting in rare cases) nearly so bad as he beforehand anticipated. In a moral point of view, the effect ought to be, to teach him good-humoured patience, freedom from selfishness, the habit of acting for himself, and of making the best of every thing, or in other words contentment. In short he should partake of the characteristic qualities of the greater number of sailors. Travelling ought also to teach him distrust; but at the same time he will discover, how many truly goodnatured people there are, with whom he never before had, or ever again will have any further communication, who yet are ready to offer him the most disinterested assistance.

* Discourse on the Study of Natural Philosophy.

APPENDIX ONE

ADMIRALTY INSTRUCTIONS FOR THE *BEAGLE* VOYAGE

VICTORIAN NAVAL officers almost always included the official Admiralty instructions in the written accounts of their voyages that were published afterwards, and Captain Robert FitzRoy was no exception. The text that follows is taken directly from the second volume of his *Narrative of the surveying voyages of HMS Adventure and Beagle*, London, 1839.

Hydrographer's Opinion – Continuation of Survey – Chain of Meridian Distances – Efficient Arrangements – Repair and raise Deck – Outfit – Boats – Lightning-Conductors – Rudder – Stove – Windlass – Chronometers – Mr Darwin – Persons on board – Changes – List of those who returned – Supplies – Admiralty Instructions – Memorandum – Hydrographer's Memorandum

WHEN IT was decided that a small vessel should be sent to Tierra del Fuego, the Hydrographer of the Admiralty was referred to for his opinion, as to what addition she might make to the yet incomplete surveys of that country, and other places which she might visit.

Captain Beaufort embraced the opportunity of expressing his anxiety for the continuance of the South American Surveys, and mentioning such objects, attainable by the *Beagle*, as he thought most desirable: and it was soon after intimated to me that the voyage might occupy several years. Desirous of adding as much as possible to a work in which I had a strong interest, and entertaining the hope that a chain of meridian distances might be carried round the world if we returned to England across the

Pacific, and by the Cape of Good Hope; I resolved to spare neither expense nor trouble in making our little Expedition as complete, with respect to material and preparation, as my means and exertions would allow, when supported by the considerate and satisfactory arrangements of the Admiralty: which were carried into effect (at that time) by the Navy Board, the Victualling Board, and the Dockyard officers at Devonport.

The *Beagle* was commissioned on the 4th of July 1831, and was immediately taken into dock to be thoroughly examined, and prepared for a long period of foreign service. As she required a new deck, and a good deal of repair about the upper works, I obtained permission to have the upper-deck raised considerably,* which afterwards proved to be of the greatest advantage to her as a sea boat, besides adding so materially to the comfort of all on board. While in dock, a sheathing of 2-inch fir plank was nailed on the vessel's bottom, over which was a coating of felt, and then new copper. This sheathing added about 15 tons to her displacement, and nearly 7 to her actual measurement. Therefore, instead of 235 tons, she might be considered about 242 tons burthen. The rudder was fitted according to the plan of Captain Lihou: a patent windlass supplied the place of a capstan: one of Frazer's stoves, with an oven attached, was taken instead of a common 'galley' fire-place; and the lightning-conductors, invented by Mr Harris, were fixed in all the masts, the bowsprit, and even in the flying jib-boom. The arrangements made in the fittings, both inside and outside, by the officers of the Dock-yard, left nothing to be desired. Our ropes, sails, and spars, were the best that could be procured; and to complete our excellent outfit, six superior boats† (two of them private property) were built expressly for us, and so contrived and stowed that they could all be carried in any weather.

Considering the limited disposable space in so very small a ship, we contrived to carry more instruments and books than one would readily suppose could be stowed away in dry and secure places; and in a part of my own cabin twenty-two chronometers were carefully placed.

* Eight inches abaft and twelve forward.
† Besides a dinghy carried astern.

Anxious that no opportunity of collecting useful information, during the voyage, should be lost; I proposed to the Hydrographer that some well-educated and scientific person should be sought for who would willingly share such accommodations as I had to offer, in order to profit by the opportunity of visiting distant countries yet little known. Captain Beaufort approved of the suggestion, and wrote to Professor Peacock, of Cambridge, who consulted with a friend, Professor Henslow, and he named Mr Charles Darwin, grandson of Dr Darwin the poet, as a young man of promising ability, extremely fond of geology, and indeed all branches of natural history. In consequence an offer was made to Mr Darwin to be my guest on board, which he accepted conditionally; permission was obtained for his embarkation, and an order given by the Admiralty that he should be borne on the ship's books for provisions. The conditions asked by Mr Darwin were, that he should be at liberty to leave the *Beagle* and retire from the Expedition when he thought proper, and that he should pay a fair share of the expenses of my table.

Knowing well that no one actively engaged in the surveying duties on which we were going to be employed, would have time – even if he had ability – to make much use of the pencil, I engaged an artist, Mr Augustus Earle, to go out in a private capacity; though not without the sanction of the Admiralty, who authorized him also to be victualled. And in order to secure the constant, yet to a certain degree mechanical attendance required by a large number of chronometers, and to be enabled to repair our instruments and keep them in order, I engaged the services of Mr George James Stebbing, eldest son of the mathematical instrument-maker at Portsmouth, as a private assistant.

The established complement of officers and men (including marines and boys) was sixty-five: but, with the supernumeraries I have mentioned, we had on board, when the *Beagle* sailed from England, seventy-four persons, namely:

Robert FitzRoy	Commander and Surveyor
John Clements Wickham	Lieutenant
Bartholomew James Sulivan	Lieutenant

Edward Main Chaffers .... Master
Robert Mac-Cormick .... Surgeon
George Rowlett .... Purser
Alexander Derbishire .... Mate
Peter Benson Stewart .... Mate
John Lort Stokes .... Mate and Assistant
                                    Surveyor
Benjamin Bynoe .... Assistant Surgeon
Arthur Mellersh .... Midshipman
Philip Gidley King .... Midshipman
Alexander Burns Usborne .... Master's Assistant
Charles Musters .... Volunteer 1st Class
Jonathan May .... Carpenter
Edward H. Hellyer .... Clerk

Acting boatswain: sergeant of marines and seven privates: thirty-four seamen and six boys.

On the List of supernumeraries were –

Charles Darwin .... Naturalist
Augustus Earle .... Draughtsman
George James Stebbing .... Instrument Maker

Richard Matthews and three Fuegians: my own steward: and Mr Darwin's servant.

Some changes occurred in the course of the five years' voyage, which it may be well to mention in this place.

In April 1832, Mr Mac-Cormick and Mr Derbishire returned to England. Mr Bynoe was appointed to act as Surgeon. Mr Mellersh received a Mate's warrant; and Mr Johnson joined the *Beagle* as Midshipman. In May Mr Musters fell a victim to fever, caught in the harbour of Rio de Janeiro: Mr Forsyth took his place.

Mr Earle suffered so much from continual ill health, that he could not remain on board the *Beagle* after August 1832; but he lived at Monte Video several months previously to his return to England. The disappointment caused by losing his services was diminished by meeting Mr Martens at Monte Video, and engaging him to embark with me as my draughtsman.

In March 1833, Mr Hellyer was drowned at the Falkland Islands, in attempting to get a bird he had shot. In September 1833, Mr Kent joined as Assistant Surgeon. In June 1834, Mr Rowlett died, at sea, of a complaint under which he had laboured for years: and the vacancy caused by his lamented decease was filled by Mr Dring.

Mr Martens left me, at Valparaiso, in 1834; and Mr King remained with his father, at Sydney, in Australia, in February 1836. After these changes, and at our return to England in October 1836, the list stood thus —

Robert FitzRoy	Captain and Surveyor
John Clements Wickham	Lieutenant
Bartholomew James Sulivan	Lieutenant
Edward Main Chaffers	Master
Benjamin Bynoe	Surgeon (Acting)
John Edward Dring	Purser (Acting)
Peter Benson Stewart	Mate
John Lort Stokes	Mate and Assistant Surveyor
Arthur Mellersh	Mate
Charles Richardson Johnson	Mate
William Kent	Assistant Surgeon
Charles Forsyth	Midshipman
Alexander Burns Usborne	Master's Assistant
Thomas Sorrell	Boatswain (Acting)
Jonathan May	Carpenter

And on the List of supernumeraries were Mr Darwin: George J. Stebbing: my steward: and Mr Darwin's servant.

Our complement of seamen, marines, and boys was complete at our return, and generally during the voyage; because, although many changes happened, we had always a choice of volunteers to fill vacant places.

Many of the crew had sailed with me in the previous voyage of the *Beagle*; and there were a few officers, as well as some marines and seamen, who had served in the *Beagle*, or *Adventure*, during the whole of the former voyage. These determined admirers of

Tierra del Fuego were, Lieutenant Wickham, Mr Bynoe, Mr Stokes, Mr Mellersh, and Mr King; the boatswain, carpenter, and sergeant; four private marines, my coxswain, and some seamen.

I must not omit to mention that among our provisions were various antiscorbutics – such as pickles, dried apples, and lemon juice – of the best quality, and in as great abundance as we could stow away; we had also on board a very large quantity of Kilner and Moorsom's preserved meat, vegetables, and soup: and from the Medical Department we received an ample supply of anti-septics, and articles useful for preserving specimens of natural history.

Not only the heads of departments exerted themselves for the sake of our health and safety, but the officers subordinate to them appeared to take a personal interest in the *Beagle*; for which I and those with me felt, and must always feel, most grateful.

Perhaps no vessel ever quitted her own country with a better or more ample supply (in proportion to her probable necessities) of every kind of useful provision and stores than the little ship of whose wanderings I am now about to give a brief and very imperfect narrative; and, therefore, if she succeeded in effecting any of the objects of her mission, with comparative ease and expedition, let the complete manner in which she was pre-pared for her voyage, by the Dock-yard at Devonport, be fully remembered.

On the 15th of November I received my instructions from the Lords Commissioners of the Admiralty.

Instructions

By the Commissioners for executing the office of Lord High Admiral of the United Kingdom of Great Britain and Ireland, &c.

You are hereby required and directed to put to sea, in the vessel you command, so soon as she shall be in every respect ready, and to proceed in her, with all convenient expedition, successively to Madeira or Teneriffe; the Cape de Verd Islands; Fernando Noronha; and the South American station; to perform the oper-ations, and execute the surveys, pointed out in the accompanying memorandum, which has been drawn up under our direction by the Hydrographer of this office; observing and following, in the prosecution of the said surveys, and in your other operations, the directions and suggestions contained in the said memorandum.

You are to consider yourself under the command of Rear-Admiral Sir Thomas Baker, Commander-in-chief of his Majesty's ships on the South American station, whilst you are within the limits of that station, in execution of the services above-mentioned; and in addition to the directions conveyed to you in the memorandum, on the subject of your supplies of provisions, we have signified to the Rear-Admiral our desire that, whenever the occasion offers, you should receive from him and the officers of his squadron, any assistance, in stores and provisions, of which you may stand in need.

But during the whole time of your continuing on the above duties, you are (notwithstanding the 16th article of the 4th section of the 6th chapter, page 78, of the General Printed Instructions) to send reports, by every opportunity, to our Secretary, of your proceedings, and of the progress you make.

Having completed the surveys, which you are directed to execute on the South American station, you are to proceed to perform the several further operations set forth in the Hydrographer's memorandum, in the course therein pointed out; and having so done, you are to return, in the vessel you command, to Spithead, and report your arrival to our Secretary, for our information and further directions.

In the event of any unfortunate accident happening to yourself, the officer on whom the command of the *Beagle* may in consequence devolve, is hereby required and directed to complete, as far as in him lies, that part of the survey on which the vessel may be then engaged, but not to proceed to a new step in the voyage; as, for instance, if at that time carrying on the coast survey on the western side of South America, he is not to cross the Pacific, but to return to England by Rio de Janeiro and the Atlantic.

[Memorandum]

A considerable difference still exists in the longitude of Rio de Janeiro, as determined by Captains King, Beechey, and Foster, on the one hand, and Captain W. F. Owen, Baron Roussin, and the Portuguese astronomers, on the other; and as all our meridian distances in South America are measured from thence, it becomes a matter of importance to decide between these conflicting authorities. Few vessels will have ever left this country with a better set of chronometers, both public and private, than the *Beagle*; and if her voyage be made in short stages, in order to detect the changes which take place in all chronometers during a continuous increase of temperature, it will probably enable us to reduce that difference within limits too small to be of much import in our future conclusions.

With this view, the run to Rio de Janeiro may be conveniently divided into four parts:

1st. Touching at Madeira, the exact position of which has been admitted by all parties. Having obtained a four days' rate there, or, if the weather and the exposed anchorage will not permit, at Teneriffe, the *Beagle* should, 2ndly, proceed with the least possible delay to Port Praya, in the Cape de Verd Islands, not only to establish a fresh four days' rate; but that point being the pivot on which all Captain Owen's longitudes turn, no pains should be spared in verifying the position he has assumed for it. From thence, 3rdly, she should make the best of her way across the Line to Fernando Noronha. This island, indeed, lies somewhat to the westward of her track, and may retard her progress a little; yet a series of chronometric observations there is essential to the object in view, because it forms the third nearly equal division of the whole run, and because it was the point of junction of Commander Foster's double line of longitudes. If two or three days' delay at either of these two last stations will enable him to obtain satisfactory occultations, and moon culminating observations, which are likely to be seen in this country, the increased certainty of the results will well atone for that loss of time. The Commander will, of course, be careful to adopt, in all those stations, the precise spot of the former observations, with which his are to be compared. The Governor of Fernando Noronha was peculiarly obliging to Commander Foster, and gave up part of his own house for the pendulum experiments. There will be no occasion now for trespassing so heavily on his kindness; but the difference of longitude between that station and Commander FitzRoy's must be well measured.

However desirable it may be that the *Beagle* should reach Rio de Janeiro as soon as possible, yet the great importance of knowing the true position of the Abrolhos Banks, and the certainty that they extend much further out than the limits assigned to them by Baron Roussin, will warrant the sacrifice of a few days, if other circumstances should enable her to beat down about the meridian of 36° W from the latitude of 16° S. The deep sea-line should be kept in motion; and if soundings be obtained, the bank should be pursued both ways, out to the edge, and in to that part already known.

Its actual extent to the eastward, and its connection with the shoals being thus ascertained, its further investigation may be left to more convenient opportunities.

At Rio de Janeiro, the time necessary for watering, &c. will, no doubt, be employed by the commander in every species of observation that can assist in deciding the longitude of Villegagnon Island.

It is understood that a French Expedition is now engaged in the examination of the coast between St Catherine's and the Rio de la Plata; it would therefore be a waste of means to interfere with that interval; and Commander FitzRoy should be directed to proceed to Monte Video, and to rate his chronometers in the same situation occupied by Captain King.

To the southward of the Rio de la Plata, the real work of the survey will begin. Of that great extent of coast which reaches from Cape St Antonio to St George's bay, we only know that it is inaccurately placed, and that it contains some large rivers, which rise at the other side of the continent, and some good harbours, which are undoubtedly worth a minute examination. Much of it, however, from the casual accounts of the Spaniards, seems to offer but little interest either to navigation or commerce, and will scarcely require more than to have its direction laid down correctly, and its prominent points fixed. It should nevertheless be borne in mind there, and in other places, that the more hopeless and forbidding any long line of coast may be, the more precious becomes the discovery of a port which affords safe anchorage and wholesome refreshments.

The portions of the coast which seem to require particular examination are –

1st. From Monte Hermoso to the Rio Colorado, including the large inlet of Bahia Blanco, of which there are three manuscripts in this office that differ in every thing but in name.

2ndly. The gulf of Todos los Santos, which is studded in the Spanish charts with innumerable islands and shoals. It is said to have an excellent harbour on its southern side, which should be verified; but a minute survey of such an Archipelago would be a useless consumption of time, and it will therefore be found sufficient to give the outer line of the dangers, and to connect that line with the regular soundings in the offing.

3rdly. The Rio Negro is stated to be a river of large capacity, with settlements 50 miles from its mouth, and ought to be partially reconnoitred as far as it is navigable.

4thly. The gulf of San Matias should be examined, especially its two harbours, San Antonio and San José, and a narrow inlet on the eastern side of the peninsula, which, if easy of access, appears to be admirably situated: and –

5thly. From the Bahia Nueva to Cape Blanco, including the Gulf of St George, the coast is of various degrees of interest, and will accordingly require to have more or less time bestowed on its different parts. The position of Cape Blanco should be determined, as there appears to be an error of some miles in its latitude, as well as much doubt about the places of two shoals which are marked near it in the Spanish charts.

From Cape Blanco to the Strait of Magalhaens, the coast has been partially corrected by Captain King; and Port Desire, having been carefully placed by him, will afford a good place for rating the chronometers, and an opportunity for exploring the river.

Port San Julian, with its bar and wide river, should be surveyed, as well as any parts of that interval which were not visited in the last expedition.

The above are the principal points of research between the Rio de la Plata and the Strait. They have been consecutively mentioned in order to bring them into one point of view; but that part of this service would perhaps be advantageously postponed till after the *Beagle*'s first return from the southward; and, generally speaking, it would be unwise to lay down here a specific route from which no deviation would be permitted. Where so many unforeseen circumstances may disturb the best-concerted arrangements, and where so much depends on climates and seasons with which we are not yet intimately acquainted, the most that can be safely done is to state the various objects of the voyage, and to rely on the Commander's known zeal and prudence to effect them in the most convenient order.

Applying this principle to what is yet to be done in the Strait, and in the intricate group of islands which forms the Tierra del Fuego, the following list will show our chief desiderata.

Captain King, in his directions, alludes to a reef of half a mile in length, off Cape Virgins, and in his chart he makes a 7 fathoms' channel outside that reef; and still further out, 5 fathoms with overfalls. Sarmiento places 50 fathoms at 10 miles ESE from that Cape; 13 fathoms at 19 miles; and, at 21 miles in the same direction, only 4 fathoms, besides a very extensive bank projecting from Tierra del Fuego, between which and the above shoals Malaspina passed in 13 fathoms. In short, there is conclusive evidence of there being more banks than one that obstruct the entrance to the Strait, and undoubtedly their thorough examination ought to be one of the most important objects of the Expedition; inasmuch, as a safe approach to either straits or harbours is of more consequence to determine than the details inside.

None of the above authors describe the nature of these shoals, whether rock or sand; it will be interesting to note with accuracy the slope, or regularity, of the depths, in their different faces, the quality of their various materials, and the disposition of the coarse or fine parts, as well as of what species of rock in the neighbourhood they seem to be the detritus; for it is probable that the place of their deposition is connected with the very singular tides which seem to circulate in the eastern end of the Strait.

Beginning at Cape Orange, the whole north-eastern coast of Tierra del Fuego as far as Cape San Diego should be surveyed, including the outer edge of the extensive shoals that project from its northern extreme, and setting at rest the question of the Sebastian Channel.

On the southern side of this great collection of islands, the Beagle Channel and Whale-boat Sound should be finished, and any other places which the Commander's local knowledge may point out as being requisite to complete his former survey, and sufficiently interesting in themselves to warrant the time they will cost; such as some apparently useful ports to the westward of Cape False, and the north side of Wakefield Channel, all of which are said to be frequented by the sealers.

In the north-western part it is possible that other breaks may be found interrupting the continuity of Sta Ines Island, and communicating from the Southern Ocean with the Strait; these should be fully or cursorily examined, according to their appearance and promise; and though it would be a very useless waste of time to pursue in detail the infinite number of bays, openings, and roads, that teem on the western side of that island, yet no good harbour should be omitted. It cannot be repeated too often that the more inhospitable the region, the more valuable is a known port of refuge.

In the western division of the Strait, from Cape Pillar to Cape Froward, there are a few openings which may perhaps be further explored, on the chance of their leading out to sea; a few positions which may require to be reviewed; and a few ports which were only slightly looked into during Captain King's laborious and excellent survey, and which may now be completed, if likely to augment the resources of ships occupied in those dreary regions.

In the eastern division of the Strait there is rather more work to be done, as the Fuegian shore from Admiralty Sound to Cape Orange has not been touched. Along with this part of the service, the Islands of Saints Martha and Magdalena, and the channel to the eastward of Elizabeth Island, will come in for examination; and there is no part of the Strait which requires to be more accurately laid down and distinctly described, from the narrowness of the channels and the transverse direction of the tides. Sweepstakes Foreland may prove to be an Island; if so, there may be found an useful outlet to the long lee-shore that extends from Cape Monmouth; and if not, perhaps some safe ports might be discovered in that interval for vessels caught there in strong westerly gales.

It is not likely that, for the purposes of either war or commerce, a much more detailed account will be necessary of those two singular inland seas, Otway and Skyring Waters, unless they should be found to communicate with one of the sounds on the western coast, or with the western part of the Strait. The general opinion in the former Expedition was certainly against such a communication, and the phenomena of the tides is also against it; still the thing is possible, and it becomes an interesting geographical question, which a detached boat in fine weather will readily solve.

These several operations may probably be completed in the summer of 1833–34, including two trips to Monte Video for refreshments; but before we finally quit the eastern coast of South America, it is necessary to advert to our present ignorance of the Falkland Islands, however often they have been visited. The time that would be occupied by a rigorous survey of this group of islands would be very disproportionate to its value; but as they are the frequent resort of whalers, and as it is of immense consequence to a vessel that has lost her masts, anchors, or a large part of her crew, to have a precise knowledge of the port to which she is obliged to fly, it would well deserve some sacrifice of time to have the most advantageous harbours and their approaches well laid down, and connected by a general sketch or running survey. Clear directions for recognizing and entering these ports should accompany these plans; and as most contradictory statements have been made of the refreshments to be obtained at the east and west great islands, an authentic report on that subject by the Commander will be of real utility.

There is reason to believe that deep soundings may be traced from these islands to the main, and if regular they would be of great service in rectifying a ship's place.

Having now stated all that is most urgent to be done on this side of the South American Continent as well as in the circuit of Tierra del Fuego, the next step of the voyage will be Concepción, or Valparaiso, to one of which places the *Beagle* will have to proceed for provisions, and where Captain King satisfactorily determined the meridian distances.

The interval of coast between Valparaiso and the western

entrance of the Strait has been partly surveyed, as well as most of the deep and narrow channels formed by the islands of Hanover, Wellington, and Madre de Dios; but of the sea face of that great chain of islands which stretches from Queen Adelaide Archipelago to Campana Island, little has yet been done. It presents a most uninviting appearance, can probably afford but little benefit to the navigator, and the chief object in urging its partial examination, is to remove a blank from this great survey, which was undertaken by Great Britain from such disinterested motives, and which was executed by Captains King and FitzRoy with so much skill and zeal.

The experience gained by the latter in that climate will enable him to accomplish all that is now required in much less time than it would have occupied in the beginning of the former expedition.

At the Gulf of Peñas the last survey terminated. Of the peninsula de Tres Montes, and of the islands between that and Chiloe, a Spanish manuscript has been procured from Don Felipe Bauzá, which may greatly abridge the examination of that interval.

From thence to Valdivia, Concepción, and Valparaiso, the shore is straight, and nearly in the direction of the meridian, so that it will require no great expenditure of time to correct the outline, and to fix the positions of all the salient points. Mocha Island is supposed to be erroneously placed: and the depth, breadth, and safety of its channel are not known.

To the south of Valparaiso the port of Topocalmo and the large shoal in the offing on which an American ship was wrecked, require special examination; and according to Captain Burgess, of the *Alert*, the coast and islands near Coquimbo are very imperfectly laid down. Indeed of the whole of this coast, the only general knowledge we have is from the Spanish charts, which seem, with the exception of certain ports, to have been merely the result of a running view of the shore. Of this kind of half-knowledge we have had too much: the present state of science, which affords such ample means, seems to demand that whatever is now done should be finally done; and that coasts, which are constantly visited by English vessels, should no longer have the motley appearance of alternate error and accuracy. If, therefore, the local Governments make no objections, the survey should be continued to Coquimbo, and indefinitely to the northward, till that period arrives when the Commander must determine on quitting the shores of South America altogether. That period will depend on the time that has been already consumed, and on the previous management of his resources, reserving sufficient to ensure his obtaining a series of well-selected meridian distances in traversing the Pacific Ocean.

The track he should pursue in executing this important duty cannot well be prescribed here, without foreseeing to what part of the coast he may have pushed the survey, and at what place he may find it convenient to take in his last supplies. If he should reach Guayaquil, or even Callao, it would be desirable he should run for the Galapagos, and, if the season permits, survey that knot of islands. Felix Island, the London bank seen by the brig *Cannon*, in 1827, in 27° 6′ S 92° 16′ W, even with the water's edge, and half a mile in length; some coral islands, supposed to be 5° or 6° south of Pitcairn Island, and other spots, which have crept into the charts on doubtful authority, would all be useful objects of research if the *Beagle*'s route should fall in their vicinity. But whatever route may be adopted, it should conduct her to Tahiti, in order to verify the chronometers at Point Venus, a point which may be considered as indisputably fixed by Captain Cook's and by many concurrent observations. Except in this case, she ought to avoid as much as possible the ground examined by Captain Beechey.

From Tahiti the *Beagle* should proceed to Port Jackson touching at some of the intervening islands, in order to divide the run into judicious chronometer stages; for the observatory at Paramatta (Port Jackson) being absolutely determined in longitude, all those intervening islands will become standard points to which future casual voyagers will be able to refer their discoveries or correct their chronometers.

From Port Jackson her course will depend on the time of the year. If it be made by the southward, she might touch at Hobart Town, King George Sound, and Swan River, to determine the difference of longitude from thence to the Mauritius, avoiding the hurricane months; to Table or Simon's Bay, according to the season; to St Helena, Ascension, and home.

If she should have to quit Port Jackson about the middle of the year, her passage must be made through Torres Strait. In her way thither, if the in-shore route be adopted, there are several places whose positions it will be advantageous to determine: Moreton Bay, Port Bowen, Cape Flinders, and one of the Prince of Wales Islands; and in pursuing her way towards the Indian Ocean, unless the wind should hang to the southward, Cape Valsche or the south-west extreme of New Guinea, one of the Serwatty Chain, Coupang, or the extreme of Timor, Rotte Island, and one of the extremes of Sandalwood Island, may be easily determined without much loss of time. And, perhaps, in crossing the ocean, if circumstances are favourable, she might look at the Keeling Islands, and settle their position.

Having now enumerated the princip [1] places at which the *Beagle* should be directed to touch in her circuit of the globe, and described the leading operations which it would be desirable to effect, it remains to make some general remarks on the conduct of the whole survey.

In such multiplied employments as must fall to the share of each officer, there will be no time to waste on elaborate drawings. Plain, distinct roughs, every where accompanied by explanatory notes, and on a sufficiently large scale to show the minutiæ of whatever knowledge has been acquired, will be documents of far greater value in this office, to be reduced or referred to, than highly finished plans, where accuracy is often sacrificed to beauty.

This applies particularly to the hills, which in general cost so much labour, and are so often put in from fancy or from memory after the lapse of months, if not of years, instead of being projected while fresh in the mind, or while any inconsistencies or errors may be rectified on the spot. A few strokes of a pen will denote the extent and direction of the several slopes much more distinctly than the brush, and if not worked up to make a picture, will really cost as little or less time. The in-shore sides of the hills, which cannot be seen from any of the stations, must always be mere guess-work, and should not be shown at all.

It should be considered an essential branch of a nautical survey, to give the perpendicular height of all remarkable hills and headlands. It requires but a single angle at each station, adds much to our geographical knowledge, materially assists the draftsman, and by tables which are now printing it will afford to the seaman a ready and exact means of knowing his distance.

All charts and plans should be accompanied by views of the land; those which are to be attached to the former should be taken at such a distance as will enable a stranger to recognize the land, or to steer for a certain point; and those best suited for the plan of a port should show the marks for avoiding dangers, for taking a leading course, or choosing an advantageous berth. In all cases the angular distances and the angular altitudes of the principal objects should be inserted in degrees and minutes on each of the views, by which means they can be projected by scale, so as to correct any want of precision in the eye of the draftsman. Such views cannot be too numerous; they cost but a few moments, and are extremely satisfactory to all navigators.

Trifling as it may appear, the love of giving a multiplicity of new and unmeaning names tends to confuse our geographical knowledge. The name stamped upon a place by the first discoverer should be held sacred by the common consent of all nations; and in new discoveries it would be far more beneficial to make the name convey some idea of the nature of the place; or if it be inhabited, to adopt the native appellation, than to exhaust the catalogue of public characters or private friends at home. The officers and crews, indeed, have some claim on such distinction, which, slight as it is, helps to excite an interest in the voyage.

Constant observations on the tides, including their set, force, and duration, the distance to which they carry salt water up the rivers, their rise at the different periods of the lunation, and the extent to which they are influenced by the periodic winds, by the sea currents, or by the river freshes, form so prominent a part of every surveyor's duty, that no specific directions on this subject can be necessary. Nor is there any occasion to insist here on the equally important subject of currents; for it is only by a great accumulation of data that we can ever hope to reduce them to regular systems, or that we can detect the mode in which they are affected by change of seasons, or influenced by distant winds.

The periods and limits of the monsoons and trade winds will naturally be a continual object of the Commander's observation and study. It is true that he can only witness what occurs during his voyage; but besides collecting facts on this and the last subject, on which others can hereafter reason, it will be of immense advantage that he should endeavour to digest them with the remarks of former voyagers when on the spot.

On the western coast of South America, for instance, some skill is required in making passages at different periods, and much scattered experience has been gained by seamen who have been long occupied there; but this information has not yet been presented to the public in an intelligible form; and it seems to be the peculiar province of an officer expressly employed on a scientific mission like this, to combine that information with his own, and to render it accessible to every navigator.

The local attraction of the *Beagle* will of course have been ascertained before she leaves England; but when favourable opportunities occur, it will be satisfactory to swing her again in different latitudes, and under large differences of variation.

No day should pass at sea without a series of azimuths, and no port should be quitted without having ascertained not only the magnetic angle, but the dip, intensity, and diurnal variation. If these observations should have been well made in the same places before, we shall at once obtain the annual change; and by multiplying them in new places, we shall have the means of inferring the magnetic curves.

The Commander has been so accustomed to the management of chronometers, that there is no doubt, with proper precautions and with proper formulæ for determining their rates, that he will succeed in obtaining good results in reasonably short intervals of time and in gradual changes of temperature; but after long periods, and sudden changes of heat and cold, it will be absolutely necessary to check them by astronomical means.

Eclipses, occultations, lunar distances, and moon-culminating stars, will furnish those means in abundance: of all these, the last can be obtained with the greatest regularity and certainty; they have become part of the current business at the establishments of the Cape of Good Hope, Paramatta, and St Helena, in the southern hemisphere; probably at Madras, and in many of the European observatories and it will therefore be scarcely possible that there should not be corresponding observations for all such as he may have made.

The eclipses of Jupiter's third and fourth satellites should also be sedulously observed whenever both immersion and emersion can be seen, as the different powers of the telescopes employed by the observers do not in that case affect the results.

There are also some remarkable phenomena, which will be announced in the Nautical Almanacks, and which will occur during the *Beagle*'s voyage. Some of these will be highly interesting to astronomers, and if it would not much derange her operations, she should be taken to some convenient anchorage for the purpose of landing the instruments.

If a comet should be discovered while the *Beagle* is in port, its position should be determined every night by observing its transit over the meridian, always accompanied by the transits of the nearest known stars, and by circum-meridional altitudes, or by measuring its angular distance from three well-situated stars by a sextant. This latter process can be effected even at sea, and the mean of several observations may give very near approximations to its real position.

Meterological Registers may be of use in a variety of ways; but then they must be steadily and accurately kept. The barometer should be read off to the third place of decimals, and recorded at regular periods of the day; nine o'clock and four o'clock may be recommended as the best, as being the usual hours of its maximum and minimum. The temperature should be marked at the same time, and the extremes of the self-registering thermometer should be daily recorded; care being taken that no reflected heat should act on any of these instruments. The temperature of the sea at the surface ought to be frequently observed and compared with that of the air. An officer cruising on the east coast of South America, between the parallels of 20° and 35°, was enabled by these means to predict with singular precision the direction and strength of the current.

In this register the state of the wind and weather will, of course, be inserted; but some intelligible scale should be assumed, to indicate the force of the former, instead of the ambiguous terms 'fresh', 'moderate', &c., in using which no two people agree; and some concise method should also be employed for expressing the state of the weather. The suggestions contained in the annexed printed paper are recommended for the above purposes, and if adopted, a copy should be pasted on the first page of every volume of the log-book; and the officer of the watch should be directed to use the same terms in the columns of the log-board.

The circularly formed Coral Islands in the Pacific occasionally afford excellent land-locked harbours, with a sufficient entrance, and would be well adapted to any nice astronomical observations which might require to be carried on in undisturbed tranquillity. While these are quietly proceeding, and the chronometers rating, a very interesting inquiry might be instituted respecting the formation of these coral reefs.

An exact geological map of the whole island should be constructed, showing its form, the greatest height to which the solid coral has risen, as well as that to which the fragments appear to have been forced. The slope of its sides should be carefully measured in different places, and particularly on the external face, by a series of soundings, at very short distances from each other, and carried out to the greatest possible depths, at times when no tide or current can affect the perpendicularity of the line. A modern and very plausible theory has been put forward, that these wonderful formations, instead of ascending from the bottom of the sea, have been raised from the summits of extinct volcanoes; and therefore the nature of the bottom at each of these soundings should be noted, and every means exerted that ingenuity can devise of discovering at what depth the coral formation begins, and of what materials the substratum on which it rests is composed. The shape, slope, and elevation of the coral knolls in the lagoon would also help the investigation; and no circumstances should be neglected which can render an account of the general structure clear and perspicuous.

A set of observations connected with the theory of the tides might likewise be carried on with peculiar propriety in one of these coral basins, provided the openings should be sufficiently wide and deep to admit the flux and reflux without material impediment. The island selected for such a purpose should be nearly midway in the ocean, and not very far from the equator.

There the tidal wave, uninfluenced by the interrupting barrier of one continent, and equally far from the reaction of the other, might be measured with very beneficial results. Delicate tide-gauges should be prepared beforehand, and immediately fixed in some snug nook, where the undulation of the sea could not reach. The rise and fall of the tide should be registered every hour, during the stay of the *Beagle*, as well as the moments (stated whether in apparent or mean time) of high and low water, as nearly as they can be obtained; and the periods at which the sea and land breezes spring up and fail should likewise be noted, with their effects on the tide, if they can be detected. A boat should be detached, on each tide, to some distance from the island, in order to ascertain the strength and direction of the stream; and all these operations should be continued, if possible, through a whole lunation.

Compiling general and particular instructions, for the navigation of all the places which he may visit, will of course be an essential part of the Commander's duty; but he will also have innumerable opportunities of collecting a variety of auxiliary information, which, when judiciously combined with the above instructions, of a purely nautical character, will much enhance their utility to all classes of vessels. Such as the general resources on which ships may depend in different places: the chief productions that can be obtained, and the objects most anxiously desired in return: the effect of seasons, of climate, and of peculiar articles of food on the health of the crew, and many others which will readily occur to his mind, and which become of great value to a stranger.

On all the subjects touched on in these memoranda, Commander FitzRoy should be directed to draw up specific reports, and to transmit them from time to time, through their Lordship's Secretary, to the Hydrographic Office, so that if any disaster should happen to the *Beagle*, the fruits of the expedition may not be altogether lost. Besides such reports, and with the same object in view, he should keep up a detailed correspondence by every opportunity with the Hydrographer.

The narrative of every voyage in the Pacific Ocean abounds with proofs of the necessity of being unremittingly on guard against the petty treacheries or more daring attacks of the natives. It should be recollected that they are no longer the timid and unarmed creatures of former times, but that many of them now possess fire-arms and ammunition, and are skilful in the use of them. Temper and vigilance will be the best preservatives against trivial offences and misunderstandings, which too often end in fatal quarrels; and true firmness will abandon objects of small importance, where perseverance must entail the necessity of violence; for it would be a subject of deep regret that an expedition devoted to the noblest purpose, the acquisition of knowledge, should be stained by a single act of hostility.

APPENDIX TWO

ROBERT FITZROY'S 'REMARKS WITH REFERENCE TO THE DELUGE'

At the end of his own account of the voyage of the *Beagle*, Robert FitzRoy included a chapter giving his personal opinion of early nineteenth-century geology. Since FitzRoy's narrative of the *Beagle*'s surveying voyage is largely unread today (most people know only of Darwin's book on the same subject) and his scientific views are generally unknown or misrepresented, it was felt to be useful to reprint here the final version of his geological chapter, entitled, 'A very few remarks with reference to the Deluge' (*Narrative of the surveying voyages of HMS* Adventure *and* Beagle, vol. 2, chap. 28). It was perfectly normal procedure at that time for the captain of an exploratory voyage to describe the scientific results of the expedition in his official account: had FitzRoy not encouraged Darwin to write a 'naturalist's journal', he would probably have written much more on these topics than he actually did. FitzRoy's interests ranged far beyond the practical mathematical and meteorological skills necessary for his surveying project: his *Narrative* displays great interest in physical anthropology in particular, and the history, geography and languages of the various tribes and races that were encountered. He believed that all human beings were originally of only one race; they had subsequently been affected by climate, habit and food and had changed in response to their environment. Such modern-sounding views were, however, rooted in a strong belief in the literal truth of the Bible. Other well-known Christian ethnologists such as James Cowles Prichard also found that the Bible told, for them, a true story about the origins and geographic divergence of the human species, and it was on this basis that missionary societies and concerned individuals like FitzRoy endeavoured to Christianize the aboriginal populations that they met in order to bring civilization – as they saw it – to their lost brothers under the skin.

As was commonplace in early nineteenth-century Britain, FitzRoy took his literal belief in the Bible beyond ethnology into the study of geology: the origin of the earth. He frankly admitted that his time on the *Beagle* had caused him to doubt the biblical account of the formation of the earth. No doubt a great deal of his uncertainty was generated by prolonged discussion with Darwin about the meaning of the various geological formations and structures that they were seeing and the startling explanations put forward by Charles Lyell in his *Principles of Geology*. But his doubts served to convince him of the truth of the 'Mosaic account'. His essay, given here in full, shows how he attempted to accommodate natural phenomena with Genesis and indirectly provide an antidote to Darwin's descriptions, which he perceived as irreligious.

––––––––––––––

To account for offering a few remarks on a subject so important and difficult as that of the Deluge, I beg to say that reflections, arising out of facts witnessed during the *Beagle*'s voyage, have occasioned them; and, as results of that expedition, it has appeared to me that they are neither irrelevant to the narrative, nor likely to be altogether uninteresting to young men in the navy.

I suffered much anxiety in former years from a disposition to doubt, if not disbelieve, the inspired History written by Moses. I knew so little of that record, or of the intimate manner in which the Old Testament is connected with the New, that I fancied some events there related might be mythological or fabulous, while I sincerely believed the truth of others; a wavering between opinions, which could only be productive of an unsettled, and therefore unhappy, state of mind. Some young men, I am well aware, are in a similar condition, while many others are content to set aside all reflection, and do as the world does; or rather, as those do among whom they generally live. Natural affection and respect for good parents, relations, and elders, never can lead a young man astray; but there is, perhaps, no guide more fallible or dangerous than the common custom of those inexperienced persons who associate together, chiefly for lack of fixed occupation; and whose principal object is to drive away self-examination, or prolonged thought, by a continual succession of idle amusement, or vivid excitement.

Wholesome and necessary as amusement and recreation are, both for mind and body, every one knows how insipid, even painful their excess becomes; and external evidence shows but too plainly where the happiness, the blessings, and the comfort men might enjoy, have by themselves been slighted, or destroyed, from forgetting the line between using, and abusing; and by turning a deaf ear to the reflection that they are but 'tenants at will'.

Much of my own uneasiness was caused by reading works written by men of Voltaire's school; and by those of geologists who contradict, by implication, if not in plain terms, the authenticity of the Scriptures; before I had any acquaintance with the volume which they so incautiously impugn. For geology, as a useful branch of science,* I have as high a respect as for any other young branch of the tree of knowledge, which has yet to undergo the trial of experience; and no doubt exists in my own breast that every such additional branch, if proved by time to be sound and healthy, will contribute its share of nourishment and vigour to the tree which sprung from an immortal root. For men who, like myself formerly, are willingly ignorant of the Bible, and doubt its divine inspiration, I can only have one feeling – sincere sorrow.

Few have time, as well as inclination, to go far into both sides of any question; but truth can hardly be drawn out of the well unless some exertion be made, in examining each argument, or in selecting a well-tried and experienced guide. It is idle to say, as I have heard asserted, that such works as those above-mentioned do little harm; experience proves the contrary; of which I am made painfully aware, not only by my own conscience, but by conversation with friends.

While led away by sceptical ideas, and knowing extremely little of the Bible, one of my remarks to a friend, on crossing vast plains composed of rolled stones bedded in diluvial detritus some hundred feet in depth, was 'this could never have been effected by

* By which word I mean 'Knowledge', in its most comprehensive signification.

a forty days' flood,' – an expression plainly indicative of the turn of mind, and ignorance of Scripture. I was quite willing to disbelieve what I thought to be the Mosaic account, upon the evidence of a hasty glance, though knowing next to nothing of the record I doubted: and I mention this particularly, because I have conversed with persons fond of geology, yet knowing no more of the Bible than I knew at that time. Thus much I feel it necessary to say, in accounting for my own approach to a subject in which all men feel deeply interested; and which has therefore been so well treated of, that these remarks would be useless, were it not that they may reach the eyes of young sailors, who have not always access to works of authority.

The Mosaic account of the Creation is so intimately connected with that of the Deluge that I must ask my young reader (whom alone I presume to address on this subject) to turn to the first chapter of Genesis, and refer to a few verses with me. We soon find a remarkable fact, which shows to my mind that the knowledge of Moses was super-human: his declaration in an early age that light was created before the sun and moon, which must till then have appeared to be the sources of light. In the fourth verse it is stated that 'God divided the light from the darkness.' This may have been effected by a rotation of the earth on its axis, turning each side in succession to the light; otherwise, had the earth remained stationary, light must have been destroyed to admit darkness, and there must have been repeated creations of light. The light was called day – 'and the evening and the morning were the first day'. Of course there could have been no morning previous to the creation of light; and the first portion of time, consonant to our present expressions, would have been that which elapsed between light and darkness, or evening. The length of a day being determined by the rotation of the earth on its axis; turning round once, so as to make an evening and a morning to each spot on the globe; the time occupied by that rotation is a natural object of interest. In the 12th verse it is said that grass, herbs, and trees, were brought forth; in the 14th and 16th, that lights were made to divide the day from the night; and that the greater light was to rule the day. It is known that neither trees, herbs, nor grass can exist long without the light and heat of the sun, therefore the rotation of the earth between the third day, when vegetation was produced, and the fourth, could not have been very different, in velocity, from its present rotation. Some men, of rare abilities, have thought that the 'days' of creation were indefinite periods, notwithstanding the statement in verse 14, which affirms that the lights in the firmament of heaven were to divide the day from the night; and to 'be for signs, and for seasons, and for days, and years'. In this one verse do we not see that the day was less than a year (signs and seasons, days and years); for had the day there meant been more than a year would not the words have been differently placed, namely – signs and seasons, years and days? Can we think that day means one space of time in the former part, and another space of time in the latter part of that one verse? Another indication that the word day, used in the first chapter of Genesis, does not mean a period much, if at all, longer than our present day, is – that it is spoken of as alternating with night. Although the word day is used in other chapters of the Bible, even so soon as the 4th verse of the 2nd chapter of Genesis, to express a period, or space of time longer than our present day, the word night is never so applied; hence, as the earth turns uniformly on its axis, and, so far as we can reason from analogy, must have turned uniformly, while turning at all, the word night in the 5th verse interprets the length of a day.

Some have laid stress upon the declaration that a thousand years are with the Lord as one day: but what is the context?* To lengthen the day to a thousand years, on account of this and a similar expression, is not more reasonable than it would be to reduce it to a night-watch. What is a watch in the night when passed? – next to nothing: so are a thousand years with the Almighty. These considerations tend to show how, without Chaldee or Hebrew learning, a man, with a common English education, may convince himself of a fact which has lately been so much controverted.†

* 'A thousand years in thy sight are but as yesterday when it is past, and as a watch in the night', Ps. xc. ver. 4.

† I may, however, here remark to my young sailor friends, that the Jews, who perform their worship in Hebrew, and are naturally at least as much interested in the Old Testament as any people, use and prefer our authorized English translation.

Partly referring to such indefinite periods as we have been discussing – and partly to reasoning unaided by revelation – some geologists have said that there were successive creations, at intervals of vast duration. They have imagined an age in which only the 'so-called' lowest orders of animated creatures existed,* an age of fishes, an age of reptiles, an age of mammalia, and an age in which man appeared: statements which have obtained much attention. Fossil fishes and shell mollusca† have been found in coal measures, and in subjacent formations: how could this have happened if vegetables had been produced first; then swept away and converted into coal, and that afterwards the lower orders of animals had appeared? We know that the fossil plants of the coal formations are similar in structure to vegetables now growing on the earth, which cannot flourish without warmth, and the light of the sun. Vegetation was produced on the third day, the sun on the fourth. If the third day was an age, how was the vegetable world nourished? But anomalies such as these appear to be endless in most geological theories: I will leave them for the present and continue my course.

In the 16th verse it is said that 'God made two great lights; the greater light to rule the day, the lesser light to rule the night: the stars also;' that is, he made the stars also.

It is not stated here that the Almighty made all the stars at that time; nor can I, after consulting very able men, find any passage of such an import. That all the stars dependant upon, or connected with, our solar system, namely, the planets and their satellites,

---

* In classing one order of creatures above or below others, we may perhaps consider them as they appear to our apprehension, in comparison with others, but we must beware of thinking them more or less imperfect. Every creature is perfectly adapted to the condition and locality for which it is designed, and absolutely perfect (speaking generally). Some that are intended to live in the dark, or some that are to exist under pressure, may at first appear to us imperfect, perhaps shapeless, unsightly objects: but, after examination into their natural history, our hasty remark is succeeded by expressions of astonishment at such wonderful arrangements of Providence as are shown – even in the most shapeless sea slug.

Multitudes of creatures exist now, especially in the sea, quite as apparently imperfect as those of the so-called lowest order of animated creation, whose impressions are found in solid rocks. There may also be animals in deep waters that could not exist except under pressure.

† Rhind, Keith, Lyell, &c.

were then created, seems to be evident from the fact of their revolving round our sun; but farther than this, it is not thought necessary (may we not presume) for man to know; therefore it is not revealed to us. In the ancient book of Job, the creation of the world is thus alluded to. 'Who laid the cornerstone thereof, when the morning stars sang together, and all the sons of God shouted for joy.'* But the earth was finished, and vegetation produced, before the creation of stars mentioned in Gen. i. 16; therefore, unless the 'cornerstone' alludes to man, it may be inferred that there were stars in existence besides those made on the fourth day. Of course, the 'singing of the stars' is a figurative expression; but as we do not meet with any similar metaphor in the Bible, unconnected with some object that we know exists; we may infer that stars existed when the allegorical, or mystical cornerstone was thus laid.†

In verses 29 and 30, the food for man and beast is mentioned, and with reference to the Deluge this should be borne in mind. It may be said that the teeth of some animals are so formed as to be fit only for grazing, or browzing; that beasts of prey have teeth adapted to tearing and gnawing and that man requires meat; but we must remember that dogs and wild beasts thrive upon a vegetable diet, and that some men never touch meat, even in the present state of the world: very different probably from its condition before the flood, as may be concluded from the inferior duration of human life.

The second and third verses of chap. ii. recall to mind the wonderful fact that the seventh day has been a marked division of time from the earliest period of historical record.‡

It is now well known that all nations, and almost all tribes of the human race, preserve traditions of a great flood in which nearly all

---

* Job. xxxviii: 6, 7.
† Much must depend upon the limit attached to the meaning of the word Heaven in the 1st Chapter of Genesis, and Heavens in the 2nd; viewed in connection with verses 1, 6, 7, 8, 9, 14, 15, 17, and 20 of Chap. i; 1, 4, of Chap. ii; verses 5, 7, 10, 13, of 2 Peter iii; and very many other passages; not omitting the Lord's prayer.
‡ We find it ordained in Gen. ii: 3; alluded to by Noah, chap. viii: ver. 10, 12; and afterwards observed regularly, down to the present time

men were destroyed:* and it is also established as a fact, that nearly all parts of the earth, hitherto examined, bear witness to their having been at some time covered by the ocean. Instead of ascribing these effects to the universal deluge, many geologists say that the earth is in a continual, though gradual state of change; that in consequence of this general mobility, places now far above the sea were once beneath it; that districts, or countries, may have been inundated in one quarter, and other regions elsewhere, but that an universal deluge never could have happened. This is implied plainly enough, if not asserted, in several geological works.

In the *Beagle*'s examination of the southern parts of South America, I had opportunities of observing immense tracts of land composed, solely, of fossil shells, bones, and an earth which looked like dried sandy mud: extensive ranges of country where no solid rock could be found, only rolled or shingle stones, embedded to a great depth in earth, exactly like that described above; and a wide district, at least 50 miles across, covered with lava of which the surface was nearly horizontal. (San José, San Julian, Santa Cruz.)

I brought to England many specimens of these shells, which, although taken from within a few feet of the surface of the land, were found to have been pressed together, crushed, and penetrated by mud, in a manner that never could have been caused by the weight of earth then lying above them, because, though solid, it could neither have mashed the shells, nor worked into their inmost recesses. It seems evident to me that those shells have undergone enormous pressure beneath an ocean, when they were surrounded with mud.† But previous to such pressure, the shells

* Sharon Turner, Harcourt, &c.
† On this subject, the pressure of an ocean, Mr Lyell remarks, (*Elements of Geology*, 1838, pp. 7, 8, 9). 'When sand and mud sink to the bottom of a deep sea the particles are not pressed down by the enormous weight of the incumbent ocean; for the water, which becomes mingled with the sand and mud, resists pressure with a force equal to that of the column of fluid above . . . Nevertheless if the materials of a stratum remain in a yielding state, and do not set or solidify, they will be gradually squeezed down by the weight of other materials successively heaped upon them, just as soft clay or loose sand on which a house is built may give way. By such downward pressure particles of clay, sand, and marl may become packed into a smaller space, and be made to cohere together

footnote continued overleaf

must have grown naturally somewhere: certainly not at the bottom of an ocean; because they are shells of a comparatively delicate structure which are usually found within a few feet of low water; some at least of the number being identical with living species.

If the square miles of solid land in which those myriads of shells are now embedded, had been upheaved (as geologists say), either

footnote continued

permanently . . . But the action of heat at various depths is probably the most powerful of all causes in hardening sedimentary strata.'

In reflecting upon these passages it appears to me that Mr Lyell has supposed what may not always take place in a deep sea, namely – that sand and mud sink to the bottom.

Whenever particles of sand and mud are at the bottom, they must be lower than contiguous particles of water, or they could not be at the bottom; therefore those particles of sand and mud have water above, while resting upon some other substance below. Pressure there can be none, excepting of some earthy particles upon others, while the specific gravity of the sand and mud exceeds that of the displaced fluid. But, if the depth of water be increased, and its specific gravity at the bottom augmented, the sand and mud at the bottom must rise, if they do not cohere together, and to the surface on which they lie; in which case the increasing weight and density of water would tend to compress and make them cohere still more.

The smaller kinds of sea-shells are very little heavier than sea water. This would prevent their being carried by the action of the sea to great depths, even if it were possible for them to be so rolled over rocks, sand, or mud, in which they would stick, or be buried, before they had been moved many miles from the place where they grew. These two considerations may help to account for the fact that seamen do not find impressions of shells, on the 'arming' of the lead, when sounding in very deep water, at a considerable distance from any shore where they grow. Sea-shells, I need hardly remark, grow only in comparatively shallow water. The specific gravity of oyster shell, when dead, is about twice that of sea-water (2092, 1028). Most other shells are much lighter, and but few at all heavier than the oyster.

Before ending this note I must remark that the horiziontal movement of water near the bottom, though gentle, may tend to press together and smooth any loose sand, mud that sinks, oazy clay, or fragments of shells, before many of their particles travel far. Water in rapid motion is known to hold sand as well as mud in suspension, but not shells, unless the current is very strong. To such a constant agitation of the sea, oscillating gently with each tide, we may perhaps ascribe the comparatively level and smoothened state of the bed of the ocean, where it has hitherto been sounded. Excepting near irregular, rocky land, one finds, generally speaking, no ravines, no vallies, no abrupt transitions in the bottom of the sea. For miles together there is an almost equal or gradually altering depth of water: and little similarity can be traced between the contour of the bed of a sea and the neighbouring dry land, until you are near the shore, where the sea acts differently, and irregular bottom is as frequent as it is usually dangerous for shipping.

gradually, or rapidly, shells could not be found there in their present confused and compressed state. Had the land sunk down many thousand feet with shells upon it, they might have been covered with mud, and on being afterwards upheaved again they would have appeared embedded regularly where they grew, in a matrix which, with the pressure of a superincumbent ocean, might have flattened and penetrated them: but they would not have been torn away from their roots, rolled, broken, mashed, and mixed in endless confusion, similarly to those now in my possession.

There is also another consideration: geologists who contend for the central heat of the earth assert that substances subjected to great pressure under the sea become altered: hence, in conformity with their theory, these shells could not have been long buried under a deep ocean, and afterwards raised in their pristine state. So little changed are these shells, except in form, that they appear as if they had been heaped together and squeezed in mud within a few years from the present time.

One remarkable place, easy of access, where any person can inspect these shelly remains, is Port San Julian. There, cliffs, from 10 to 100 feet high, are composed of nothing but such earth and fossils; and as those dug from the very tops of the cliffs are just as much compressed as those at any other part, it follows that they were acted upon by an immense weight not now existing. From this one simple fact may be deduced the conclusions – that Patagonia was once under the sea; that the sea grew deeper over the land in a tumultuous manner, rushing to and fro, tearing up and heaping together shells which once grew regularly or in beds: that the depth of water afterwards became so great as to squeeze or mass the earth and shells together by its enormous pressure; and that after being so forced down, the cohesion of the mass became sufficient to resist the separating power of other waves, during the subsidence of that ocean which had overwhelmed the land. If it be shown that Patagonia was under a deep sea, not in consequence of the land having sunk, but because of the water having risen, it will follow as a necessary consequence that every other portion of the globe must have been flooded to a nearly equal height, at the same time; since the tendency to equilibrium in fluids would prevent any one part of an ocean from rising much above any other part, unless sustained at a greater elevation by external force; such as the attraction of the moon, or sun; or a strong wind; or momentum derived from their agency. Hence therefore, if Patagonia was covered to a great depth, all the world was covered to a great depth; and from those shells alone my own mind is convinced, (independent of the Scripture) that this earth has undergone an universal deluge.

The immense fields of lava, spoken of in a preceding page (407), and which to an ordinary observer appear to be horizontal, are spread almost evenly over such an extent of country, that the only probable conclusion seems to be, that the lava was ejected while a deep sea covered the earth, and that tidal oscillations,* combined with immense pressure, spread and smoothed it, while in a rapidly cooling though viscous state, over the surface of the land.

The vast quantity of shingle, or rounded stones of all sizes, may be accounted for in a manner unconnected with that of water acting upon a shore; though doubtless a great proportion of the shingle we see has been rounded in that manner. Melted stone, thrown out of a volcano, and propelled through water with great velocity, might be rounded and cooled as shot are when dropped into water from a tower. In modern volcanoes we observe that some matter is thrown into the air, while other, and the greater quantity, runs over the edges of a crater, overflowing the adjacent tracts of land.

Proceeding to the west coast of South America, we find that near Concepción there are beds of marine shells at a great height above the level of the sea. These, say geologists, were once under the ocean, but, in consequence of the gradual upheaval of the land, are now far above it. They are closely compressed together, and some are broken, though of a very solid and durable nature; and being near the surface of the land are covered with only a thin stratum of earth. They are massed together in a manner totally different from any in which they could have grown, therefore the argument used in Patagonia is again applicable here. But in addition to this, there is another fact deserving attention: namely,

* *Narrative*, vol. 2: see remarks on tides in the Appendix.

that there are similar beds of similar shells, (identical with living species) about, or rather below the level of the present ocean, and at some distance from it.*

In crossing the Cordillera of the Andes Mr Darwin found petrified trees, embedded in sandstone, 6,000 or 7,000 feet above the level of the sea: and at 12,000 or 13,000 feet above the sea-level he found fossil sea-shells, limestone, sandstone, and conglomerate in which were pebbles of the 'rock with shells'. Above the sandstone in which the petrified trees were found, is 'a great bed, apparently about one thousand feet thick, of black augitic lava; and over this there are at least five grand alternations of such rocks, and aqueous sedimentary deposits, amounting in thickness to several thousand feet'.† These wonderful alternations of the consequences of fire and flood, are, to me, indubitable proofs of that tremendous catastrophe which alone could have caused them; of that awful combination of water and volcanic agency which is shadowed forth to our minds by the expression 'the fountains of the great deep were broken up, and the windows of heaven were opened.'

The upheaval of the island of Santa Maria has been quoted by geologists, from my statement; and it will be interesting to learn whether that island has remained at its new elevation, or whether, like the shore at Talcahuano,‡ it has sunk down again. If the coast in that neighbourhood has been gradually rising, it is strange that old Penco Castle should still stand so low.

In Mr Lyell's *Elements of Geology*,§ he mentions Mr Darwin having found, near Callao, 'at the altitude of eighty-five feet above the sea, pieces of cotton thread, plaited rush, and the head of a stalk of Indian corn, all of which had evidently been imbedded with the shells' (marine). 'At the same height on the neighbouring mainland, he found other signs corroborating the opinion that the ancient bed of the sea had there also been uplifted eighty-five feet, since the region was first peopled by the Peruvian race.' The

* *Narrative*, vol. 2, pp. 421, 2, 3.
† Mr Darwin's letters to Professor Henslow: printed for the Cambridge Philosophical Society – 1835.
‡ *Narrative*, vol. 2, pp. 420–1.
§ 1838, pp. 295–6.

neighbourhood of Lima has suffered from immense waves caused by earthquakes, and the relics found among the shells may have been scattered by one of those waves. The bed of shells may have been disturbed by the earthquake and its consequences, the ground may have been rent, and afterwards closed again, or the opening may have been filled up by loose earth and anything lying on it, as has taken place at Concepción. That the country near Callao, or Lima, has not been upheaved, to any sensible amount, since the last great earthquake, which was accompanied by a wave that swept over and destroyed Callao, is evident from the present position of a pillar erected soon after that event to mark the place to which the waves advanced inland.* This pillar now stands so low, that waves, such as those which ruined Talcahuano, would inevitably reach its base; again destroying the whole of Callao, still situated on a flat, very few feet above the sea, near where old Callao stood.

I have now mentioned the principal facts connected with the *Beagle*'s voyage, which I am desirous of noticing with reference to the Deluge. Want of space prevents my adding others: I have hardly room left to lay before my young readers some general considerations, arising partly out of these facts, which I hope may interest and perhaps be useful to them.

When one thinks of the Deluge, questions arise, such as 'Where did the water come from to make the flood; and where did it go to after the many months it is said to have covered the earth?' To the first the simplest answer is 'From the place whence the earth and its oceans came:' – the whole being greater than its part, it may be inferred that the source which supplied the whole could easily supply an inferior part: and, to the second question, 'part turned into earth, by combination with metallic bases; part absorbed by, and now held in the earth; and part evaporated'.† We know nothing of the state of the earth, or atmosphere surrounding it, before the Flood; therefore it is idle and unphilosophical to reason on it, without a fact to rely on. We do not know whether it moved in the same orbit; or turned on its axis in a precisely similar

* In 1746.
† Electricity may have acted a prominent part in these changes.

manner; whether it had then huge masses of ice near the poles; or whether the moon was nearer to it, or farther off. Believers in the Bible know, however, that the life of man was very much longer than it now is, a singular fact, which seems to indicate some difference in atmosphere, or food, or in some other physical influence. It is not so probable that the constitution of man was very different (because we see that human peculiarities are transmitted from father to son), as it is to suppose that there was a difference in the region where he existed. It is easy to settle such speculations by the reflection – 'It was the will of Him who is Almighty;' but as in most cases we see that secondary causes are employed to work out His will, we may imagine that the extraordinary prolongation of man's existence was effected by such means.

Connected with these questions respecting the additional quantity of water is the reflection that the amount must have been very great. This may be placed in another light. Sir John Herschel says,* 'On a globe of sixteen inches in diameter such a mountain (five miles high) would be represented by a protuberance of no more than one hundredth part of an inch, which is about the thickness of ordinary drawing paper. Now as there is no entire continent, or even any very extensive tract of land, known, whose general elevation above the sea is any thing like half this quantity, it follows, that if we would construct a correct model of our earth, with its seas, continents, and mountains, on a globe sixteen inches in diameter, the whole of the land, with the exception of a few prominent points and ridges, must be comprised on it within the thickness of thin writing paper; and the highest hills would be represented by the smallest visible grains of sand.' Such being the case, a coat of varnish would represent the diluvial addition of water; and how small an addition to the mass does it appear!

Let us now refer briefly to the recorded account of the Flood. Without recapitulating dates and events, I will at once advert to the ark: an immense vessel,† constructed of very durable wood,‡

* Treatise on Astronomy, Cabinet Cyclopaedia, page 22.
† Sharon Turner, Harcourt, Burnett, &c.
‡ Some of our English ships have lasted more than a century.

and well stored with vegetable provision for all that it contained. Some cavillers have objected to the heterogeneous mixture of animals embarked; on the ground that they could not have been assembled; and would have destroyed one another. We may reply: He who made, could surely manage. But, without direct miraculous interposition (though we should never forget that man is a miracle, that this world is a miracle, that the universe is a miracle), imagine the effect that would be produced on the animal creation by the approach of such a war of elements. Do we not now find animals terrified by an earthquake – birds shunning the scene of violence – dogs running out of a town,* and rats forsaking a sinking ship? What overcoming terror would possess the animated beings on an island, if it were found to be rapidly sinking while worse than tropical torrents, aggregated water-spouts, thunder and lightning, earthquakes and volcanic erup-tions united to dismay, if not to paralyse, the stoutest human heart: yet such probably would be but a faint similitude of the real deluge. Those who have themselves witnessed the war of ele-ments, in some regions of our globe, are perhaps more able to conceive an idea, however inadequate, of such a time, than persons who have scarcely travelled beyond Europe, or made more than ordinary sea voyages. Happily for man, hurricanes or typhoons occur but rarely: earthquakes, on a great scale; their overwhelming waves; and devastating eruptions of volcanoes, still less often. That the approach of a general calamity would have affected animals, what we now see leads us to infer, and that many would have fled to the ark, is only in accordance with the wonderful instinct they are gifted with for self-preservation. Proud man would, in all probability, have despised the huge construction of Noah, and laughed to scorn the idea that the mountains could be covered, even when he saw the waters rising. Thither, in his moral blindness, he would have fled, with num-bers of animals that were excluded from the ark, or did not go to it; for we do not see all animals, even of one kind, equally instinctive. As the creatures approached the ark, might it not have been easy to admit some, perhaps the young and the small, while

---

* Concepcion and Talcahuano. *Narrative*, vol. 2, pp. 403, 5.

the old and the large were excluded?* As we do not know what was the connection or partition of land, before the deluge; how the creatures were distributed; or, what was the difference of climate between one region and another; we cannot say that any particular kind could not have been near the ark because of crossing the sea, or having far to travel.

There is abundant proof that animals have changed their habits, shape, coat, colour, or size, in consequence of migration, or transportation to different climates; therefore we cannot tell, from what is now seen, what alterations have taken place since their second dispersion.

Many able men† have pointed out how water penetrating to metallic bases, may cause volcanic eruptions; how matter thrown up, and materials torn or washed off the earth may have combined, mechanically as well as chemically; how gases may have assisted the transformations: how creatures may have been instantaneously overwhelmed, or gradually entombed; how lime may have been one among many powerful agents; how seeds, and spawn, and the germs of insects may have been preserved; and why, among such multitudes of fossil remains as we now find,

---

* The small number of enormous animals that have existed since the Deluge, may be a consequence of this shutting out of all but a very few. We are not told how many creatures died in the ark; some of those least useful to man may have gone; but, even if none died, the few that quitted the ark could hardly have long withstood the rapid increase of enemies, unless their increase had been proportionally quick. Whether Job had himself seen, or only heard of, the leviathan and the behemoth, does not appear; but that those monsters were the megalosaurus and the iguanadon there seems to be little doubt (Burnett, p. 67.) Excepting the serpent in Africa, which opposed the passage of Regulus and a Roman army, I am not aware whether profane history mentions any well-authenticated instance of such enormous reptiles; but I cannot look at our representations of dragons, wyverns, griffins, &c. without thinking that, at least, tradition must have handed down the memory of some such monsters; even if a stray one here and there did not actually live in the earlier historical ages: pterodactyles, plesiosauri, ichthyosauri, &c. are too like them, in general figure, to admit of this idea being treated as altogether chimerical. Tradition, no doubt exaggerated by imagination, may have handed down the fact of such creatures having once existed: indeed the casual finding of a skeleton might confirm reports, if not originate them.

† Davy, Sharon Turner, Fairholme, Burnett, Granville Penn, Sumner, Young, Rhind, Lyell, Cockburn, &c.

only in a few places are there remains of man incontrovertibly fossil.*

Still there are some points but lightly touched, or unnoticed, by any person whose works bearing particularly on this subject I have yet seen. One is the rapidity with which certain substances combine under water, and form stone; such, for instance,† as those used in Roman cement; another is the possibility of fragile substances, such as shells, small creatures, leaves, corallines, branches, &c., being enveloped in a muddy matrix, while floating at various depths, according to their specific gravities; and the precipitation (chemically speaking) or consolidation, or simple deposition of such cohering masses.‡

The similarity of coal to asphalte inclines one to suspect an identity of origin; and that coal, in a fluid state, enveloped quantities of vegetable matter – was for some time agitated by the continual tides and tidal currents of the diluvial ocean, and afterwards hardened by cooling, by pressure, or by chemical change; if not by all three. We find the impressions of leaves, stems, and branches – and even large woody trunks embedded in coal: but that the matrix, in which the leaves were enveloped and subjected to pressure, was not triturated vegetable matter is probable, because the casts of delicate vegetable substances found in it show few, if any, signs of friction or maceration. The impressions are as beautifully perfect as those of shells in fossils where the shell itself has disappeared. Might we not as well say that limestone was formed out of decomposed or pulverized shells, as assert that coal was formed out of the luxuriant herbage, the ferns and the palms, of a former state of the world?

Asphalte is at first buoyant; that trees and other vegetable productions are so I need not remark; but coal sinks in water, and asphalte may be altered chemically so as to sink like coal. Experiments on the asphalte of the famous lake at Trinidad have proved

---

* These fossil remains of man are not only mixed with those of animals, or fish; but in some cases they are buried at a distance beneath fossil bones of animals. See Fairholme on the Mosaic Deluge, pp. 41–52; Miers's Chile, vol. i p. 455, &c.

† Lyell, Elements of Geology, 1838, p. 75–6.

‡ The simplest experiments with pulverized, or numerous minute substances in water, show that they attract one another mutually, and then cohere.

that there is so very close an analogy between that substance and coal, that a gas, exactly resembling coal gas, and burning equally well; a bituminous oil; a substance like coal-tar; and a residuum, similar to coke; result from its distillation.

Electricity may have been a powerful agent in crystallization; in the rapid deposition of strata;* in the formation of mineral veins;† in earthquakes and volcanoes; in the formation or decomposition of water; and in other ways of which we are yet, and perhaps ever shall be, totally ignorant.

Successive strata may have been rapidly deposited by tidal oscillations and currents, aiding chemical or mechanical combinations.

The depth to which bodies would sink in an ocean several miles deep has not been proved, and there is reason to think that it is much less than people generally imagine. An eminent man has said that a knowledge of 'the depression of the bed of the ocean below the surface, over all its extent, is attainable (with whatever difficulty and however slowly) by direct soundings;'‡ and, in consequence of a conversation on this subject with him in 1836, he wrote to me, suggesting a mode which might be tried. I consulted with a friend as to the possibility of success, and his letter,§ taken

---

* Crosse.
† Fox.
‡ Treatise on Astronomy, by Sir John Herschel, Cabinet Cyclopedia, p. 154.
§ 'I return Sir John Herschel's letter on deep-sea sounding. Anything from him is sure to be interesting and instructive; but there is a circumstance unnoticed in his communication which might obstruct the descent of a sounding apparatus to very great depth.

'Mr Perkins found that at a depth of only 3,000 feet, sea-water was compressed 1/27th of its bulk at the surface (Lib^y of Useful Knowledge, vol. 1. Art. Hydrostatics). Hydrostatic pressure has usually been estimated from depth alone, assuming that the density of the fluid was uniform; such, however, cannot be the case in an elastic fluid like water, for at great depths, being in a compressed state, it is more dense than at the surface.

'In estimating the amount of hydrostatic pressure at great depths, we should know the vertical height of the column and mean density of the fluid; and since density increases with depth, by reason of superincumbent pressure, the water at great depths must be enormously compressed, and, consequently, in a very dense state. Let us now inquire how increasing density (from compression alone) might affect an apparatus sent down by a weight, in order to reach the bottom, presuming that the solids composing the float and sinker were incompressible, and retained their form and magnitude during the operation.

footnote continued overleaf

in connection with the facts related by Scoresby;* with what has been found by those who have sounded to great depths; and with my own practical experience in sounding – has induced me to think that man never will reach the lowest depths of the deepest oceans by any method his ingenuity may contrive; because the water increases in density with the depth, in a ratio perhaps more than arithmetical. Every seaman knows that in sounding at great depths very heavy leads must be used with ordinary lines, or very thin lines with ordinary leads; the object being the same – that of overcoming the augmenting buoyancy of the line by a weight unusually heavy. But line, such as is used for sounding, is not buoyant at the surface of the sea; a coil of it thrown overboard sinks directly. Then what is it that causes any weight attached to a sounding-line to sink slower and more slowly, after the first few hundred fathoms, the deeper it penetrates; if not the increased resistance to sinking, found by the weight and line? 'Friction, caused by passing through the water,' I may be told. Can that friction be compared with the augmented tendency to sink that would be given by the continually increasing weight of line, if the water did not increase in density?

The pressure of the column of water over any weight, after it

* Scoresby's Arctic Regions.

---

footnote continued

'Let bees'-wax be a float, and cast-iron a sinker, and let each, for illustration, be one cubic foot in dimensions. Let it be possible that at some depth water may be compressed into one-fourth of its bulk at the surface, and still retain the properties of a fluid; let it also be granted that a solid will swim if specifically lighter than the contiguous fluid, and sink if heavier than an equal volume of the fluid. The specific gravity of bees'-wax is stated to be 964; that of cast iron, 7248; and that of sea-water at the surface, 1028: hence the buoyancy of wax immersed in sea-water at the surface, may be called 64, and the tendency of cast-iron to sink, from the same surface, 6220. Deducting 64, we have 6156 as the whole tendency of the mass (wax attached to iron) to sink from the surface. Let us now suppose that the machine has attained a depth where the water is compressed into a four-fold density, represented by 4112 for a cubic foot; and we have 3148 for the tendency of the wax to float, but only 3136 for the tendency of the iron to sink: and the inclination to ascend rather than descend, might be represented by 12. Thus we see that an apparatus may not be certain of arriving at the bottom of an ocean: as in an opposite manner, a balloon may not reach the highest regions of the atmosphere. Either machine could only attain a position where there would be no tendency either to descend or ascend.

Plymouth, 24th Feb. 1837.   'WILLIAM WALKER.'

has been sunk some 100 fathoms, is shown by the time and exertion required to haul it up again. The operation of sounding in very deep water, with any considerable weight, occupies several hours, and a great number of men. That water is elastic has been proved by Canton's experiments as well as others: but there are familiar illustrations of this fact visible in ricochet shot, in 'ducks and drakes', in the splashing of water, and in the rebounding of rain-drops from water. Being elastic, and the lower strata being under enormous pressure, it follows that those strata of water must be more dense than the body above them. No one doubts that the lower regions of the atmosphere are denser than the higher; yet air is but a rarer and much more elastic fluid than water. That which takes place in air, to a great extent, may be expected to occur in a very diminished degree with water. If it were not so, why should stones be blown up, casks violently burst, or rocks suddenly torn asunder by the application of the principle usually described as the hydrostatic paradox? If the water were not highly compressed before the explosion takes place, would there not be a gradual yielding, a tearing asunder by degrees, instead of a sudden and violent bursting?

The object of this digression is to show that although bodies which are not buoyant may sink to a considerable depth, it does not follow that they must sink to the bottom. Each separate thing may sink a certain distance, in proportion to its specific gravity, and there remain. The greatest depths ever reached by heavy weights, attached to lines, do not exceed a mile and a half; a small distance, probably, compared with the depth of the diluvial flood.

Although metals, stone, rock, or coal may have sunk deeply in the waters, other substances, such as earth, mud, bones, animal and human remains, &c., may have been held at various depths until decomposed by water; or combined and consolidated by volcanic gases, or electric currents. In this manner the preservation of delicate corallines, shells, skeletons of animals, &c., may be accounted for. Suspended in water, surrounded by earth in a dissolved state, combined by chemical agency, deposited on land, and consolidated by pressure, by volcanic or by latent heat, they may have become fossils. Thick-skinned animals may have floated longest, because their hides would have buoyed them up for a greater length of time,* hence their remains should be found near, or upon the surface of the ground, in some cases water-worn, in others uninjured, according as they had been strewed among shingle, or deposited in a yielding mass. That bones were not rolled about much among the stones in which they are found,† is evident from the fact that bones, if so rolled among them, would soon be ground to powder. It is clear that, however much the bones may have been water-worn before deposition on land, both they, and the adjacent shingle, must have been deposited there nearly about the same time.

Tripoli stone, and other substances composed chiefly, if not entirely, of microscopic insects, may have been formed by the accumulation and cohesion of myriads of such minute creatures, swept together off the land, like swarms of locusts, aggregated by the rolling of the waves, agglutinated, deposited on the land, and afterwards heavily compressed. Or they may have been insects bred in water; such as those which Mr Darwin calculated to amount to 'one hundred thousand in a square inch of surface'; while the sea was streaked with them for a great distance.‡ Microscopic objects such as these may have been killed by some gas rising from a volcano beneath; then drawn together by mutual attraction, rolled over and over, and landed among other recent compositions. In what other way could such a mass of these animalcules be heaped together?

There are also effects of existing causes which authors have only mentioned by name, in reference to the Deluge, without explaining that the effects alluded to would have been enormously increased at that time: I mean the tides. In the Appendix to this volume is a short statement of the manner in which tides may act – upon the principles of the ocean oscillating in its bed; and of tides being caused, partly by the water being elevated by the moon and sun, partly by a westward momentum given to it by their attraction, and partly by the oscillation caused by the return of the fluid after being elevated. If this globe were covered with water to

---

* When 'blown' after putrefaction began.
† Those, for instance, of Blanco Bay, *Narrative*, vol. 2, p. 112.
‡ Darwin's Letters to Professor Henslow, printed for private distribution among the members of the Cambridge Philosophical Society, in 1835.

the height of a few miles above the present level of the ocean, three more particular effects would take place: an enormous pressure upon the previously existing ocean, and on all low land; a diminished gravity in the uppermost waters, resulting from their removal from the earth's centre; and immense tides, in consequence of the increased depth of the mass, the diminished weight of the upper fluid, and the augmentation of the moon's attraction. As the waters increased on the earth, the tides would also increase, and vast waves would rush against the sides of the mountains, stripping off all lighter covering, and blowing up,* or tearing down, enormous masses of rock. Similar effects would take place as the diluvial ocean decreased, until it became bounded by its proper limits. Such oscillations I conceive to be alluded to by the words 'going and returning',† and by the expression, 'they go up by the mountains; they go down by the valleys';‡ which exactly describes the rushing of enormous waves against high land. When a wave strikes against a rock, it dashes up every projection that opposes it; but – its impetus at an end – down the water runs again through cavities and hollows: such, on a grand scale, would be the effect of a diluvial wave urged against a mountain-side.

In such a war of waters, earth, and fire, a buoyant, closed-in vessel – without masts, rudder, or external 'hamper' to hold wind, or catch a sea – might have floated uninjured; and the fewer openings, of any description, in her cover, or sides, the better for her security. Seeing nothing of the conflict around might have diminished the excessive terror which must have been felt by those that were within, except the confiding Chief. We do not find that the largest or highest 'swell' injures a good 'sea-boat', when in deep water, and far from land: the foaming 'breakers' alone destroy. But, after all, such conjectures as these are vain, we cannot now know how far miraculous interposition extended – how far secondary causes were employed.

The landing of the ark on a mountain of middle height appears remarkable; because the climate of the highest, on which we might naturally suppose the ark rested, did we not know to the

* By the extraordinary power, or principle, called the hydrostatic paradox.
† Gen. viii: v. 5, marginal translation.
‡ Psalm civ: ver. 8.

contrary, might have been insupportable during the time that Noah waited for the recess of the waters.* Reasoning from existing circumstances, the temperature of the surface of the ocean would have been nearly that of the contiguous air: but after the waters had receded, high mountain tops would have gradually acquired their present frozen state.

Here the reflection arises – when did icebergs begin to appear? Was not the climate equable and temperate all over the world for some time after the Deluge, in consequence of the slow drying and warming of tropical regions, and gradual formation of ice near the poles? Such a condition of climate would have favoured the distribution of animals. Those who oppose the idea that animals migrated to various quarters of the globe, surely do not reflect that the swallow, the wild swan, the wild goose, the wild horse, the Norway rat, and numerous other creatures, now migrate periodically in search of food or a better climate. Similar instinct may have taught animals to wander then, till they reached the places suited to them;† and there the same instinct would retain them. Want of proper food, or climate, or the attacks of enemies,‡ may have destroyed stragglers who did not migrate; therefore, when we find no kangaroos in Europe, it is no proof that kangaroos did not once exist there. Elks are now found in North America – we know they were formerly in Europe – is that race here now? During the first few hundred years after the flood, extraordinary changes may have taken place in the geography of the world, in consequence of the drying and altering of various portions; also from the effects of volcanic eruptions and earthquakes, occasioned perhaps by electric action on newly-exposed land, as well as by other causes. Many places, now islands, may

---

* The Deluge began in the 600th year of Noah's life, in the second month, and seventeenth day of the month (Gen. vii: 11), and Noah quitted the ark in the 601st year, in the second month, and twenty-seventh day of the month (Gen: viii: 14), making a period of twelve months and ten days. Noah waited in the ark nearly five months after it grounded on Ararat.

† We see abundant evidence that either living creatures are adapted to particular climates and localities, or that climates and localities are adapted to particular creatures; which latter, it has been proved by many authors, are altered by any material change of the former.

‡ It should be remembered, that man was allowed to eat flesh after the flood. Gen. ix. ver. 3.

have been united to a continent for a considerable period after the deluge; much land may have sunk down, much may have risen up, in various parts of the world. Such changes are said to be going on even now, though on a small scale (Lyell, Darwin, &c.); what may they not have been during the first few centuries after the flood? Volcanic eruptions, such as those of the Galapagos, Andes, Etna, Auvergne, Indian Islands, &c., were then perhaps in such activity as they have never shown since.

What the division in the earth was, in the days of Peleg, does not distinctly appear: but it could only have been a separation from the true faith; a partition of territory among men; or some mighty convulsion, some rending or contraction, as it were, of the earth, which was so general as to have occasioned a marked and unqualified record, as of an event well known to all.

Many philosophers think that the world has a central region of surpassing heat, and that the greater part of the interior of the globe is in a state of incandescence, if not of fusion. That small portion which they call the crust of the earth is supposed to be the only cooled part; and they differ merely as to the degree of fluidity in the central region. I take it for granted that they have duly estimated the moon's tendency to cause tides in a fluid mass, within her influence: if there were no crust, of course she would cause such an effect, but a well-hardened case, we must suppose, can resist any such movement in the central fluid mass. Upon the principle of the arch, it would be easier to imagine resistance to pressure from without than from within; but the case or crust of our globe must be so solid that it neither yields nor vibrates to an internal expansive force.

This theory, however, is unsupported by any satisfactory evidence. Men of character and attainments have advocated it, although resting on conjecture: but when we look back along the roll of history, and discover so few philosophers who have not greatly erred, although famed in their day, it is natural to pause, and not acquiesce hastily in mere human assertion unsubstantiated by proof. Boring the ground, or examining the temperature of the bottom of a deep mine, affords no estimate for that of the central regions: Sir John Herschel says,* that 'the deepest mine

* Treatise on Astronomy; Cabinet Cyclopaedia, p. 22, Art. 30.

existing does not penetrate half a mile below the surface; a scratch or pinhole duly representing it on the surface of such a globe, (sixteen inches in diameter), would be imperceptible without a magnifier.' As our globe is about 8,000 miles in diameter, and external influence may be supposed to penetrate some distance, we can draw but unsatisfactory conclusions from experiments at depths not nearly so great even as that to which the ocean descends, and made chiefly in temperate or cold climates.

Having no pretension to more knowledge than any observant seaman may acquire in the course of a few years active employment afloat, it would be as vain as presumptuous in me, were I to offer any conjecture about the central mass of the earth. Perhaps, at a future day, when the nature of aerolites; the agency of electricity; and of electric communication through the superficial, if not through the interior regions of the globe, are better known, other opinions, respecting this wonderful world which we inhabit, may be formed by philosophers.

I have now fulfilled my intention of endeavouring to be useful, in however small a degree, to young persons of my own profession. If the few remarks laid before them, in this and the preceding chapter, at all increase their interest in the subjects spoken of, and tend, even in the least, to warn them against assenting hastily to new theories – while they induce a closer examination into the Record of truth – my object in writing them will be fully attained.

BIOGRAPHICAL GUIDE

ALISON, Robert Edward
English resident of Valparaiso who helped Charles Darwin with his geological researches in South America.

AZARA, Félix de (1742–1821)
Spanish geographer and explorer who wrote *Voyages dans l'Amérique méridionale*, Paris, 1809.

BANKS, Joseph (1743–1820)
Travelled with James Cook on the *Endeavour* expedition to the South Seas and Australia (1768–71). President of the Royal Society of London from 1778.

BEATSON, Alexander (1759–1833)
Governor of St Helena (1808–13). Author of *Tracts relative to the island of St Helena*, London, 1816.

BEAUFORT, Francis (1774–1857)
Hydrographer to the British Admiralty (1829–55). Introduced a scale to describe wind force. Instrumental in Charles Darwin's introduction to Robert FitzRoy.

BEECHEY, Frederick William (1796–1856)
Naval officer and explorer. Travelled on the *Hecla* expedition to the Arctic under the command of W. E. Parry. Commanded the *Blossom* expedition to the Pacific (1825–8) and the *Sulphur* expedition to South America (1835–6). Author of *Narrative of a voyage to the Pacific and Beering's Strait*, London, 1831.

BELL, Thomas (1792–1880)
Dental surgeon and comparative anatomist. Identified Charles Darwin's reptile specimens. Published many works on zoology including *History of British quadrupeds*, London, 1837.

BENNETT, Edward Turner (1797–1836)
Zoologist and first secretary of the Zoological Society of London. Identified the fish specimens from F. W. Beechey's Voyage to the Pacific.

BONPLAND, Aimé (1773–1858)
Travelled with F. W. H. A. von Humboldt in Central and South America (1799–1804). Remained in Buenos Aires and made a further

expedition up the river Paraná. Imprisoned for nine years in Paraguay.

BORY DE SAINT-VINCENT, Jean Baptiste Georges Marie (1778–1846)
Naturalist and explorer. Editor of the *Dictionnaire classique de l'histoire naturelle*, Paris, 1822–31. Wrote *Voyage dans les quatres principales îles des mers d'Afrique*, Paris, 1803.

BROWN, Robert (1773–1858)
Scottish botanist who travelled with Matthew Flinders on the *Investigator* expedition to Australia (1801–5). Librarian, curator and friend of Sir Joseph Banks.

BUCKLAND, William (1784–1856)
Eminent geologist, canon of Christ Church, Oxford, and dean of Westminster from 1845.

BUFFON, Georges-Louis Leclerc, Comte de (1707–1788)
Prominent naturalist and philosopher. Author of the great *Histoire Naturelle*, Paris, 1749–1804.

BURCHELL, William John (1782–1863)
Explorer and naturalist who travelled in the interior of South Africa (1810–15) and through South America (1825–9). His account of the American voyage was never published but his materials were freely available to naturalists in London.

BYNOE, Benjamin (1804–1865)
Assistant naval surgeon on the *Beagle* voyage (1831–6).

BYRON, John (1723–1786)
Commanded the *Dolphin* circumnavigation of the globe (1764–6). Author of *Narrative . . . of the loss of the* Wager, London, 1768.

CALDCLEUGH, Alexander (d. 1858)
English resident in Valparaiso.

CHAMISSO, Adelbert von (1781–1838)
Travelled with Otto von Kotzebue on a Russian expedition to explore the Pacific (1815–18). Director of the botanic gardens in Berlin from 1833.

COOK, James (1728–1779)
Navigator. Commanded the *Endeavour* expedition to the South Seas and Australia (1768–71) and, with the *Resolution* and *Adventure*, to Australia (1772–5). Killed on Hawaii.

CORFIELD, Richard Henry (1804–1897)
Schoolfellow of Charles Darwin. Resident in Valparaiso.

CUMING, Hugh (1791–1865)
Naturalist and conchologist. Resided in Valparaiso (1819–26). Travelled in Pacific (1826–7) and Philippines (1835–9). His collections were distributed to museums after his return to London in 1839.

CUVIER, Georges (1769–1832)

Eminent comparative anatomist and palaeontologist. Author of numerous works, including *Recherches sur les ossemens fossiles des quadrupèdes*, Paris, 1812 and *Discours sur les révolutions du globe*, Paris, 1812.

DAMPIER, William (1652–1715)

Buccaneer and naval officer. Travelled in the West Indies and in South American waters. Author of *Voyage round the world*, London, 1697. Commanded the *Roebuck* expedition to the southern hemisphere (1698–1700), described in *Voyage to New Holland*, London, 1703–9.

DANIELL, John Frederic (1790–1845)

Physicist and meteorologist. Author of *Meteorological essays*, London, 1823.

DE LA BECHE, Henry Thomas (1796–1855)

Geologist. Founder and director of the Geological Survey of Great Britain from 1835.

DILLON, Peter (1785?–1847)

Author of *Narrative . . . of a voyage in the south seas*, London, 1829.

EARLE, Augustus (1793–1838)

Artist on the *Beagle* (1831–2) until poor health forced him to leave the voyage at Montevideo. Author of *A narrative of a nine months' residence in New Zealand*, London, 1832.

EHRENBERG, Christian Gottfried (1795–1876)

Microscopist who published many studies on infusoria.

ELLIS, William (1794–1872)

Missionary in the South Sea islands from 1817. Another of the celebrated *Polynesian researches*, London, 1829.

FALKNER (OR FALCONER), Thomas (1707–1784)

Jesuit missionary who travelled in Paraguay (1732–9) and in Argentina to Tierra del Fuego from 1740. Resided in England from 1768; author of *A description of Patagonia*, Hereford, 1774.

FITZROY, Robert (1805–1865)

Promoted to captain on the *Beagle*'s first voyage to South America under the command of Phillip Parker King (1826–30). Commander of the *Beagle* during a second expedition (1831–6) on which Charles Darwin sailed. Author of *Narrative of the surveying voyages of HMS Adventure and Beagle*, London, 1839. Director of the Meteorological Office from 1855. Committed suicide.

FLINDERS, Matthew (1774–1814)

Commanded the *Investigator* expedition to Australia (1801–5) during which he was imprisoned by the French in Mauritius (1803–10). Author of *A voyage to Terra Australis*, London, 1814.

FORSTER, Johann Reinhold (1729–1798)

Travelled with James Cook on the second expedition to Australia (1772–5). Author of *Observations made during a voyage round the world*, London, 1778.

FRÉZIER, Amédée-François (1682–1773)

Commanded an expedition to South American waters and lived for three years in Peru and Chile (1712–15). Author of *Relation du voyage de la mer du Sud*, Paris, 1716.

GAY, Claude (1800–1873)

Professor of physics and chemistry in the French college at Santiago, Chile, from 1828. Travelled widely in South America until his return to Paris in 1842. Author of *Historia fisica y politica de Chile*, Paris, 1844–71.

GOULD, John (1804–1881)

Taxonomist and artist. Published numerous papers on the classification of birds and fine illustrated works, including *The birds of Australia*, London, 1840–8. Identified Charles Darwin's bird specimens.

HALL, Basil (1788–1844)

Naval officer who travelled a great deal in South America and China. Author of *Account of a voyage of discovery to the west coast of Corea*, London, 1818 and *Extracts from a journal written on the coasts of Chili, Peru and Mexico*, Edinburgh, 1824.

HEAD, Francis Bond (1793–1875)

Manager of the Rio Plata Mining Association from 1825 who travelled widely in South America. Author of *Rough notes taken during some rapid journeys across the Pampas and . . . Andes*, London, 1826. Lieutenant-governor of Upper Canada from 1835.

HENSLOW, John Stevens (1796–1861)

Professor of Mineralogy at Cambridge University, 1822–7; Professor of Botany from 1827. Rector of Hitcham, Suffolk from 1837. Charles Darwin's friend and mentor.

HERSCHEL, John Frederick William (1792–1871)

Celebrated natural philosopher and astronomer. Directed an astronomical expedition to Cape Town (1833–8). Author of *A preliminary discourse on the study of natural philosophy*, London, 1830.

HOLMAN, James (1786–1857)

Traveller and author of *A voyage round the world*, London, 1834–5.

HUMBOLDT, Friedrich Wilhelm Heinrich Alexander von (1769–1859)

Travelled in Central and South America (1799–1804) with Aimé Bonpland. Author of a celebrated *Personal Narrative* (trans. London,

1814–29) and *Voyage aux régions équinoxiales du nouveau continent*, Paris, 1805–34.

KING, Phillip Parker (1793–1856)

Naval officer. Commander of HMS *Adventure* and *Beagle*'s first expedition to South America (1826–30). Settled in Australia in 1830. His account of the voyage of the *Beagle* and *Adventure* was edited by Robert FitzRoy and published in FitzRoy's *Narrative*, London, 1839.

KIRBY, William (1759–1850)

Entomologist. Author of *Introduction to entomology*, with William Spence, London, 1815–26 and the Bridgewater treatise, *The habits and instincts of animals*, London, 1835.

KOTZEBUE, Otto von (1787–1846)

Naval officer. Accompanied A. I. Krusenstern on the Russian expedition to the South Pacific (1803–6). Commanded the *Rurick* expedition into the South Seas and Bering's Straits (1815–18) and a later circumnavigation (1823–6). Author of *A voyage of discovery*, London, 1821, and *A new voyage round the world*, London, 1830.

LA BILLARDIÈRE, Jacques Julien de (1755–1834)

Commanded the French expedition to search for La Pérouse (1791–4). Imprisoned by the Dutch in Java until released in 1795. Author of *Relation du voyage à la recherche de La Pérouse*, Paris, 1800, and various works on Australian botany.

LA PÉROUSE, Jean François de Galaup, Comte de (1741–1788)

Naval officer who commanded the *L'Astrolabe* expedition to the Pacific (1785–88). Shipwrecked and killed on the reefs of Vanikoro in the Santa Cruz group near Australia.

LATREILLE, Pierre André (1762–1833)

Taxonomist and Professor of Entomology at the Muséum d'Histoire Naturelle in Paris.

LESSON, René-Primevère (1794–1849)

Travelled on the *Coquille* expedition to the southern hemisphere commanded by Louis Isidore Duperrey and J. S. C. Dumont d'Urville (1822–5). Described the scientific results of the voyage in several books, published 1827–35.

LUMB, Edward

British merchant in Buenos Aires. Helped Charles Darwin locate and purchase fossil bones in 1834.

LYELL, Charles (1797–1875)

Prominent geologist and author of *Principles of Geology*, London, 1830–33. Advocated a 'uniformitarian' view of nature. Charles Darwin's friend and mentor after the *Beagle* voyage.

MCCORMICK, Robert (1800–1890)

Naturalist and naval surgeon. Surgeon in the *Beagle* (1831–2) until his departure from the voyage at Rio de Janeiro, Brazil. Participated in an expedition to search for Sir John Franklin in 1852.

MAGELLAN, Ferdinand (c. 1480–1521)

Portuguese navigator who completed the first circumnavigation of the world (1519–22).

MARTENS, Conrad (1801–1878)

Artist. Replaced Augustus Earle as artist in the *Beagle* (1833–4). Left the voyage and settled in Australia in 1835.

MOLINA, Juan Ignacio (1740–1829)

Jesuit priest who resided in Concepcion, Chile until 1768. Author of *Compendio della Storia geografica, naturale, e civile del regno del Chile*, Bologna, 1776.

NARBROUGH, John (1640–1688)

Admiral. Commander of a squadron that saw action in the Mediterranean (1674–7) and in Algiers (1677–9). Author of *An account of several late voyages and discoveries to the South and North*, London, 1694.

ORBIGNY, Alcide d' (1802–1857)

Travelled in South America on a collecting expedition for the Muséum d'Histoire Naturelle in Paris (1826–34). Published his results in *Voyage dans l'Amérique méridionale*, Paris, 1834–47.

OWEN, Richard (1804–1892)

Prominent comparative anatomist and palaeontologist. Conservator and later professor at the Museum of the Royal College of Surgeons in London. Identified Charles Darwin's mammalian fossil specimens.

OWEN, William Fitzwilliam (1774–1857)

Naval officer and author of *Narrative of voyages to explore the shores of Africa, Arabia and Madagascar*, London, 1833.

PARISH, Woodbine (1796–1882)

Diplomat and naturalist. Chargé d'affaires at Buenos Aires (1825–32). Pursued palaeontological research in Brazil and published *Buenos Ayres and the provinces of Rio de la Plata*, London, 1839.

PENTLAND, Joseph Barclay (1797–1873)

Traveller in South America. Consul-general in Bolivia (1836–9). Surveyed Bolivian Andes with Woodbine Parish (1826–7).

PERNETY, Antoine Joseph (1716–1801)

Accompanied de Bougainville to South America (1763). Author of *Journal historique d'un voyage fait aux îles Malouines*, Paris, 1769.

QUOY, Jean René Constant (1790–1869)

Travelled on the *Uranie* expedition to the southern hemisphere under the command of Louis Claude de Freycinet (1817–20) and on the *Astrolabe* expedition under J. S. C. Dumont d'Urville (1826–9). Described the zoological collections of the *Uranie* voyage (Paris, 1824) and of the *Astrolabe* (Paris, 1830–32).

REID, John (1809–1849)

Anatomist and Professor of Anatomy at St Andrew's University from 1841. Author of *Physiological, anatomical and pathological researches*, Edinburgh, 1848.

ROSAS, Don Juan Manuel Ortez (1793–1877)

Governor of Buenos Aires (1829–32) who led a campaign against the Indians in order to gain more territory (1833–5). Governor for a second time (1835–52).

ROSS, James Clark (1800–1862)

Authority on polar navigation. Commanded the *Erebus* and *Terror* expedition to the Antarctic (1839–43) and a search party for Sir John Franklin in the Arctic (1848). Author of *A voyage of discovery and research in the southern and Antarctic regions*, London, 1847.

SCORESBY, William (1789–1857)

Master-mariner and naturalist, engaged in whale fisheries in the Arctic. Author of *Account of the Arctic regions and northern whale fishery*, London, 1820.

SMITH, Andrew (1797–1872)

Army doctor who resided at the Cape of Good Hope (1821–37). Travelled extensively in the interior of South Africa. Author of *Illustrations of the zoology of South Africa*, London, 1838–47. An authority on Africa after his return to England in 1837.

SOWERBY, George Brettingham, the elder (1788–1854)

Conchologist and artist. Identified some of Charles Darwin's shell specimens.

STOKES, John Lort (1812–1885)

Naval officer. Lieutenant on the *Beagle* (1831–6). Friend of Charles (1826–30) and as mate and assistant surveyor during the second voyage (1831–6).

STOKES, Pringle (d. 1828)

Captain of the *Beagle* during the first voyage to South America under Phillip Parker King (1826–30). Committed suicide in 1828.

STURT, Charles (1795–1869)

Explorer and administrator who made several expeditions into the interior of Australia (1827–45). Author of *Narrative of an expedition into Central Australia*, London, 1849.

SULIVAN, Bartholomew James (1810–1890)

Naval officer. Lieutenant on the *Beagle* (1831–6). Friend of Charles
Darwin during the voyage and afterwards.

SWAINSON, William (1789–1855)

Zoologist and popular science writer. Advocate of a 'quinarian' system
of classification. Author of *Zoological illustrations*, London, 1820–23,
and *Exotic Conchology*, London, 1821–2.

ULLOA Y DE LA TORRE GIRAL, Antonio de (1716–1795)

Accompanied a French scientific expedition to America (1736–45).
Author of *Relación histórica del viaje a la América meridional*, Madrid,
1748.

WATERHOUSE, George Robert (1810–1888)

Naturalist. Curator of the Zoological Society (1836–43). Described
Charles Darwin's insect specimens from the *Beagle* voyage.

WILLIAMS, Henry (1792–1867)

Naval officer, then missionary to New Zealand from 1822. Established
a mission at Paihia and Waimate. Involved with the unrest over the
treaty of Waitangi.

YORK MINSTER

One of the Fuegians brought to England by Robert FitzRoy in 1830
and returned to Tierra del Fuego in 1833.

Discover more about our forthcoming books through Penguin's FREE newspaper...

Penguin
Quarterly

It's packed with:

- exciting features
- author interviews
- previews & reviews
- books from your favourite films & TV series
- exclusive competitions & much, much more...

Write off for your free copy today to:
Dept JC
Penguin Books Ltd
FREEPOST
West Drayton
Middlesex
UB7 0BR
NO STAMP REQUIRED

READ MORE IN PENGUIN

In every corner of the world, on every subject under the sun, Penguin represents quality and variety – the very best in publishing today.

For complete information about books available from Penguin – including Puffins, Penguin Classics and Arkana – and how to order them, write to us at the appropriate address below. Please note that for copyright reasons the selection of books varies from country to country.

In the United Kingdom: Please write to *Dept. JC, Penguin Books Ltd, FREEPOST, West Drayton, Middlesex UB7 0BR*

If you have any difficulty in obtaining a title, please send your order with the correct money, plus ten per cent for postage and packaging, to *PO Box No. 11, West Drayton, Middlesex UB7 0BR*

In the United States: Please write to *Penguin USA Inc., 375 Hudson Street, New York, NY 10014*

In Canada: Please write to *Penguin Books Canada Ltd, 10 Alcorn Avenue, Suite 300, Toronto, Ontario M4V 3B2*

In Australia: Please write to *Penguin Books Australia Ltd, 487 Maroondah Highway, Ringwood, Victoria 3134*

In New Zealand: Please write to *Penguin Books (NZ) Ltd, 182–190 Wairau Road, Private Bag, Takapuna, Auckland 9*

In India: Please write to *Penguin Books India Pvt Ltd, 706 Eros Apartments, 56 Nehru Place, New Delhi 110 019*

In the Netherlands: Please write to *Penguin Books Netherlands B.V., Keizersgracht 231 NL–1016 DV Amsterdam*

In Germany: Please write to *Penguin Books Deutschland GmbH, Friedrichstrasse 10–12, W–6000 Frankfurt/Main 1*

In Spain: Please write to *Penguin Books S. A., C. San Bernardo 117–6° E–28015 Madrid*

In Italy: Please write to *Penguin Italia s.r.l., Via Felice Casati 20, I–20124 Milano*

In France: Please write to *Penguin France S. A., 17 rue Lejeune, F–31000 Toulouse*

In Japan: Please write to *Penguin Books Japan, Ishikiribashi Building, 2–5–4, Suido, Bunkyo-ku, Tokyo 112*

In Greece: Please write to *Penguin Hellas Ltd, Dimocritou 3, GR–106 71 Athens*

In South Africa: Please write to *Longman Penguin Southern Africa (Pty) Ltd, Private Bag X08, Bertsham 2013*

READ MORE IN PENGUIN

A CHOICE OF CLASSICS

Netochka Nezvanova Fyodor Dostoyevsky

Dostoyevsky's first book tells the story of 'Nameless Nobody' and introduces many of the themes and issues which dominate his great masterpieces.

Selections from the Carmina Burana
A verse translation by David Parlett

The famous songs from the *Carmina Burana* (made into an oratorio by Carl Orff) tell of lecherous monks and corrupt clerics, drinkers and gamblers, and the fleeting pleasures of youth.

Fear and Trembling Søren Kierkegaard

A profound meditation on the nature of faith and submission to God's will, which examines with startling originality the story of Abraham and Isaac.

Selected Prose Charles Lamb

Lamb's famous essays (under the strange pseudonym of Elia) on anything and everything have long been celebrated for their apparently innocent charm. This major new edition allows readers to discover the darker and more interesting aspects of Lamb.

The Picture of Dorian Gray Oscar Wilde

Wilde's superb and macabre novel, one of his supreme works, is reprinted here with a masterly Introduction and valuable Notes by Peter Ackroyd.

Frankenstein Mary Shelley

In recounting this chilling tragedy Mary Shelley demonstrates both the corruption of an innocent creature by an immoral society and the dangers of playing God with science.